The Student's Companion

Second Edition

Laboratory Techniques for Organic Chemistry

Standard Scale and Microscale

John W. Lehman

Professor Emeritus,
Lake Superior State University

Prentice Hall, Upper Saddle River, New Jersey 07458

Editor-in-Chief, Science: *Nicole Folchetti*
Senior Editor: *Andrew Gifillan*
Associate Editor: *Jennifer Hart*
Assistant Managing Editor, Science: *Gina M. Cheselka*
Project Manager, Production: *Ashley M. Booth*
Cover Designer: *John Christiana*
Art Editor: *Connie Long*

© 2008 Pearson Education, Inc.
Pearson Prentice Hall
Pearson Education, Inc.
Upper Saddle River, NJ 07458

Printed in the United States of America

ISBN 0-13-159381-1

Pearson Education Ltd., *London*
Pearson Education Australia Pty. Ltd., *Sydney*
Pearson Education Singapore, Pte. Ltd.
Pearson Education North Asia Ltd., *Hong Kong*
Pearson Education Canada, Inc., *Toronto*
Pearson Educación de Mexico, S.A. de C.V.
Pearson Education—Japan, *Tokyo*
Pearson Education Malaysia, Pte. Ltd.
Pearson Education, *Upper Saddle River, New Jersey*

The Student's Lab Companion *is dedicated to the many students whose achievements and misadventures in my organic chemistry laboratory provided the substance and inspiration for the book.*

About the Author

John W. Lehman received his Ph.D. in chemistry from the University of Colorado in Boulder. He has taught chemistry for 35 years at Lake Superior State University, a small university in Michigan's scenic Upper Peninsula, where teaching—rather than research—is the most important activity. In recognition of his teaching skills, he received the State of Michigan's Teaching Excellence Award in 1990. In 2001 he funded a chemistry scholarship to help bring outstanding students into the chemistry program at Lake Superior State University. His groundbreaking lab text, *Operational Organic Chemistry*, was first published in 1981, and he has written three additional books for the organic chemistry laboratory.

Contents

Appendixes and Bibliography   322

Preface

To the Instructor

The Student's Lab Companion describes all of the laboratory operations that are likely to be used in a typical organic chemistry lab course. It is intended primarily for use in lab courses for which the experiments are either provided by the instructor or obtained from some other source that does not include adequate descriptions of lab techniques. The book can also be used by professional chemists or advanced students engaged in organic synthesis or chemistry research, and as a general reference to organic laboratory techniques.

This book is based on the author's lab textbook, *Multiscale Operational Organic Chemistry: A Problem-Solving Approach to the Laboratory Course.* It includes the entire "Operations" section of that book, with some additions and revisions. The operation descriptions apply to the use of both standard scale glassware, such as that provided by an organic chemistry lab kit with 19/22 standard-taper joints, and microscale glassware, such as that provided by a microscale lab kit with 14/10 standard-taper joints and threaded connectors.

The Student's Lab Companion also contains information about lab safety and chemical hazards, a "Guide to Success in the Organic Chemistry Lab" designed to allay students' trepidation about the lab experience and help them perform better, instructions for writing lab reports and keeping lab notebooks, descriptions of stoichiometric calculations, tables of properties of organic compounds for qualitative analysis, and more. A section on scientific methodology is intended to help students "think like scientists" as they carry out their assigned lab experiments. Appendix VII provides an introduction to the chemical literature and is followed by a comprehensive Bibliography that includes a number of useful resources on the practice of organic chemistry. The Bibliography also cites general interest books and articles that can help students write reports and research papers.

Preface to the Second Edition

During some 35 years of teaching chemistry, I've witnessed nearly everything that can go wrong in an organic chemistry lab. To give students the benefit of my observations, I've added new "troubleshooting" sections, entitled *When Things Go Wrong*, to most of the operations. If a student combines the wrong layers during an extraction, obtains no product upon cooling a recrystallization solution, or records an infrared spectrum with no absorption bands, referring to the troubleshooting section should lead the student to a solution, or at least to an understanding of what went wrong.

The serious damage that can be produced by the release of hazardous chemicals into the environment and the unsustainable rate at which we consume chemical and energy resources have convinced me that chemistry students should learn how to work with chemicals in an environmentally responsible way. This edition therefore contains a new introductory section on "Chemistry and the Environment," which includes a discussion of the principals of green chemistry.

Each category of operations in the book now begins with a brief summary of the operations and their uses. Some operations have been reorganized and a new one on excluding air from reaction mixtures has been added. I have added new information about some instrumental techniques, such as the use of ATRs and disposable cards in infrared spectrometry, and made substantive revisions of many operations such as the one on flash chromatography. Other features of the second edition include a revised and updated Bibliography, a new section on finding chemical safety information, the inclusion of calculations for solution preparation in Appendix IV, and directions for improvising synthetic procedures and scaling down reactions in Appendix V .

I would like to express my gratitude to Nicole Folchetti, Andrew Gilfillan, and Jennifer Hart, who provided encouragement and assistance during the preparation of this edition, and to the staff at Prentice Hall for their contributions to its production. The spectra in this book are reproduced from the spectral libraries of the Sigma–Aldrich Corporation, whose generosity is gratefully acknowledged.

All textbooks can benefit from comments and criticism by their adopters. I welcome correspondence from users of this book to point out errors. suggest improvements, or share their own experiences in the organic chemistry lab.

John W. Lehman
jlehman@lssu.edu

Introduction

To the Student

In the dictionary, an *operation* is defined as a process or series of acts performed to effect a certain purpose or result. For example, the *recrystallization* operation, which is used to purify a solid substance, involves at least four separate steps: (1) *heating* the solid in a boiling solvent until it dissolves, (2) *cooling* the resulting solution until the solid crystallizes, (3) *filtering* the crystals to separate them from the solvent, and (4) *washing* the crystals to remove residual impurities. Such operations are often called *techniques,* but the latter term is also used to refer to a single step, such as filtration, in a complex operation.

This book describes all of the operations (designated by "OP" followed by the operation number) you need to know to successfully complete your organic chemistry laboratory course. It can also be used to help you carry out an undergraduate research project involving organic synthesis or analysis and for most other purposes that require the use of organic chemistry techniques in the laboratory. It covers the use of both *standard scale* equipment, for preparing relatively large amounts of product (usually about 1–10 g), and *microscale* equipment, for preparing smaller amounts of product (usually less than 1 g). You will find the operation number and page location of each operation in the Table of Contents.

The experiments you will be performing will be provided separately by your instructor or obtained from some other source. Before you come to the lab to work on an experiment, you should always read or review the operations you will be performing during that experiment. Otherwise, you will waste valuable lab time looking up information needed to carry out the operations, and you may not be able to finish an experiment in the time allotted.

Your instructor may tell you what operations will be used in a particular experiment. If not, you should read through the procedure carefully to find out what operations you will be using. In most cases, the operations should be apparent from the procedure, as shown by the operation names and numbers (in parentheses) following the italicized instructions or amounts in this example:

> *Weigh* (OP-4, Weighing) 5.00 g of eucalyptus oil into a clean, dry, round-bottom flask and add 10 mL (OP-5, Measuring Volume) of anhydrous ethyl ether. Then *stir in* (OP-10, Mixing) 1.80 g (OP-4, Weighing) of powdered maleic anhydride. *Heat* (OP-7, Heating) the reaction mixture under gentle *reflux* (OP-7c, Heating under Reflux) for 45 min. At the end of the reaction period, *transfer* (OP-6, Making Transfers) the mixture to a small Erlenmeyer flask while it is still warm and let it stand until crystals form. Then *cool* (OP-8, Cooling) the flask in an ice-water bath for 5 min. Collect the product by *vacuum filtration* (OP-16, Vacuum Filtration) and *recrystallize* (OP-28, Recrystallization) it from dry methanol. *Dry* (OP-26, Drying Solids) the product, *measure its mass* (OP-4, Weighing) and *melting point* (OP-33, Melting Point), and then record its *infrared spectrum* (OP-39, Infrared Spectrometry).

After you have used an operation once or twice, you should not have to reread the entire description the next time you use it, but you should at least read the summary and review the "General Directions" section (if provided) to refresh your memory. Eventually, you should not need to refer to the operation description at all, unless you are applying the operation in a new situation or unexpected problems arise. If you do run into problems while carrying out an operation, read *When Things Go Wrong*, a troubleshooting guide included with most of the operations.

Appendix I contains illustrations of standard scale and microscale laboratory equipment. Appendixes II and III tell you how to keep a lab notebook and write lab reports. Appendix IV will help you perform calculations used in carrying out and writing up experiments. Appendix V tells you how to develop experimental plans, write flow diagrams for experiments, improvise procedures for organic syntheses, and scale down reactions. Appendix VI contains tables of properties for qualitative organic analysis—the identification of organic compounds. Appendix VII is a guide to the chemical literature that describes important works in organic chemistry, directs you to information about them in the Bibliography, and in some cases, tells you how to use them. The Bibliography lists a large number of useful works about the practice of organic chemistry, ranging from enormous multivolume sets such as *Chemical Abstracts* to short papers from the *Journal of Chemical Education*. References to entries listed in the Bibliography are made throughout this book in the form [Bibliography, F20], where the letter refers to a category and the number to a location within that category; for example, F20 is the 20th book listed under category F, Spectrometry. If you are required to write a research paper for your organic chemistry course, looking through relevant books or articles cited in Category L and elsewhere in the Bibliography should give you ideas and provide useful information.

At some time before the first organic chemistry laboratory period, you should read the next section, "A Guide to Success in the Organic Chemistry Lab." It will tell you what to expect in an organic chemistry lab, how to work efficiently so that you don't fall behind, and how to get along in the lab. The following section, "Scientific Methodology," is intended to help you "think like a scientist," which in turn will help you do better work and get more out of your laboratory experience.

Citations to journal articles are given in the form J. Chem. Educ. **2002**, *79, 721, where the abbreviated journal name (the* Journal of Chemical Education *in this example) is followed by the date, volume number, and page number.*

A Guide to Success in the Organic Chemistry Lab

What to Expect in Your Organic Chemistry Lab Course

Students are sometimes apprehensive about having to work in an organic chemistry lab. They may have heard that organic chemistry is difficult or that the lab is a dangerous place because of the hazardous nature of organic chemicals. While some students find the subject matter of organic chemistry difficult, most of them have little trouble completing the laboratory experiments successfully. In fact, many students enjoy the lab far more than the lecture course, and they generally receive higher grades in lab than in lecture.

It is true that some organic chemicals are quite hazardous and can cause serious injury, or even death, if not handled properly. But if the chemistry lab were a very dangerous place to work, you would expect professional chemists to have short life spans. In fact, chemists tend to live longer than professionals in most other fields; in one listing of occupational risks, chemists rank near the bottom (lowest risk), right between school administrators and ticket agents. In my 35 years of teaching organic chemistry, I've observed only a few lab accidents that caused significant injury, most of them involving broken glass. Only a handful of students were injured by contact with chemicals, and none of them suffered permanent injury. That is not to say that you can afford to be careless in the lab, because the potential for a serious accident is always there. But if you follow the safety rules listed in the next section and take reasonable precautions when handling chemicals, you should have little reason for concern about lab accidents.

Getting Started

By the end of the first week of your organic chemistry lab, you should have read the "Laboratory Safety" section of this book and any other safety rules or information provided by your instructor. Before you begin working in the laboratory, your instructor should review the safety rules and tell you what safety supplies, such as safety goggles and protective gloves and aprons, you must have in order to work in the lab. During the first laboratory period, the instructor will show you where the safety equipment is located and tell you how to use it. As you locate each item, check it against the following list and make a note of its location. (Your instructor may suggest additions or changes to the list.)

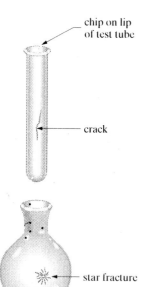

chip on lip
of test tube

crack

star fracture

Figure IN1 Glassware defects

- Fire extinguishers
- Fire blanket
- Safety shower
- Eyewash fountain
- First aid supplies
- Spill cleanup supplies

You should also learn the locations of chemicals, consumable supplies (such as filter paper and boiling chips), waste containers, and various items of equipment such as balances and drying ovens.

Your instructor will assign you a locker and provide a list of locker supplies. You will then need to check into the laboratory. This usually involves getting a locker key or combination lock and checking your locker for missing or damaged items. You will find illustrations of typical locker supplies in Appendix I at the back of this book. If you find any glassware items with chips, cracks, or star fractures (see Figure IN1), you should have them replaced; they may cause cuts, break on heating, or shatter under stress. If necessary, clean up any dirty glassware in your locker (see OP-1) and organize it neatly at this time.

Working Efficiently

Because of wide variations in individual working rates, it is usually not possible to schedule experiments so that everyone can finish in the allotted time. If all labs were geared to the slowest student, the objectives of the course could not be accomplished in the limited time available. If you fall behind in the lab, you may need to put in extra hours outside your scheduled laboratory period to complete the course. The following suggestions should help you work more efficiently and finish each experiment on time.

1. *Be prepared to start the experiment the moment you reach your work area.* Don't waste the precious minutes at the start of a laboratory period doing calculations, reading the experiment, washing glassware, or carrying out other activities that should have been done at the end of the previous period or during the intervening time. The first half hour of any lab period is the most important—if you use it to collect the necessary materials, set up the apparatus, and get the initial operation (reflux, distillation, etc.) under way, you should have no trouble completing the experiment on time.

2. *Organize your time efficiently.* Schedule a time each week to read the experiment and operation descriptions and complete any prelab assignment—an hour before the lab period begins is too late! Plan ahead so that you know approximately what you will be doing at each stage of the experiment. A written experimental plan, prepared as described in Appendix V, is invaluable for this purpose.

3. *Organize your work area.* Before performing any operation, arrange all of the equipment and supplies you will need during the operation neatly on your bench top, in the approximate order in which they will be used. Place small objects (such as spatulas) and items that might be contaminated by contact with the bench top on a paper towel, laboratory tissue, or mat. After you use each item, move it to an out-of-the-way location where it can be cleaned and then returned to its proper location when time permits; for example, put dirty glassware in a washing trough in the sink. Keep your locker well organized, placing each item in the same location after use so that you can immediately find the equipment you need. This will also help you identify missing items, so that you can hunt for them before you leave the lab; otherwise they will probably disappear before the next lab period and you may be charged for them.

Getting Along in the Laboratory

You will get along much better in the organic chemistry laboratory if you can maintain peace and harmony with your coworkers—or at least keep from aggravating them—and stay on good terms with your instructor. Following these common-sense rules will help you do that:

1. *Leave all chemicals where you found them.* You will understand the reason for this rule once you experience the frustration of hunting high and low for a reagent, only to find it at another student's station in a far corner of the lab. Containers should be taken to the reagent bottles to be filled; reagent bottles should never be taken to your lab station.

2. *Take only what you need.* Liquids and solutions should ordinarily be obtained using bottle-top dispensers, graduated cylinders, automatic pipets, or other measuring devices, so that you will take no more than you expect to use for a given operation. Solids can be weighed out directly from stock bottles.

3. *Prevent contamination of chemicals.* Do not use your own pipet or dropper to remove liquids from stock bottles and do not return unused chemicals to stock bottles. Be sure to close all bottles tightly after use — particularly those containing anhydrous chemicals and drying agents.

4. *If you must use a burner, inform your neighbors,* unless they are already using burners. This will allow them to cover any containers of flammable solvents and take other necessary precautions. In some circumstances, you may have to use a different heat source, move your operation to a safe location (such as a fume hood), or find something else to do while flammable solvents are in use.

5. *Return all community equipment to the designated locations.* This may include ring stands, steam baths, lab kits, clamps, condenser tubing, and other items that aren't in your own locker. Since such items will be needed by students in other lab sections, they should always be returned to the proper storage area at the end of the period.

6. *Clean up for the next person.* Few experiences are more annoying than discovering that the lab kit you just checked out is full of dirty glassware or that your lab station is cluttered with paper towels, broken glass, and spilled chemicals. The last 15 minutes or so of every laboratory period should be set aside for cleaning up your lab station and the glassware used during the experiment. Put things away so that your workstation is uncluttered. Clean the bench top with a towel or a wet sponge; remove rubber tubing, other supplies, and debris from the sink; and thoroughly wash any dirty glassware that is to be returned to the stockroom, as well as that from your locker. Clean up any spills and broken glassware immediately. If you spill a corrosive or toxic chemical, such as sulfuric acid or aniline, inform the instructor before you attempt to clean it up.

7. *Heed Gumperson's Second Law*, which advises you to maximize the labor and minimize the oratory while in the **labor**atory. This doesn't mean that all conversation must come to a halt. Quiet conversation during a lull in the experimental activity is okay, but a constant stream of chatter directed at a student performing a delicate operation is distracting and can lead to an accident. For the same reason, radios, tape players, CD players, and other audio devices should not be brought into the laboratory. Your instructor may *not* dissolve your Eminem CDs in sulfuric acid, but you can never be sure of that.

Scientific Methodology

If you are taking an organic chemistry course, you are probably planning a career in some field of science or technology, or a field that is based on scientific knowledge and principles, such as medicine. To succeed in such a field, you must learn to think and work like a scientist. Scientists follow certain basic principles that are often lumped together under the expression "the scientific method." In fact, there is no universal scientific method that all scientists follow rigorously. But most of them go through at least some of the following steps when dealing with a scientific problem:

- Define the problem
- Plan a course of action
- Gather evidence
- Evaluate the evidence
- Develop a hypothesis
- Test the hypothesis
- Reach a conclusion
- Report the results

Defining the Problem.   Most people think of a "problem" in a negative sense, as in "We've got a problem here," or "What's your problem?" To a scientist, a problem is not a perceived difficulty but an *opportunity* for exploring and learning more about some aspect of the physical world. A problem may be inherent in an assigned task, or it may arise from anything that the scientist is curious about, such as an unexplained phenomenon or an unexpected observation. A problem is often defined in the form of a question: What is the identity of the liquid my instructor gave me? What is the mercury concentration in Lake Michigan salmon? How do fireflies generate light?

Planning a Course of Action. A scientist must plan his or her own course of action for solving a scientific problem. This often requires that the scientist carry out a literature search to glean information and data relating to the problem, decide which experimental methods and instruments to use, and develop a detailed procedure to be followed. In most lab courses, however, the procedure is given in a lab textbook or provided by the instructor. In some courses, you may be allowed to develop your own procedures — with your instructor's permission.

Gathering Evidence. Scientists will gather as much evidence as they feel is needed to solve a problem and convince other scientists that their solution is correct. Evidence is gathered by making careful *observations* and *measurements.*

To make valid observations, you must be *objective,* reporting only what you actually saw and not what you expected to see. A wildlife biologist who expects wild chimpanzees to behave just like chimpanzees at the zoo is not likely to make any important discoveries about chimpanzees! Keep the following points in mind when you make observations:

1. Don't confuse an *observation* with an *inference.* An observation is whatever you perceive with your senses (sight, smell, touch, taste, or hearing) during an event. An inference is a guess about the *cause* of the event. Writing "The solution turned brown upon addition of 0.1 M $KMnO_4$" records an observation. Writing "The solution must have contained an alkene because it turned brown upon addition of 0.1 M $KMnO_4$" is an inference.
2. Be prepared to be surprised. If you observe something you didn't expect, don't simply disregard the observation or report what you thought you should have seen. Consider an unexpected observation as an opportunity to learn something you didn't know before — something that might lead to a new discovery.
3. Write down your observations as you make them or shortly afterward. If you wait too long, you are likely to leave out important details.
4. Record your observations clearly, completely, and systematically. For example, if you are doing repetitive measurements or carrying out the same test on a series of samples, record your observations in a table.

Making accurate measurements, such as determining the mass or melting point of the product of a chemical synthesis, requires a certain amount of skill and know-how. You can obtain such skills by, for example, watching your instructor demonstrate the operation of an instrument and then practicing on the instrument until you obtain consistent and accurate results *before* you use it to make a measurement you intend to report. If you are not sure how to use an instrument properly, ask the instructor to show you.

Evaluating the Evidence. Evaluating the evidence involves assessing the reliability of your experimental results and looking for clues among your results that may lead you to a solution to the problem. For example, inspecting the infrared spectrum of an unknown liquid for evidence of a specific functional group and comparing its boiling point with the boiling points of known compounds may help you solve the problem, "What is the identity of the liquid my instructor gave me?" Solving such problems often requires clear and logical thinking; in other cases a more creative, intuitive approach can be valuable. In either case, all of your reasoning and intuition may be fruitless if your experimental results are unreliable. The validation of experimental results requires first that you ask yourself whether the results make sense physically. If you obtain a melting point that is much lower than the literature value, a product mass that is higher than the theoretical yield for a synthesis, or any other result that seems suspect, then you need to find out whether the result is, in fact, erroneous. You should review everything you did that led to that result, using notes from your lab notebook to jog your memory as necessary. Perhaps you only need to repeat a melting point or dry a product longer, but in any case you should find out what you did wrong and correct it. In some kinds of experiments, validation of results may also require a statistical analysis of experimental data. In that case, your instructor will provide directions.

Developing and Testing Hypotheses.   A hypothesis can be regarded as an "educated guess" about the cause of some phenomenon or the outcome of an experiment. Hypotheses can help us see the significance of an object or event that would otherwise mean little. For example, the movements of the planets seemed erratic and mysterious before Copernicus formulated his hypothesis that the earth revolves around the sun. A hypothesis must be *testable* to have validity; that is, it must be formulated in such a way that experiments can be devised whose outcome might prove the hypothesis *wrong*. John Dalton's hypothesis that atoms are indivisible was proven wrong after it was shown that bombarding uranium atoms with neutrons caused them to split into smaller atoms. But Dalton's more fundamental hypothesis that all matter is made of atoms has been tested repeatedly over the years and has never been proven wrong, so it is generally accepted as true.

In a typical organic chemistry lab course, it is unlikely that you will have an opportunity to formulate and test a general hypothesis about the nature of matter. But you may be able to propose one or more *working hypotheses* for most of the experiments assigned. A working hypothesis can be a prediction about the outcome of an experiment based on the information available, which for most chemistry lab courses is provided in the experiment write-up. For example, if a preparation can yield one of several possible products, your working hypothesis could predict which product you consider to be the most likely one. Such a hypothesis can be proposed and revised at any time during the course of an experiment.

One drawback of a working hypothesis is that you may become so attached to it that you will overlook or ignore evidence that contradicts it. For this reason, some scientists prefer to explore a problem without any preconceived ideas about the outcome—which is not always easy to do. Whenever you propose a working hypothesis, you should be ready to revise or abandon it without regret if the evidence does not support it.

Reaching Conclusions.   If a hypothesis passes all of the tests you carry out, then you are ready to state a conclusion, such as "The liquid my instructor gave me is benzaldehyde." This doesn't necessarily mean that your conclusion is correct. You may have not carried out enough tests or carried them out accurately enough to justify your conclusion. But if you have performed an experiment carefully and reasoned logically, you will probably arrive at a valid conclusion.

Reporting Results.   You should report all results of an experiment as clearly, completely, and unambiguously as possible. Write up your results in good English using complete sentences and correct spelling. Be as specific as you can, avoiding such generalities as "My yield was lower than expected due to human error." Label any tables clearly, giving names or standard abbreviations for all physical properties and the units in which they are measured.

If an experiment requires that you graph your data, use accurately ruled graph paper, never notebook paper or paper you have ruled yourself. Plot the dependent variable on the *y*-axis and the independent variable on the *x*-axis. Label both axes with the physical quantities being graphed and their units, if any. Select appropriate uniform scale intervals so that your data points extend most of the way up and across the graph paper. If the relationship you are graphing is linear, draw the straight line that best fits the data points.

Applying Scientific Methodology

In most organic chemistry lab courses, the procedure for each experiment is provided and the expected outcome of the experiment is described. How can you apply scientific methodology in such a case? Consider a typical organic chemistry experiment involving the reaction of isopentyl alcohol with acetic acid in the presence of a catalyst such as phosphoric acid. By reading the experiment, you learn that the expected product of the reaction is isopentyl acetate (also known as "banana oil"). With a little thought, you can formulate a statement of the problem such as "When isopentyl alcohol is heated with acetic acid in the presence of phosphoric acid, will there be a reaction that produces pure isopentyl acetate?" If you wish, you can formulate one or more working hypotheses, which could include predictions about the yield or purity of the product. Such a hypothesis might be, for example, "Heating isopentyl alcohol with acetic acid in the presence of phosphoric acid will yield isopentyl acetate with a purity of at least 95%, by mass."

The course of action should be mapped out for you in the experimental procedure. As you read the procedure, consider the kind of evidence you can gather by measurement or observation. During the experiment itself, make careful observations of everything that might provide evidence relating to the problem. What changes occur when the chemicals are combined and heated together? Do the changes indicate that a chemical reaction is occurring? Do they give any clue about the possible identity of the product? What physical properties of the product can you observe (such as odor) or measure (such as boiling point) during the experiment, and what bearing do they have on the identity and purity of the product? If you analyze the product by gas chromatography or infrared spectrometry, what do your results tell you about the identity and purity of the product? After you have completed the experiment you can evaluate the evidence, decide whether the results confirm or disprove your hypothesis, and arrive at a conclusion that answers the question proposed in your problem statement. Appendixes II and III suggest ways of recording and reporting your experimental results. Your instructor will let you know what kind of report he or she prefers.

Laboratory
Safety

Safety Standards

In the United States, a Federal Laboratory Safety Standard designated as OSHA 29 CFR 1910.1450 requires that any laboratory in which hazardous chemicals are handled should have a written safety plan called a *chemical hygiene plan*. Your institution's chemical hygiene plan should describe safe operating procedures to be followed in carrying out various laboratory operations, and emergency response procedures to be followed in case of accident. According to the OSHA standard, laboratory instructors are required to see that students know and follow established safety rules, have access to and know how to use appropriate emergency equipment, and are aware of hazards associated with specific experiments. The lab instructor alone cannot prevent laboratory accidents, however. You also have a responsibility to follow safe laboratory practices while performing experiments and to be ready to respond in case of accident.

Protecting Yourself

Just as construction workers protect themselves from accidents by wearing hard hats and steel-toed boots, people who work with chemicals should wear appropriate clothing and personal protective equipment (such as safety goggles) that reduce the likelihood of injury in case of an accident.

Eye protection is essential at all times; that is the law in many states and should be the rule in every chemistry laboratory. Safety glasses provide only limited protection because they have no side shields, so it is best to wear safety *goggles* that protect your eyes from chemical splashes and flying particles coming from any direction.

In any chemistry lab you should wear clothing that is substantial enough and covers enough of your body to offer some protection against accidental chemical spills and flying glass or other particles. Long-sleeved shirts or blouses and long pants or dresses are recommended, especially when made of denim or other heavy materials. Some synthetic fabrics can be dissolved by chemicals such as acetone, and could melt in contact with a flame or other heat source. Wear shoes that can protect you from spilled chemicals and broken glass, not open sandals or cloth-topped athletic shoes. To protect your clothing (and yourself) from chemical spills, it is a good idea to wear a lab jacket or lab apron.

Always wear appropriate gloves when handling caustic chemicals, which can burn the skin, or toxic chemicals that can be absorbed through the skin. No single type of glove protects against all chemicals, but neoprene gloves offer good to excellent protection against many commonly used chemicals, and disposable nitrile gloves are adequate for use in most undergraduate labs. Latex gloves are not recommended because some people are allergic to latex and because they are permeable to many hazardous chemicals. In 1997 a Dartmouth College chemistry professor died after spilling a few drops of dimethylmercury on her latex gloves!

Preventing Laboratory Accidents

Most organic lab courses are completed without incident, apart from minor cuts or burns, and serious accidents are rare. Nevertheless, the potential for a serious accident always exists. To reduce the likelihood of an accident, *you must learn the following safety rules and observe them at all times.* Additional safety rules or revisions of these rules may be provided by your instructor.

1. *Wear approved eye protection in the laboratory at all times.* Even when you are not working with hazardous materials, another student's actions could endanger your eyes, so never remove your safety goggles or safety glasses until you leave the lab. Do not wear contact lenses in the laboratory, because chemicals splashed into an eye may get underneath a contact lens and cause damage before the lens can be removed. Learn the location of the eyewash fountain nearest to you during the first laboratory session, and learn how to use it.

2. *Never smoke in the laboratory or use open flames in operations involving low-boiling flammable solvents.* Anyone who smokes in an organic chemistry laboratory is subject to immediate expulsion. Before you light a burner or even strike a match, inform your neighbors of your intention to use a flame. If anyone nearby is using flammable solvents, either wait until they are finished or move to a safer location, such as a fume hood. Diethyl ether and petroleum ether are extremely flammable, but other common solvents, such as acetone and ethanol, can be dangerous as well. When ventilation is inadequate, the vapors of diethyl ether and other highly volatile liquids can travel a long way; lighting a burner at one end of a lab bench with an open bottle of ether at its other end has been known to start an ether fire. Learn the location and operation of the fire extinguishers, fire blankets, and safety showers at the first laboratory session.

3. *Consider all chemicals to be hazardous and minimize your exposure to them.* Never taste chemicals, do not inhale the vapors of volatile chemicals or the dust of finely divided solids, and prevent contact between chemicals and your skin, eyes, and clothing. Many chemicals can cause poisoning by ingestion, inhalation, or absorption through the skin. Strong acids and bases, bromine, thionyl chloride, and other corrosive materials can produce severe burns and require special precautions, such as wearing gloves and lab aprons. Some chemicals cause severe allergic reactions, and others may be carcinogenic (tending to cause cancer) or teratogenic (tending to cause birth defects) by inhalation, ingestion (swallowing), or skin absorption.

To prevent accidental *ingestion* of toxic chemicals, do not bring food or drink into the laboratory or use mouth suction for pipetting, and wash your hands thoroughly after handling any chemical. To prevent *inhalation* of toxic or carcinogenic chemicals, work under an efficient fume hood or use a gas trap (see OP-14) to keep chemical fumes out of the laboratory atmosphere. To prevent *contact* with corrosive or toxic chemicals, wear appropriate gloves and a lab apron or lab jacket. Clean up chemical spills immediately, using a neutralizing agent and plenty of water for acids and bases, and an absorbent for solvents. In case of a major spill, or if the chemical spilled is very corrosive or toxic, notify your instructor before you try to clean it up.

4. *Exercise great care when working with glass and when inserting or removing thermometers and glass tubing.* Among the most common injuries in a chemistry lab are cuts from broken glass and burns from touching hot glass. Protect your hands with gloves or a towel when inserting glass tubes or thermometers into stoppers or thermometer adapters and when removing them. Grasp the glass close to the stopper or thermometer adapter and gently twist it in or out. Remember that hot glass remains hot for some time; so after you have fire-polished a glass rod or complet-ed another glass-working operation, give the glass plenty of time to cool before you touch it. Refer to OP-3 for more information about safe glass-working procedures.

5. *Wear appropriate clothing in the laboratory.* Wear clothing that is sub-stantial enough to offer some protection against accidental chemical spills, and shoes that can protect you from spilled chemicals and broken glass. Human hair is very flammable, so wear a hair net while using a burner if you have long hair.

6. *Dispose of chemicals properly.* For reasons of safety and environmental protection, most organic chemicals should not be washed down the drain. Except when your instructor or an experiment's directions indicate other-wise, place used organic chemicals and solutions in designated waste con-tainers. Some water-soluble chemicals and aqueous solutions can be safely washed down the drain, but consult your instructor if there is any question as to the best method for disposing of a particular chemical or solution. See the section on "Disposal of Hazardous Wastes" for additional information.

7. *Never work alone in the laboratory or perform unauthorized experiments.* If you wish to work in the laboratory when no formal lab period is sched-uled, you must obtain written permission from the instructor and be cer-tain that others will be present while you are working.

Reacting to Accidents: First Aid

If you have or witness a serious accident involving poisoning or injury, report it to the instructor immediately. Serious accidents should be treated by a competent physician, but applying some basic first aid procedures before a physician arrives can help minimize any damage.

If you *have* an accident that requires quick action to prevent perma-nent injury, take the appropriate action as described here, if you can, and see that the instructor is informed of the accident. If you *witness* an acci-dent, call the instructor immediately and leave the first aid to him or her, unless (1) no instructor or assistant is in the laboratory area, (2) the victim requires immediate attention because of stopped breathing, heavy bleed-ing, or any other life-threatening condition, or (3) you have had appropriate emergency response training.

If an accident victim stops breathing or goes into shock as a result of any kind of accident, standard procedures for artificial respiration and treating shock should be applied. Descriptions of these procedures can be found in Chapter 2 of *The CRC Handbook of Laboratory Safety*, 4th ed. [Bibliography, C1].

Eye Injuries. If any chemical enters your eyes, *immediately* flush them with water from an eyewash fountain while holding your eyelids open. If you are wearing contact lenses, remove them first. Continue irrigation for at least 15 minutes (or until a nurse or a physician arrives), then have your eyes examined by a physician. If foreign bodies such as glass particles are propelled into your eye, seek immediate medical attention. Removal of such particles is a job for a specialist.

Chemical Burns. If a corrosive chemical is spilled on your skin or cloth-ing, remove any contaminated clothing and *immediately* flush the affected area with a large amount of water until the chemical is completely removed, using a safety shower if the area of injury is extensive or if it is inaccessible to washing from a tap. Speed and thoroughness in washing are the most important factors in reducing the extent of injury. Dry the area gently with a clean, soft towel. If you (or another victim) experience pain or if the skin is red or swollen, immerse the injured area in cold water or apply cold, wet dressings. Do not use neutralizing solutions, ointments, or greases on chemical burns unless they are specifically called for in a first aid procedure. Unless the skin is only reddened over a small area, a chemical burn should be examined by a nurse or physician. First aid procedures for burns caused by specific chemicals, such as bromine, are given in the *Sigma–Aldrich Library of Regulatory and Safety Data* [Bibliography, C4] and in the *First Aid Manual for Chemical Accidents* [Bibliography, C3].

If a chemical burn is very extensive or severe, the victim should lie down with the head and chest a little lower than the rest of the body. If the victim is conscious and is able to swallow, he or she should be provided with plenty of nonalcoholic liquid to drink (water, tea, coffee, etc.) until a physician or an ambulance arrives.

Thermal Burns. If you are burned by hot glass, another hot object, or a flame, try to determine the type and extent of the burn and then take the appropriate action as described next.

- *First-degree burn.* The skin is reddened but there are no blisters or broken skin. Immerse the affected area in clean, cold water or apply ice to reduce the pain and facilitate healing.
- *Second-degree burn.* Blisters are raised. Immerse the burned area in clean, cold water or apply ice, then cover the area with sterile gauze or another clean dressing. If legs or arms are burned, keep them elevated above the trunk of the body. Never puncture blisters raised by a second-degree burn.
- *Third-degree burn.* The skin is broken and underlying tissue is damaged. Place a thick sterile dressing or clean cloth over the affected area and have the burn examined by a nurse or physician as soon as possible. Do not remove burned clothing, immerse the burned area in cold water, or cover the burn with greasy ointment. If legs or arms are burned, keep them elevated above the trunk of the body.

In case of an extensive thermal burn, the burned area should be covered with the cleanest available cloth material and the victim should lie down, with the head and chest lower than the rest of the body, until a physician or an ambulance arrives. If the injured person is conscious and is able to swallow, he or she should be provided with plenty of nonalcoholic liquid to drink (water, tea, coffee, etc.).

Bleeding, Cuts, and Abrasions. In case of a cut or abrasion that does not involve heavy bleeding, cleanse the wound and the surrounding skin with soap and lukewarm water, applying it by wiping away from the wound. Try to remove imbedded glass shards, if there are any, by using tweezers if necessary. Hold a sterile gauze pad over the wound until bleeding stops. If the damage is confined to the skin, apply an antibacterial cream such as Neosporin to prevent infection. Place a fresh gauze pad over the wound and secure it loosely with a triangular or rolled bandage. Replace the pad and bandage as necessary with clean, dry ones. Avoid contact between the wound and the mouth, fingers, handkerchiefs, or other unsterile objects. If the wound is deep or extensive, it should be treated further by a nurse or physician.

In case of a major wound that involves heavy bleeding, *immediately* apply pressure directly over the wound with a cloth pad (such as a clean handkerchief or other clean cloth) or sterile dressing, pressing firmly with one or both hands to reduce the bleeding as much as possible. Then call for assistance. A compression bandage should be applied on top of the original dressing if profuse bleeding continues. (Removing the original dressing may delay clotting.) If no compression bandage is available, you can roll a clean cork in sterile gauze and bind it to apply pressure directly over the wound. The victim should lie down with the bleeding part higher than the heart, and the dressing should be held in place with heavy gauze or other cloth strips. A physician or an ambulance should be called as soon as possible, and the victim should be kept warm with a blanket or coat. If the injured person is conscious and is able to swallow, he or she should be provided with plenty of nonalcoholic liquid to drink (water, tea, coffee, etc.) until a physician arrives.

Poisoning. If, after contact with, inhalation of, or accidental ingestion of a chemical, you experience a burning sensation in the throat, discoloration of the lips or mouth, stomach cramps, nausea and vomiting, or confusion, *immediately* seek treatment for chemical poisoning.

If a poison has been *ingested* (swallowed), an ambulance and a poison control center should be called immediately. If a victim of poisoning is conscious, loosen tight clothing around the neck and waist, have the victim rinse his or her mouth several times with cold water and spit it out, and then give the victim one to two cups of water or milk to drink. Unless otherwise advised by a poison control center, induce vomiting by tickling the back of the throat or by giving two tablespoons (~30 mL) of ipecac syrup followed by a cup of water. Vomiting should not be induced if the victim is unconscious, is in convulsions, or has severe pain and burning sensations in the mouth or throat, or if the poison is a petroleum product or a strong acid or alkali. When vomiting begins, lower the victim's head to prevent vomit from reentering the mouth, and collect a sample of the vomit for possible analysis. If convulsions occur, remove any objects that might cause injury, or move the victim away from such objects. Watch out for and remove any obstructions in the victim's mouth (including dentures). If necessary, insert a soft pad between the victim's teeth to protect the tongue from being bitten. If the victim has stopped breathing, clear the airway and administer artificial respiration. When there is time, the poison should be identified (if possible) and an appropriate antidote should be given (see the *Merck Index*, 9th ed., pp. MISC-22ff., or call a poison control center). If the poison cannot be identified and a medical professional is not present, a heaping tablespoon (15 g) of a universal antidote can be given in a half glass of warm water. The universal antidote can be prepared by combining two parts of activated charcoal with one part of magnesium oxide and one part of tannic acid. If it is known, a sample of the poison should be saved for the physician.

If a poison has been *inhaled*, the victim must be taken to fresh air and a physician should be called immediately. Loosen any tight clothing around the neck and waist, and use a tongue depressor or other device, if necessary, to keep the victim's airway open. Administer manual (not mouth-to-mouth) artificial respiration if the victim has stopped breathing. Keep the victim warm and as quiet as possible until a physician arrives. If the poison is a highly toxic gas, such as hydrogen cyanide, hydrogen sulfide, or phosgene, the persons attempting to rescue the victim should wear self-contained respirators while they are in contact with the vapors.

In case of *skin contact* with a toxic substance, follow the same general procedure as that for chemical burns. The *Sigma–Aldrich Library of Regulatory and Safety Data* [Bibliography, C4] or another appropriate source should be consulted for specific procedures to be used for injury by certain substances.

Reacting to Accidents: Fire

Fires are unlikely in most modern organic chemical laboratories, which use flameless heat sources for nearly all operations. However, burners may be used for special applications, such as bending glass, and some chemicals may ignite on a hot surface, such as the top of a hot plate.

In case of fire, your first response should be to *get away* as quickly as possible and let the instructor deal with the fire. However, if a fire is small and confined to a container such as a flask or beaker, you may be able to put it out by placing a watch glass over the mouth of the container. If no instructor is in the laboratory, obtain a fire extinguisher of the appropriate type and attempt to put out the fire by aiming the extinguisher at the base of the fire, while maintaining a safe distance. Most labs should be equipped with class BC or ABC dry-chemical extinguishers, which are effective against solvent and electrical fires. A class D fire, which involves burning metals or metal hydrides such as sodium metal and lithium aluminum hydride, can be extinguished with an appropriate class D fire extinguisher or by smothering the fire in dry sand, sodium chloride, or sodium carbonate. If a fire is too large to be put out by a fire extinguisher, sound the nearest fire alarm and evacuate the area.

If your hair or clothing catches on fire, *do not panic. Walk* (don't run) directly to the nearest fire blanket or safety shower and attempt to extinguish the fire. Do not wrap yourself in a fire blanket while you are standing, as that may direct the flames around your face; instead, *drop* to the floor and *roll* as you wrap the blanket around your body. To use a safety shower, get directly under the shower head and pull the chain. If another person's hair or clothing has caught fire, try to prevent panic as you lead him or her to a safety shower or fire blanket.

Chemical Hazards

For your own health and safety, it is essential that you *exercise caution while handling chemicals* and *minimize your exposure to them*. Most academic chemistry departments have policies and procedures for dealing with hazardous chemicals, which should be incorporated into the institution's chemical hygiene plan. Your instructor will inform you of any departmental or institutional rules relating to the safe handling and disposal of hazardous chemicals. Some of the operations in this book describe specific chemical hazards and handling precautions under the heading "Safety Notes."

There are several different kinds of chemical hazards. A chemical may be toxic and therefore capable of causing illness or death when ingested, inhaled, or allowed to contact the skin. Certain chemicals can burn when exposed to a spark, an open flame, or a high temperature. A few chemicals can explode when heated, subjected to shock, or mixed with certain organic materials. Some chemicals may react spontaneously or when combined with other chemicals, generating heat that could cause a fire.

The severity of any hazards associated with a chemical is sometimes indicated by a labeling system, such as the one established by the National Fire Protection Association (NFPA). The NFPA system rates health, flammability, and reactivity hazards on a scale of 0 to 4, where 0 refers to the absence of hazard and 4 refers to the most severe potential hazard associated with a chemical. For example, exposure to a material with a health hazard rating of 1 could cause irritation but only minor injury, even with no medical treatment, while brief exposure to a material with a health hazard rating of 4 could cause death or major injury, even with prompt treatment. These hazard numbers are typically displayed on three quadrants of a diamond-shaped diagram. The quadrants are color-coded, with red referring to a fire hazard, blue to a health hazard, and yellow to a chemical reactivity hazard. For example, the NFPA symbol for diethyl ether shows that it is extremely flammable (4) but presents only a moderate health risk (2) and is relatively unreactive (1).

A chemical with a high *fire* hazard number should obviously be kept away from ignition sources, including flames, sparks, and hot surfaces. For example, diethyl ether will ignite if it is spilled on a hot plate at a temperature of 160°C or higher.

A chemical with a high *reactivity* hazard number should be handled with great care, following any precautions described in the experiment or in another source of safety information, such as the Material Safety Data Sheet (MSDS, described below) for the chemical. Such chemicals may ignite or explode if subjected to shock or brought in contact with metal or other materials that may catalyze their decomposition.

The value of a *health* hazard number may suggest the appropriate response to inhalation of or contact with a chemical. If you inhale significant amounts of a chemical's vapor or the dust of a solid chemical, you should go to a window or other area where you can breathe fresh air, unless the chemical's health hazard number is 0. In case of skin contact with a chemical having a health hazard number other than 0, you should wash the area of contact with soap and water. If you get a chemical in your eyes, follow the procedures described under the heading "Eye Injuries" in the previous section on first aid. If a chemical has a high health hazard number, or if inhalation or contact is prolonged, see your instructor to determine whether treatment is required.

The fourth quadrant of an NFPA diamond may contain a letter W with a line through it for chemicals that react with water, a letter P for chemicals that may polymerize spontaneously, or the letters OXY to indicate an oxidant.

Water-sensitive chemicals react violently and exothermically upon contact with water, often generating toxic fumes. Some water-sensitive chemicals generate toxic fumes even when exposed to moist air. Such chemicals should be kept in tightly closed containers that are opened only for transfers and closed immediately afterward. They should be used under fume hoods at some distance from any source of water.

red
blue ⟨ 4 2 ✕ 1 ⟩ yellow
diethyl ether

Chemicals that polymerize readily may sometimes undergo spontaneous, rapid polymerization, which generates heat. If this reaction occurs in a closed container such as a capped bottle, it could cause the container to shatter violently. Most chemicals of this type are stabilized by the addition of an antioxidant or other stabilizer, so they are unlikely to react unless the stabilizer has been removed.

Chemicals designated as strong oxidants must not be allowed to contact other chemicals, except for those specified in an experimental procedure. In fact, you should never mix *any* chemicals together unless directed to, since mixing incompatible chemicals may result in the generation of toxic gases, fire, or an explosion.

Carcinogens. Some chemicals used in a typical organic chemistry lab are suspected *carcinogens*—agents suspected of causing cancer. With the exception of a few chemicals seldom encountered in an undergraduate lab, such chemicals present little risk of cancer to students if used as directed. The carcinogenic activity of a compound is generally established by animal tests in which high doses of the chemical are administered by various routes for prolonged periods. For example, phenacetin taken orally has been found to cause cancer in laboratory animals, but it is highly unlikely that anyone will ingest (swallow) enough phenacetin during a laboratory experiment to incur a risk of cancer. Chromium(VI) compounds have been shown to cause cancer of the lungs, nasal cavity, and sinuses in humans, but most of the persons at risk are industrial workers who have been continuously exposed to the dust of chromium compounds in the workplace, not students using chromic acid solutions a drop or so at a time.

It is only prudent to take appropriate precautions when handling any potential carcinogen, such as wearing protective gloves and clothing and working under a hood. Some halogenated hydrocarbons are suspected of causing cancer if inhaled, so avoid breathing the vapors of dichloromethane, chloroform, and other chlorinated solvents.

Teratogens. *If you are pregnant or think you may be pregnant, inform your instructor before you enter the organic chemistry lab for the first time.* Some organic chemicals are known or suspected *teratogens*, meaning that they may harm a developing fetus. As is the case for carcinogens, handling a chemical designated as a teratogen does not necessarily represent a danger to the fetus. Ethanol (ethyl alcohol) is a teratogen when ingested, but people are far more likely to ingest ethanol in a bar or restaurant, or at home, than in a chemistry laboratory. Nevertheless, some chemicals that are routinely used in organic chemistry laboratories may represent a significant danger to a developing fetus. Therefore, it is important to contact your instructor, the laboratory coordinator, or a designated chemical hygiene officer to discuss your options. The best option may be to take the laboratory course after your child is delivered. If you prefer to remain in the lab course, you should obtain the consent of your physician and then make arrangements with your instructor to minimize your exposure to potential teratogens. Such arrangements might include substituting less hazardous chemicals for teratogenic ones, performing alternative experiments that do not require the use of teratogens, or using gloves, protective clothing, and a hood when handling any teratogenic chemical.

Finding and Using Chemical Safety Information

Some basic hazard information on a number of organic compounds can be found in *The Merck Index* [Bibliography, A11]. More detailed information can be found in the *Sigma–Aldrich Library of Regulatory and Safety Data* [Bibliography, C4], which compiles safety and regulatory information for more than 20,000 chemicals. Other useful sources of chemical hazard information include *Sax's Dangerous Properties of Industrial Materials* [Bibliography, C5], *Hazards in the Chemical Laboratory* [Bibliography, C6], and *Bretherick's Handbook of Reactive Chemical Hazards* [Bibliography, C8]. You can find hazard information online using the the NIOSH web site at www.cdc.gov/niosh/npg.

The labels on the containers in which chemicals are originally received must include appropriate hazard warnings and may provide additional hazard information along with handling precautions and emergency management procedures. To reduce waste and prevent contamination, chemicals are ordinarily transferred to stock bottles before being used in the chemistry laboratory, but some hazard warnings may be included on the stock bottle labels.

Most organic chemistry lab texts include hazard information and safe-handling procedures for hazardous chemicals used in their experiments. For example, the author's textbook *Microscale Operational Organic Chemistry* contains the following information about sodium borohydride, a reducing agent used in many chemical syntheses.

> Sodium borohydride is corrosive and can react violently with concentrated acids, oxidizing agents, and other chemicals. Aqueous sodium borohydride solutions with pH values below 10.5 have been known to decompose violently, so be sure that your reaction mixture is sufficiently alkaline. Avoid contact with $NaBH_4$, do not breathe its dust, and keep it away from other chemicals.

The most complete source of information about the hazards associated with any chemical is its *Material Safety Data Sheet (MSDS)*, which is provided by the manufacturer or vendor of the chemical and is usually available on the manufacturer's Web site. The MSDSs for chemicals used in your organic chemistry laboratory should be available from the chemistry department office, the chemical hygiene officer, or some other designated source at your institution. The MSDS for a chemical must include the following information:

- Identity of the chemical
- Physical properties and relevant chemical characteristics
- Information on potential physical hazards such as fire, explosion, reactivity, and chemical incompatibility
- Health hazards from both short- and long-term exposure
- Toxicology data such as LD_{50} values
- Significant exposure routes such as inhalation, ingestion, or contact
- Exposure limits set by OSHA and other agencies
- Cancer-causing potential
- Precautions for safe handling and storage
- Control measures for preventing accidents, such as the use of personal protection gear
- Emergency procedures in case of accidental exposure

The Sigma-Aldrich Company's MSDS for benzoic acid is shown in Figure 1, which follows.

SIGMA-ALDRICH

MATERIAL SAFETY DATA SHEET

Date Printed: 11/20/2007
Date Updated: 01/31/2006
Version 1.12

Section 1 - Product and Company Information

Product Name	BENZOIC ACID, >=99.5%, A.C.S. REAGENT
Product Number	242381
Brand	SIAL
Company	Sigma-Aldrich
Address	3050 Spruce Street
	SAINT LOUIS MO 63103 US
Technical Phone:	800-325-5832
Fax:	800-325-5052
Emergency Phone:	314-776-6555

Section 2 - Composition/Information on Ingredient

Substance Name	CAS #	SARA 313
BENZOIC ACID	65-85-0	No

Formula	C7H6O2
Synonyms	Acide benzoique (French) * Acido benzoico (Italian) * Benzenecarboxylic acid * Benzeneformic acid * Benzenemethanoic acid * Benzoate * Benzoesaeure (German) * Carboxybenzene * Dracylic acid * E 210 * HA 1 (acid) * Kyselina benzoova (Czech) * Phenylcarboxylic acid * Phenylformic acid * Retarder BA * Retardex * Salvo liquid * Salvo powder * Tenn-Plas
RTECS Number:	DG0875000

Section 3 - Hazards Identification

EMERGENCY OVERVIEW
  Harmful.
  Harmful if swallowed. Irritating to eyes, respiratory system and skin.

HMIS RATING
  HEALTH: 2
  FLAMMABILITY: 1
  REACTIVITY: 0

NFPA RATING
  HEALTH: 2
  FLAMMABILITY: 1
  REACTIVITY: 0

For additional information on toxicity, please refer to Section 11.

Section 4 - First Aid Measures

ORAL EXPOSURE
If swallowed, wash out mouth with water provided person is conscious. Call a physician.

INHALATION EXPOSURE
If inhaled, remove to fresh air. If not breathing give artificial respiration. If breathing is difficult, give oxygen.

DERMAL EXPOSURE
In case of skin contact, flush with copious amounts of water for at least 15 minutes. Remove contaminated clothing and shoes. Call a physician.

EYE EXPOSURE
In case of contact with eyes, flush with copious amounts of water for at least 15 minutes. Assure adequate flushing by separating the eyelids with fingers. Call a physician.

Section 5 - Fire Fighting Measures

FLASH POINT
250 °F 121 °C Method: closed cup

AUTOIGNITION TEMP
572 °C

FLAMMABILITY
N/A

EXTINGUISHING MEDIA
Suitable: Water spray. Carbon dioxide, dry chemical powder, or appropriate foam.

FIREFIGHTING
Protective Equipment: Wear self-contained breathing apparatus and protective clothing to prevent contact with skin and eyes.
Specific Hazard(s): Emits toxic fumes under fire conditions.

Section 6 - Accidental Release Measures

PROCEDURE TO BE FOLLOWED IN CASE OF LEAK OR SPILL
Evacuate area.

PROCEDURE(S) OF PERSONAL PRECAUTION(S)
Wear respirator, chemical safety goggles, rubber boots, and heavy rubber gloves.

METHODS FOR CLEANING UP
Sweep up, place in a bag and hold for waste disposal. Avoid raising dust. Ventilate area and wash spill site after material pickup is complete.

Section 7 - Handling and Storage

HANDLING
User Exposure: Do not breathe dust. Avoid contact with eyes, skin, and clothing. Avoid prolonged or repeated exposure.

STORAGE
Suitable: Keep tightly closed.

Section 8 - Exposure Controls / PPE

ENGINEERING CONTROLS
Safety shower and eye bath. Mechanical exhaust required.

PERSONAL PROTECTIVE EQUIPMENT
Respiratory: Use respirators and components tested and approved under appropriate government standards such as NIOSH (US) or CEN (EU). Where risk assessment shows air-purifying respirators are appropriate use a dust mask type N95 (US) or type P1 (EN 143) respirator.
Hand: Compatible chemical-resistant gloves.
Eye: Chemical safety goggles.

GENERAL HYGIENE MEASURES
Wash thoroughly after handling.

EXPOSURE LIMITS

Country	Source	Type	Value
USA	ACGIH	TLV	$10 \ mg/m^3$
Remarks: inhalable particulate			
USA	ACGIH	TLV	$3 \ mg/m^3$
Remarks: respirable dust in air			
USA	OSHA.	PEL	$15 \ mg/m^3$
Remarks: total dust			
USA	OSHA.	PEL	$5 \ mg/m^3$
Remarks: respirable dust			

Section 9 - Physical/Chemical Properties

Appearance	Physical State: Solid
	Color: White
	Form: Fine crystals

Property	Value	At Temperature or Pressure
Molecular Weight	122.12 AMU	
pH	N/A	
BP/BP Range	248.9 °C	760 mmHg
MP/MP Range	121.0 - 125.0 °C	
Freezing Point	N/A	
Vapor Pressure	10 mmHg	132 °C
Vapor Density	4.21 g/l	
Saturated Vapor Conc.	N/A	
SG/Density	1.32 g/cm3	
Bulk Density	N/A	
Odor Threshold	N/A	
Volatile%	N/A	
VOC Content	N/A	
Water Content	N/A	
Solvent Content	N/A	
Evaporation Rate	N/A	
Viscosity	N/A	
Surface Tension	N/A	
Partition Coefficient	N/A	
Decomposition Temp.	N/A	

Flash Point	250 °F 1 21 °C	Method: closed cup
Explosion Limits	N/A	
Flammability	N/A	
Autoignition Temp	572 °C	
Refractive Index	N/A	
Optical Rotation	N/A	
Miscellaneous Data	N/A	
Solubility	N/A	

N/A = not available

Section 10 - Stability and Reactivity

STABILITY
Stable: Stable.
Materials to Avoid: Strong oxidizing agents, Strong bases, Strong reducing agents.

HAZARDOUS DECOMPOSITION PRODUCTS
Hazardous Decomposition Products: Carbon monoxide, Carbon dioxide.

HAZARDOUS POLYMERIZATION
Hazardous Polymerization: Will not occur

Section 11 - Toxicological Information

ROUTE OF EXPOSURE
Skin Contact: May cause skin irritation.
Skin Absorption: May be harmful if absorbed through the skin.
Eye Contact: Causes eye irritation.
Inhalation: Material may be irritating to mucous membranes and upper respiratory tract. May be harmful if inhaled.
Ingestion: Harmful if swallowed.

SIGNS AND SYMPTOMS OF EXPOSURE
To the best of our knowledge, the chemical, physical, and toxicological properties have not been thoroughly investigated.

TOXICITY DATA
Oral
Man
500 mg/kg
LDLO

Oral
Rat
1700 mg/kg
LD50

Inhalation
Rat
> 26 mg/m3
LC50
Remarks: Sense Organs and Special Senses (Nose, Eye, Ear, and Taste) : Eye: Lacrimation. Behavioral:Somnolence (general depressed activity).

Intraperitoneal
Rat

1600 MG/KG
LD50

Intravenous
Rat
1700 MG/KG
LD50

Oral
Mouse
1940 mg/kg
LD50

Remarks: Behavioral:Somnolence (general depressed activity).
Lungs, Thorax, or Respiration:Respiratory depression.
Gastrointestinal:Other changes.

Intraperitoneal
Mouse
1460 MG/KG
LD50

Skin
Rabbit
> 10000 mg/kg
LD50

IRRITATION DATA
Eyes
Rabbit
Remarks: Mild irritation effect

Skin
Human
22 mg
3D
I
Remarks: Moderate irritation effect

Skin
Rabbit
500 mg
24H
Remarks: Mild irritation effect

Eyes
Rabbit
100 mg
Remarks: Severe irritation effect

CHRONIC EXPOSURE - MUTAGEN
Species: Human
Dose: 5 MMOL/L
Cell Type: lymphocyte
Mutation test: DNA inhibition

Section 12 - Ecological Information

No data available.

Section 13 - Disposal Considerations

APPROPRIATE METHOD OF DISPOSAL OF SUBSTANCE OR PREPA-
RATION

Contact a licensed professional waste disposal service to dispose of this material.
Dissolve or mix the material with a combustible solvent and burn in a chemical
incinerator equipped with an afterburner and scrubber. Observe all federal, state,
and local environmental regulations.

Section 14 - Transport Information

DOT

Proper Shipping Name: Environmentally hazardous substances, solid, n.o.s.
UN#: 3077
Class: 9
Packing Group: Packing Group III
Hazard Label: Class 9
PIH: Not PIH

IATA

Non-Hazardous for Air Transport: Non-hazardous for air transport.

Section 15 - Regulatory Information

EU ADDITIONAL CLASSIFICATION

Symbol of Danger: Xn
Indication of Danger: Harmful.
R: 22-36
Risk Statements: Harmful if swallowed. Irritating to eyes.
S: 26
Safety Statements: In case of contact with eyes, rinse immediately with plenty of
water and seek medical advice.

US CLASSIFICATION AND LABEL TEXT

Indication of Danger: Harmful.
Risk Statements: Harmful if swallowed. Irritating to eyes, respiratory system and
skin.
Safety Statements: In case of contact with eyes, rinse immediately with plenty of
water and seek medical advice.

UNITED STATES REGULATORY INFORMATION

SARA LISTED: No
TSCA INVENTORY ITEM: Yes

CANADA REGULATORY INFORMATION

WHMIS Classification: This product has been classified in accordance with the
hazard criteria of the CPR, and the MSDS contains all the information required
by the CPR.
DSL: Yes
NDSL: No

Section 16 - Other Information

DISCLAIMER

For R&D use only. Not for drug, household or other uses.

WARRANTY

The above information is believed to be correct but does not purport to be all in-clusive and shall be used only as a guide. The information in this document is based on the present state of our knowledge and is applicable to the product with regard to appropriate safety precautions. It does not represent any guarantee of the properties of the product. Sigma-Aldrich Inc., shall not be held liable for any damage resulting from handling or from contact with the above product. See re-verse side of invoice or packing slip for additional terms and conditions of sale. Copyright 2007 Sigma-Aldrich Co. License granted to make unlimited paper copies for internal use only.

Figure 1   Material Safety Data Sheet for acetic acid. (Reprinted with permission from Aldrich Chemical Co., Inc., Milwaukee, WI)

Chemical manufacturers, for reasons related to legal liability, tend to list every possible mishap that might result from the use of their chemicals, no matter how unlikely. Because MSDSs are intended primarily for chemi-cal and industrial workers, whose exposure to chemicals may be intense and prolonged, they often recommend protective apparatus and measures that may not be necessary when chemicals are used in small quantities for short periods of time. This makes it difficult to determine which hazards associat-ed with the chemicals used in a given organic chemistry experiment are truly significant. Nevertheless, the MSDS for a given chemical can be very useful if the information it provides is used appropriately.

Chemistry and the Environment

Disposal of Hazardous Wastes

Some of the chemicals and solutions generated in organic chemistry labs are regarded as hazardous wastes, which are regulated by various state and federal agencies. Institutions that generate hazardous wastes are required to collect and dispose of the wastes in ways that pose minimal potential harm to human health and to the environment. Some hazardous wastes can be purified and reused, or converted to nonhazardous materials. Other hazardous wastes are placed in landfills, incinerated, or otherwise disposed of by commercial waste disposal firms.

Except when precluded by local regulations, moderate quantities of many common chemicals can be safely and acceptably disposed of down a laboratory drain, if certain procedures are followed. Small quantities (less than 100 g) of most water-soluble organic compounds, including water-soluble alcohols, aldehydes, amides, amines, carboxylic acids, esters, ethers, and ketones, can be disposed of in this way. The organic compound should be mixed or flushed with at least 100 volumes of excess water. Dilute aqueous solutions of inorganic compounds can be disposed of down the drain if they contain cations and anions from the following list and no other cations or anions.

Cations of these metals: Al, Ca, Cu, Fe, Li, Mg, Mo, Pd, K, Na, Sn, Zn

Anions: HSO_3^-, BO_3^{3-}, $B_4O_7^{2-}$, Br^-, CO_3^{2-}, Cl^-, OCN^-,

OH^-, I^-, O^{2-}, PO_4^{3-}, SO_4^{2-}, SO_3^{2-}

Some less common cations and anions, such as Cs^+ and SCN^-, are also permissible.

Substances that should *not* be poured down the drain include the following:

- Water-soluble organic compounds that are highly flammable or boil below 50°C
- Hydrocarbons, halogenated hydrocarbons, and other water-insoluble organic compounds
- Flammable or explosive solids, liquids, or gases
- Phenols and other taste- or odor-producing substances
- Wastes containing poisons in toxic concentrations
- Corrosive wastes capable of damaging the sewer system

These and any other chemical wastes designated by your instructor should be placed in specially labeled waste containers, which should be provided in the laboratory.

Green Chemistry

While chemistry has enriched our lives by providing a vast number of useful consumer products, some chemicals can create havoc when released into the environment. In recent years many kinds of chemical pollution have made the headlines: the chlorofluorocarbons that endanger Earth's protective ozone shield, the methyl isocyanate gas that caused 3800 deaths in India in 1984, the PCBs and mercury compounds that make some fish dangerous to eat, among others. So it is important that organizations and individuals that produce and use chemicals, whether they are manufacturers such as Dow Chemical or organic chemistry students, consider the environmental consequences of their actions.

The US Environmental Protection Agency coined the term *Green Chemistry* in the early 1990s to help encourage new technologies that could reduce or eliminate the use and generation of hazardous substances in chemical processes. More generally, Green Chemistry simply means doing chemistry in an environmentally responsible manner. As Kermit the Frog has complained, "It's not easy being green," but some green chemical processes are simpler and less costly than the ones they replace. And being green always pays off by helping to insure cleaner air, cleaner water, and a safer environment in general.

The Twelve Principles of Green Chemistry. P. T. Anastas and J. C. Warner formulated twelve principles of Green Chemistry that appeared in their book *Green Chemistry: Theory and Practice* [Bibliography, L4]. Those principles are paraphrased and illustrated here.

1. *It is easier to prevent waste than to dispose of it afterward.* Waste materials must be cleaned up or treated, which costs money and creates a need for more landfills. Even after disposal, the waste material usually leads to environmental contamination.

2. *A chemical synthesis should incorporate as many reactant atoms as possible into the final product.* The *atom economy* of a reaction is the percentage of the total molar mass of the starting materials that would appear in a desired product if the yield were 100%.

$$\text{atom economy} = \frac{\text{molar mass of desired product}}{\text{sum of molar masses of all reactants}}$$

For example, a new process for manufacturing the pain reliever ibuprofen has an atom economy of 0.77, meaning that 77% of the mass of the reactants should end up in the ibuprofen, assuming that all of the reactant molecules are converted to product molecules. It replaces an older 6-step process that had an atom economy of only 0.40, which means that 60% of the reactants' atoms are — in effect — wasted. The newer process not only reduces the amount of waste that has to be disposed of; it also decreases the cost of the process by reducing the amount of reactants needed.

The *reaction efficiency* of a chemical reaction is a good measure of its "greenness."

$$\text{reaction efficiency} = \text{atom economy} \times \text{percent yield}$$

For example, the following synthesis of ethyl acetate has an atom economy of 0.83.

$$\underset{\text{MW} = 60}{CH_3\overset{\displaystyle O}{\overset{\displaystyle \|}{C}}OH} + \underset{\text{MW} = 46}{CH_3CH_2OH} \underset{\text{catalyst}}{\rightleftharpoons} \underset{\text{MW} = 88}{CH_3\overset{\displaystyle O}{\overset{\displaystyle \|}{C}}OCH_2CH_3} + \underset{\text{MW} = 18}{H_2O}$$

$$\text{atom economy} = \frac{88}{(60 + 46)} = 0.83$$

But if the yield is 65%, the reaction efficiency is only 54% (0.83 × 65%). So chemical processes need to be designed to produce high yields as well as high atom economies.

3. *A chemical synthesis should use and generate the least toxic substances possible.* For example, adipic acid is an important chemical used for manufacturing polyurethane, nylon, and other commercial products. It is usually made by oxidizing cyclohexane, which is made from benzene, a toxic and carcinogenic product of petroleum refining.

benzene   cyclohexane   adipic acid

This process won the Presidential Green Chemistry Challenge Award, described below, in 1998.

A newer process, developed by Karen M. Draths and John W. Frost of Michigan State University, uses the simple sugar glucose as the starting material.

glucose   cis,cis-muconic acid   adipic acid

Glucose is nontoxic and can be derived from cellulose and other plant materials.

4. *The toxicity of chemical products should be minimized.* Sometimes a relatively nontoxic product can fulfill the same function as a more toxic one. For example, methylene chloride (dichloromethane) is a toxic and possibly carcinogenic liquid that softens or dissolves many kinds of plastic and has therefore been used to weld plastic parts together. Recently a California company cooperated with the International Reciprocal Trade Association (IRTA) in seeking a safer alternative to methylene chloride and discovered that acetone, a much less toxic solvent, worked just as well. So reducing the toxicity of chemical products requires not only actions taken by chemical manufacturers to modify their product line, but also actions taken by end users that will reduce the demand for the more toxic chemicals.

5. *The use of solvents, separation agents, and other auxiliary substances in a synthesis should be minimized, and any such substances used should be safe.* Many of the solvents used in chemical processes,

such as benzene and chloroform, present serious health risks. Whenever possible they should be eliminated entirely or replaced by safer solvents such as water, liquid carbon dioxide, or ionic liquids. Ionic liquids are low-melting ionic organic compounds that usually have long names, such as 1-ethyl-3-methylimidazolium bis(pentafluoroethylsulfonyl)imide. They make excellent reaction solvents because they are nonflammable, very stable, and can dissolve a wide range of both organic and inorganic compounds. Their main liability is their high cost. For example, the Aldrich Chemical Company sells 50 g of 1-butyl-4-methylpyridinium tetrafluoroborate for about $200, so high-volume industrial use of these solvents is unlikely unless increased demand brings the prices down.

6. *The energy requirements of chemical processes should be minimized.* Using less energy reduces both the environmental and economic cost of manufacturing chemicals. Energy requirements can be lowered by, for example, designing reactions so that they can be carried out at ambient temperatures, eliminating the energy cost of heating the reactants. Using catalysts and microwave generators can reduce energy requirements, not only by lowering reaction temperatures, but also by speeding up reactions so that energy input is needed for shorter time periods.

7. *Renewable resources should be used to produce starting materials.* Many chemicals can be derived from plants and other renewable sources rather than from nonrenewable resources such as petroleum, coal, and natural gas. For example, ethanol is a common alcohol that is used in many chemical processes. Most synthetic ethanol is made from ethylene, which is derived from petroleum, but ethanol can also be produced economically by fermentation of corn plants and other agricultural products.

8. *Unnecessary derivitization should be avoided.* Preparing derivatives of reactants during a chemical process requires additional chemicals and generates more waste.

9. *Using catalysts is better than using reagents in stoichiometric amounts.* Catalysts are used in relatively small quantities, and they are reusable because they accelerate reactions without being consumed. Catalysts should be as selective as possible, so that they promote the formation of the desired product without also promoting the formation of by-products.

10. *Chemical products should be designed to degrade when released into the environment after use.* Products that are biodegradable reduce the need for landfills and are less likely to build up to hazardous levels in the environment. For example, packing "peanuts" made of starch—unlike Styrofoam packing materials—dissolve in water and are easily digested by organisms after they are discarded.

11. *Chemical processes should be monitored in real time to reduce or prevent the formation of hazardous substances.* This requires the development of analytical procedures designed to measure the composition of reactants and products throughout a synthetic process, and not just to analyze the product after the process is completed. It also requires

that there be some way to stop or slow down a process when problems arise.

12. *The substances used in a chemical process should be chosen to minimize the potential for accidents.* Using less hazardous materials reduces the danger to workers and helps prevent release of hazardous chemicals into the environment in case of an accident.

Applying Green Chemistry in the Organic Chemistry Laboratory. Although Green Chemistry is usually thought to apply mainly to industrial processes, many of its principles can be applied in an academic laboratory setting as well. Here are some ways that you or your instructor might make your organic chemistry lab course greener.

Use smaller amounts of chemicals. Many organic chemistry labs provide special microscale kits that make it possible to synthesize compounds using only fractions of a gram of starting material rather than several grams. Even when standard scale lab equipment is used, reactions can often be successfully scaled down by reducing the amounts of all reactants, solvents, and other chemicals proportionately (See Appendix V).

Replace toxic reagents by less toxic ones. In the past, most oxidations of secondary alcohols to ketones in academic chemistry labs were carried out using chromium(VI) oxide (CrO_3), a toxic, cancer-causing chemical that is a serious environmental contaminant. Now most reactions of this sort are carried out with ordinary laundry bleach, a dilute solution of sodium hypochlorite in water.

Use reactions with a high atom economy. Consider two possible laboratory syntheses of methyl *t*-butyl ether.

$$
\mathbf{1} \quad
\underset{\substack{| \\ CH_3}}{\overset{\substack{CH_3 \\ |}}{CH_3CONa}} + CH_3I \longrightarrow
\underset{\substack{| \\ CH_3}}{\overset{\substack{CH_3 \\ |}}{CH_3COCH_3}} + NaI
$$

MW = 96 MW = 142   MW = 88   MW = 150

$$
\mathbf{2} \quad
\underset{\substack{| \\ CH_3}}{\overset{\substack{CH_2 \\ \|}}{CH_3C}} + CH_3OH \xrightarrow{H_2SO_4}
\underset{\substack{| \\ CH_3}}{\overset{\substack{CH_3 \\ |}}{CH_3COCH_3}}
$$

MW = 56 MW = 32   MW = 88

Reaction **1**, which involves the well-known Williamson synthesis of ethers, has an atom economy of only 0.37, while the reaction **2**, an alkene addition reaction, has an atom economy of 1.00, or 100%!

$$\text{Reaction 1: atom economy} = \frac{88}{(96 + 142)} = 0.37$$

$$\text{Reaction 2: atom economy} = \frac{88}{(56 + 32)} = 1.00$$

All addition reactions have atom economies of 100%, because all of the reactant atoms are incorporated into the product. Substitution reactions, such as reaction **1**, tend to have relatively low atom economies.

Reduce the use of organic solvents. Water is a much less hazardous reaction solvent than common organic solvents such as dichloromethane or diethyl ether, but many organic reactants don't dissolve in it. This limitation can sometimes be overcome by using a special catalyst (called a phase-transfer catalyst) to facilitate the transfer of chemicals across the boundary between an aqueous solution and an organic substance, or simply by stirring vigorously to increase the surface area of a water-insoluble organic reactant that is in contact with an aqueous solution. See Experiment 24 for a discussion of phase-transfer catalysis.

Organic solvents are often used to extract (see OP-18) organic products from aqueous reaction mixtures, even if the product is relatively insoluble in water. By "salting out" (see OP-18b) such a product with sodium chloride or another salt so that it forms a distinct layer, it may be possible to separate the product directly from the aqueous layer in good yield, eliminating the need for an extraction solvent.

Use renewable resources. Some reactions can be carried out using chemicals derived from plants rather than chemicals synthesized from petroleum products and other nonrenewable resources. For example, aspirin is synthesized commercially from benzene, but in a chemistry lab it can be synthesized from methyl salicylate, the major ingredient in wintergreen oil. Unfortunately most methyl salicylate is also synthesized from non-renewable petroleum products, but wintergreen oil from natural sources such as sweet birch trees can be obtained from various suppliers.

Reduce the energy requirements of a reaction. Some reactions can be carried out in the chemistry lab at a lower temperature, or even at ambient temperature, by using a more reactive reagent or an appropriate catalyst. Other reactants can be carried out in a microwave oven rather than by direct heating. Microwave radiation is an efficient heat source because it can be tuned to heat up some substances (especially water) and not others.

Minimize waste, especially hazardous waste. One way of reducing waste is to use only what you need. Measure out the stoichiometric quantity of a reactant and no more. Don't measure out 10 mL of a solvent if you only need to use 8 mL, because the excess shouldn't be returned to the stock bottle and will end up as waste. Just being careful to avoid unnecessary losses helps to reduce waste, and it may improve your lab grade by enabling you to report higher product yields.

Hazardous wastes can sometimes be converted to less hazardous substances in the laboratory. For example, acidic wastes can be converted to less hazardous salts by neutralizing them with sodium carbonate. Your instructor may provide directions for treating certain hazardous wastes before they are transferred to waste containers.

Laboratory Operations

This section describes all of the operations you should need to know in order to successfully complete an organic chemistry laboratory course or carry out a research project involving organic synthesis or analysis. Each operation is designated by a number, which is listed in large type on the top right-hand corner of every odd-numbered page and can be located quickly by thumbing through the pages. The operations are separated into the following categories:

A. Basic Operations
B. Operations for Conducting Chemical Reactions
C. Separation Operations
D. Washing and Drying Operations
E. Purification Operations
F. Measuring Physical Constants
G. Instrumental Analysis

There is some overlap between the categories. For example, gas chromatography is essentially a separation method because a gas chromatograph separates the components of a sample, but this operation is used primarily for qualitative and quantitative analysis of samples, so it is placed in Section G rather than Section C.

Although the theoretical background and basic methodology for an operation are the same regardless of the scale on which the operation is performed, the equipment and procedures may be different for standard scale and microscale work. Therefore, always read the introductory material for each operation and any other parts that are not specifically designated for standard scale or microscale methods. Then read the appropriate section(s) designated by ⬙ if you will be using standard scale equipment or by ⬙ if you will be using microscale equipment. In cases for which the same methods or equipment can be used for both standard scale and microscale work, the section(s) will be designated by ⬙⬙.

A. Basic Operations

In this section you will learn about some of the basic lab operations required for doing experiments in organic chemistry. Although you may already have used some of them in a general chemistry course, you should still read the descriptions carefully because an operation may require different equipment or be performed in a different way in the organic chemistry lab. For example, many volume measurements for microscale organic chemistry are carried out using Pasteur pipets or syringes rather than graduated cylinders.

The success of a reaction may depend on how clean and moisture-free your equipment is, so you must learn how to clean and dry glassware thoroughly, as described in Operation 1 (OP-1). Most experiments in organic chemistry are conducted using specialized glassware components that are held together by ground joints or other connectors. Operation 2 tells you how to assemble and disassemble ground-joint components. For some experiments you may have to cut, bend, seal, or fire-polish glass tubing or glass rod, as described in OP-3. Nearly all experiments require that you accurately weigh a specified quantity of a limiting reactant or another chemical. Weighings can be done easily using an electronic balance, as described in OP-4. Some of the many kinds of apparatus for measuring volume are illustrated and their operation is described in OP-5. Operation 6 describes several methods for transferring solids and liquids from one container to another. The method used may depend on the amount of material being transferred.

OPERATION 1   Cleaning and Drying Glassware

Cleaning Glassware

Clean glassware is essential for good results in the organic chemistry laboratory. Even small amounts of impurities can sometimes inhibit chemical reactions, catalyze undesirable side reactions, and invalidate the results of chemical tests or rate studies. Always clean your dirty glassware at the end of each laboratory period, or as soon as possible after the glassware is used. This way, your glassware will be clean and dry for the next experiment, and you will be ready to start work when you arrive. If you wait too long to clean glassware, residues may harden and become more resistant to cleaning agents; they may also attack the glass itself, weakening it and making future cleaning more difficult. It is particularly important to wash out strong bases such as sodium hydroxide promptly, because they can etch the glass permanently and cause glass joints to "freeze" tight. When glassware has been thoroughly cleaned, water applied to its inner surface should wet the whole surface and not form droplets or leave dry patches. However, used glassware that has been scratched or etched may not wet evenly.

You can clean most glassware adequately by vigorous scrubbing with hot water and a laboratory detergent such as Alconox, using a brush of appropriate size and shape to reach otherwise inaccessible spots. A plastic trough or another suitable container can serve as a dishpan. A tapered centrifuge-tube brush can be used to clean conical vials as well as centrifuge tubes. A nylon mesh scrubber is useful for cleaning spatulas, stirring rods, beakers, and the outer surfaces of other glassware. Pipe cleaners or cotton swabs can be used to clean narrow funnel stems, eyedroppers, Hickman still side ports, and so on.

Organic residues that cannot be removed by detergent and water will often dissolve in organic solvents such as technical-grade acetone. (Never use reagent-grade solvents for washing.) For example, it is difficult if not impossible to scrub the inside of a porcelain Buchner or Hirsch funnel, but squirting a little acetone around the inside of the funnel stem and letting it drain through the porous plate should remove chemical residues that may have lodged there. Use organic solvents sparingly and recycle them after use, as they are much more costly than water. Be certain that acetone is completely removed from glassware before you return it to a lab kit, because it will dissolve a foam lab-kit liner.

After washing, always rinse glassware thoroughly with water (a final distilled-water rinse is a good idea) and check it to see if the water wets its surface evenly rather than forming separate beads of water. If it does not pass this test, scrub it some more or use a cleaning solution such as Nochromix. Note that some well-used glassware may not pass the test because of surface damage, but it may still be clean enough to use after thorough scrubbing.

Drying Glassware

Always dry glassware thoroughly if it will be used with organic reactants and solvents under nonaqueous conditions. Don't waste time drying wet glassware if it will come into contact with water or an aqueous solution during an experiment. Just let it drain for a few minutes before you use it.

The easiest (and cheapest) way to dry glassware is to let it stand overnight in a position that allows easy drainage. You can dry the outer surfaces of glassware with a soft cloth, but don't use a cloth to dry any surfaces that will be in contact with chemicals, because of the likelihood of contamination. If a piece of glassware is needed shortly after washing, drain it briefly to remove excess water. Then rinse it with one or two small portions of wash acetone and dry it in a stream of clean, dry air or put it in a drying oven for a few minutes. Compressed air from an air line may contain pump oil, moisture, and dirt, so do not use it directly from the line for drying. Air can be cleaned and dried as described in OP-27.

Glassware that is to be used for a moisture-sensitive reaction (such as a Grignard reaction) must be dried very thoroughly before use. If possible, clean the glassware during the previous lab period, let it dry overnight or longer, then dry it in an oven set at about 110°C for 20 to 30 minutes. Assemble the apparatus and attach one or more drying tubes (see OP-12) as soon as possible after oven-drying; otherwise, moisture will condense inside it as it cools. If the glassware must be cleaned the same day as it is used, rinse it with acetone after washing and flush it with clean, dry air before you put it in the oven. You can also dry glassware by passing a "cool" Bunsen burner flame over the surface of the assembled apparatus, but this practice should never be used in laboratories where volatile solvents such as diethyl ether are in use. It should be done only with the instructor's permission and according to his or her directions.

Take Care! Use tongs or heat-resistant gloves when handling hot glass.

OPERATION **2**

Using Specialized Glassware

Most specialized glassware components used in organic chemistry have rigid ground-glass joints called *standard-taper* joints. The size of a tapered joint is designated by two numbers, such as 19/22 (for typical standard scale glassware) or 14/10 (for typical microscale glassware), in which the first number is the diameter at the top of the joint and the second is the length of the taper, measured in millimeters (see Figure A1).

The glassware in a commercial organic lab kit, or its equivalent purchased as separate parts, can be used to construct apparatus for many different laboratory operations. The glassware provided in typical standard scale lab kits and Mayo–Pike style microscale lab kits is illustrated in Appendix I. Setups for the various operations are illustrated in the appropriate operation descriptions in this book. For example, to find an illustration of a setup for microscale fractional distillation, refer to the microscale section of operation OP-32.

An alternative type of microscale glassware developed by Kenneth L. Williamson uses elastomer connectors rather than ground-glass joints to hold glassware assemblies together. Williamson glassware is not described in this book.

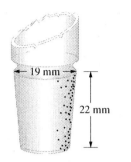

Figure A1 19/22 standard-taper joint

 ### Lubricating Joints

For some operations, such as vacuum distillation, glass joints should be lubricated with a suitable ground-glass joint grease. For most other operations, particularly with microscale equipment, lubrication of glass joints is unnecessary and may be undesirable. Your instructor should inform you if lubrication will be necessary. To lubricate a ground-glass joint, apply a thin layer of joint grease completely around the top half of the inner (male) joint. Do not lubricate the outer (female) joint. Be careful to keep grease away from the open end of the joint, where it may come into contact with and contaminate your reaction mixture or product. When you assemble the components, press the outer and inner joints together firmly with a slight twist to form a seal around the entire joint, with no gaps. Grease should never extend beyond the joint inside the apparatus.

After disassembling the apparatus, remove the grease completely by using a suitable organic solvent. You can remove petroleum-based greases with petroleum ether or hexanes, and silicone greases by thorough cleaning with dichloromethane. An inner joint can be cleaned by wrapping a small amount of cotton loosely around the end of an applicator stick, dipping it in the solvent, and wiping the joint with the moist cotton.

Take Care! Keep flames away from petroleum ether and hexanes. Avoid contact with dichloromethane and do not breathe its vapors.

 ### Assembling Standard Scale Glassware

Standard-taper joints are rigid, so a setup using standard scale glassware must be assembled carefully to avoid strain that can result in breakage. First, place the necessary clamps and rings at appropriate locations on the ring stand (use two ring stands for distillation setups). Then assemble the apparatus *from the bottom up, starting at the heat source.* Position the heat source on a ring or other support so that it can be removed easily when the heating period is over; otherwise, it may continue to heat a reaction mixture or an empty distilling flask even after it is switched off, causing a danger of breakage, tar formation, or even an explosion. Clamp the reaction flask or boiling flask securely at the proper distance from the heat source.

As you add other components, clamp them to the ring stand(s), but do not tighten the clamp jaws completely until all of the components are in place and aligned properly. Use as many clamps as are necessary to provide adequate support for all parts of the apparatus. A vertical setup, such as the one for addition under reflux [OP-11], requires at least two clamps for security because if the setup is bumped, the clamp holding the reaction flask may rotate and deposit your glassware on the lab bench, with very expensive consequences. Some vertical components, such as Claisen connecting tubes, need not be clamped if they are adequately supported by the component below. Nonvertical components, such as distillation condensers, should be clamped; otherwise, they may be jarred loose and fall. Clamping condensers and other components at an angle to a ring stand requires an adjustable clamp with a wing nut on the shaft. This wing nut is tightened after the apparatus is aligned. Distillation receivers should be supported by a ring and wire gauze or another suitable support. They should not be clamped, because they may have to be replaced quickly during a distillation.

Some joints, such as the joint that connects a condenser to a vacuum adapter, tend to separate easily, so they should be held together with joint clips or strong rubber bands. For example, you can secure a vacuum adapter to a condenser by stretching a rubber band around the tubulation on both or by snapping a joint clip around the joint rim. Condensers and vacuum adapters should never be allowed to hang unsupported, even momentarily while you are assembling the apparatus.

Position the clamps so that all parts are aligned correctly and their glass joints slide together easily. Then seat the joints firmly, with a slight twist if necessary, and tighten all the clamps. Examine the joints for gaps, then check to make sure that the apparatus is held securely by the clamp jaws and that the clamp holders are secured tightly to the ring stand(s).

Figure A2 summarizes the steps followed in assembling one kind of ground-glass apparatus. (Most of the glassware setups you will be using are less complex than the one illustrated.)

 Assembling Microscale Glassware

Microscale glassware comes in a variety of configurations. Most undergraduate microscale laboratory courses use glassware of the kind developed and tested by students at Bowdoin and Merrimack Colleges under the direction of Professors Dana W. Mayo and Ronald M. Pike. The glassware provided in a typical Mayo–Pike style microscale lab kit is illustrated in Appendix I. There are variations in the construction of this type of glassware, but the components are usually held together by threaded plastic compression caps. These microscale components are connected with standard-taper joints, just as for standard scale equipment, but a compression cap and O-ring are used to give a tight, greaseless seal.

Steps

1 Position clamps, rings.
2 Position heat source.
3 Clamp boiling flask securely.
4,5 Add Claisen adapter and connecting adapter.
6 Clamp West condenser in place.
7 Attach vacuum adapter with rubber band or spring clamp.
8 Attach receiving flask, support with ring and wire gauze.
9 Readjust all clamps to align parts.
10 Press joints together.
11 Tighten clamps.
12 Add stopper.
13 Add thermometer adapter and position thermometer.

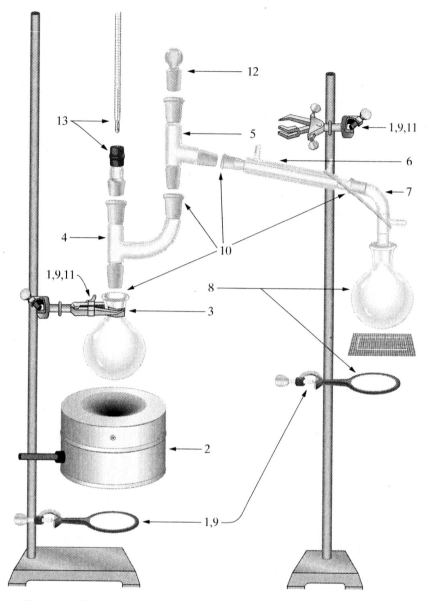

Figure A2 Steps in the assembly of standard scale ground-glass apparatus

You can assemble such a joint as illustrated in Figure A3 for a conical vial and a water-cooled condenser. First, put a compression cap, threaded side down, over the male (outer) joint of the condenser. Hold it in place as you roll an O-ring over the joint onto the clear part of the glass. Make sure that the entire O-ring is above the ground joint, then release the cap, which should be held in place by the O-ring. Now insert the male joint of the condenser into the female (inner) joint of the conical vial and screw the cap over the threads at the top of the vial, tightly enough so that the outer joint cannot be rotated around the inner joint. Be careful, because screwing it down too tightly may break the threads or cause strains in the glass that will lead to eventual breakage.

Other microscale components, such as Claisen connecting tubes, drying tubes, and Hickman stills, can be connected to conical vials or microscale

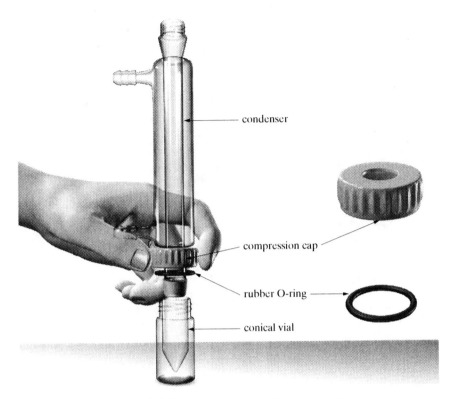

condenser

compression cap

rubber O-ring

conical vial

Figure A3 Connecting microscale components

reaction flasks to perform a wide variety of laboratory operations, such as distillation (Figure E11, OP-30) and addition under reflux (Figure B15, OP-11). All of the components are connected by compression caps, as shown by the appropriate illustrations in the operation descriptions.

Unlike standard scale apparatus, microscale apparatus can be assembled on the bench top and *then* clamped to a ring stand, often with a single microclamp, rather than being assembled from the bottom up on one or more ring stands. Since the reaction vessel (flask or conical vial) is not going to fall off the joint it is connected to, it need not be clamped. Instead, you can clamp the apparatus higher up—on a condenser, for example—to make it more stable. Be sure that the microclamp is held securely to the ring stand and that the clamp jaws are tightened securely around the apparatus so that it doesn't wobble or fall to the bench and break. If you intend to heat a reaction mixture in the flask or the conical vial, clamp the apparatus so that, if necessary (as when a reaction mixture is boiling too vigorously), you will be able to loosen the clamp holder at the ring stand and raise the entire apparatus above the heat source quickly.

 Disassembling Glassware

Disassemble (take apart) ground-joint glassware promptly after use, as joints that are left coupled for an extended period of time may freeze together and become difficult or impossible to separate without breakage. Ground joints can usually be separated by pulling the components apart with a twisting motion. For microscale glassware, unscrew the compression cap completely before you separate the joints. If a joint is frozen, you can sometimes loosen it by tapping it gently with the handle of a wooden-handled spatula or by applying steam to the joint while rotating the apparatus slowly, and then pulling the components apart with a twisting motion. If this doesn't work, see your instructor. Clean [OP-1] the glassware thoroughly and return each component to its proper location in the lab kit or to the stockroom. To reduce assembly time, caps and O-rings are sometimes left on microscale components when they are returned to their lab-kit case; this can make the components more difficult to clean thoroughly, so your instructor may prefer that you remove them.

Take Care! The glass may break, so protect your hands with heavy gloves.

OPERATION 3   Using Glass Rod and Tubing

Glass connecting tubes, stirring rods, and other simple glass items are required for certain operations in organic chemistry. Soft-glass rod and tubing can be worked easily with a Bunsen burner, but borosilicate glass (Pyrex, Kimax, etc.) requires the hotter flame provided by a Meker-type burner or an oxygen torch. To distinguish borosilicate from soft glass, dip the glass into anhydrous glycerol; most (but not all) borosilicate glass will seem to disappear in the liquid because it has nearly the same refractive index as glycerol (1.475).

 ### Cutting Glass Rod and Tubing

Glass rods and tubes are cut by scoring them at the desired location and snapping them in two. Score the rod or tube by drawing a sharp triangular file (or other glass-scoring tool) across the surface at a right angle to the axis of the tubing. Often, only a single stroke is needed to make a deep scratch in the surface; don't use the file like a saw. To cut a thin, fragile glass tube, such as a melting-point tube or the capillary tip of a Pasteur pipet, it is best to use a special glass scorer, but a sharp triangular file may work if applied carefully so as not to crush the glass. Moisten the scratch with water or saliva. Using a towel or gloves to protect your hands, place your thumbs about 1 cm apart on the side opposite the scratch and, while holding the glass firmly in both hands, press forward against the glass with your thumbs as you rotate your wrists outward (Figure A4).

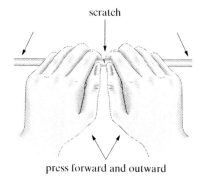

scratch

press forward and outward

Figure A4 Breaking glass tubing

 ### Working Glass Rod and Tubing

The cut ends of a glass rod or tube should always be *fire-polished* to remove sharp edges and prevent accidental cuts. To fire-polish a glass rod or tube, hold it at a 45° angle to a burner flame (see OP-7 for directions on using a burner) and rotate its cut end slowly in the flame until the edge becomes rounded and smooth (Figure A5).

To round the cut end of a glass rod, rotate the rod in a burner flame, holding it at a 45° angle with its tip at the inner blue cone of the flame. The end should be hemispherical in shape, not rounded only at the edges and flat on the bottom.

Take Care! Don't burn yourself on the hot end of a glass rod or tube or lay the glass onto combustible materials.

To flatten one end of a glass rod, rotate it with its tip at the inner blue cone of the flame until it is incandescent and very soft, but not starting to bend. Then press the softened end straight down onto a hard surface, such as the base of a ring stand. The flattened end should be well centered and about twice the diameter of the rod.

To seal one end of a glass tube, hold the tube at a 45° angle to the burner with its end just above the inner blue cone of the flame and rotate it until the soft edges come together and eventually merge. Remove the tube from the flame as soon as it is closed and immediately blow into the open end to obtain a sealed end of uniform thickness. Let the tube cool to room temperature. Then check it for leaks by connecting the open end to an aspirator or a vacuum line with a length of rubber tubing, placing the closed end in a test tube containing a small amount of dichloromethane, and turning on the vacuum. (**Take Care!** Avoid contact with dichloromethane and do not breathe its vapors.) If the tube is not properly sealed, the liquid will leak into it when you apply suction. To seal the end of a thin, fragile tube such as a melting-point capillary, rotate its open end in the *outer* edge of the flame.

To bend glass tubing, first place a flame spreader on the barrel of a Bunsen burner (or use a Meker-type burner). Hold the tubing over the burner flame parallel to the long axis of the flame spreader and rotate it constantly at a slow, even rate until it is nearly soft enough to bend under its own weight (see Figure A6). (The flame will turn yellow as the glass begins to soften.) Remove the hot tubing from the flame and immediately bend it to the desired shape with a firm, even motion and a minimum of force (if much force is required, the glass is not soft enough). Bend it in a vertical plane, with the ends up and the bend at the bottom; the bend should follow a smooth curve with no constrictions.

Figure A5 Fire polishing

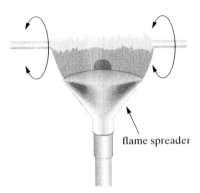

Figure A6 Bending tubing

 Inserting Glass Items into Stoppers

Improper insertion of glass tubes and thermometers through stoppers is one of the most frequent causes of laboratory accidents. The resulting cuts and puncture wounds can be very severe, requiring medical treatment and sometimes causing the victim to go into shock. Thermometers are particularly easy to break, especially at the scored immersion line.

Safety Notes

Sometimes you may have to insert a glass tube through a hole bored in a cork or rubber stopper. To bore a hole in a solid stopper, obtain a *sharp* cork borer that is slightly smaller in diameter than the object to be inserted in the stopper. (If the borer is dull, use a special cork-borer sharpener to sharpen it.) Lubricate its cutting edge with a small amount of glycerine, then *twist* it through the stopper using a minimum of force (do not try to "punch" out the hole). Rotate the borer and stopper in opposite directions, checking the alignment frequently to make sure that the borer is going in straight. When the borer is about halfway through, twist it out and start boring from the opposite end of the stopper until the holes meet. You can remove the plug left inside the cork borer with a rod that comes with a set of cork borers.

Cork borer

Take Care! Don't grasp the tube too far from the stopper. The glass may break and lacerate your hand.

To insert a glass tube into a rubber stopper, first lubricate the hole lightly with glycerol or another suitable lubricant; water may work if the hole is not too tight. You can use a cotton swab or an applicator stick to apply the lubricant evenly. Protect your hands with gloves or a towel, then grasp the tube close to the stopper and twist it through the hole with firm, steady pressure. Apply force directly along the axis of the tube, as any sideways force may cause it to break. Using excessive force or forcing the tube through a hole too small for it can also cause it to break. After the tube is correctly positioned, rinse off any glycerol with water. Follow the same directions to insert a thermometer through a rubber stopper or a standard scale thermometer adapter cap. A microscale thermometer adapter uses an O-ring rather than a rubber cap to hold the thermometer in place. Never try to insert a thermometer into an assembled microscale thermometer adapter; instead, roll the O-ring onto the thermometer and then secure the thermometer in the adapter as described in OP-9.

To remove a glass tube from a stopper, lubricate the part of the glass that will pass through the stopper with water or glycerol, protect your hands with gloves or a towel, and twist the tube out with a firm, continuous motion. Hold the tube close to the stopper or cap and avoid applying any sideways force that could cause it to break. If you can't remove the tube by this method, obtain a cork borer of a size that will just fit around it, lubricate the borer's cutting edge, and twist it gently through the stopper until the tube can be pulled out easily. (Follow the same directions to remove a thermometer from a standard scale thermometer adapter cap.)

OPERATION 4   Weighing

Most modern chemistry laboratories are equipped with electronic balances that display the mass directly, without any preliminary adjustments (Figure A7). If you will be using a different type of balance, your instructor will demonstrate its operation. For microscale experiments, chemicals should be weighed on balances that measure to at least the nearest milligram (0.001 g). For standard scale experiments, balances that measure to the nearest centigram (0.01 g) are acceptable for most purposes, but milligram balances are preferable. Most products obtained from a preparation are transferred to vials or other small containers, which should be *tared*—weighed empty—and then reweighed after the product has been added. As a rule, the container should be weighed with its cap and label on and this *tare mass* should be recorded.

A balance is a precision instrument that can easily be damaged by contaminants, so avoid spilling chemicals on the balance pan or on the balance itself. If spillage does occur, *clean it up immediately*. If you spill a liquid or corrosive solid on any part of the balance, notify your instructor as well. Before you leave the balance area, replace the caps on all reagent bottles, return them to their proper locations if you obtained them elsewhere, and see that the area around the balance is clean and orderly.

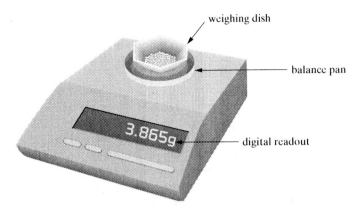

weighing dish

balance pan

3.8659g

digital readout

Figure A7 An electronic balance

Weighing Solids

Solids can be weighed in glass containers (such as vials or beakers), in aluminum or plastic weighing dishes, or on glazed weighing papers. In standard scale work, solid reactants are usually weighed on glazed paper or in weighing dishes and then transferred to the reaction flask. In microscale work, it is best to weigh solid reactants directly in the reaction vessel to avoid losses in transfer. A round-bottom flask or pear-shaped flask should be supported in a beaker or on a cork ring for weighing. Hygroscopic solids, those that absorb moisture from the atmosphere, should be weighed in screw-cap vials or other containers that can be capped immediately after the solid is added. Filter paper and other absorbent papers should not be used for weighing, since a few particles will always remain in the fibers of the paper.

To weigh a sample of a solid that is in a tared container, such as a preweighed screw-cap vial, set the digital readout to zero by pressing the appropriate button, then place the container on the balance pan. Be sure that the draft shield (if there is one) is in place, then read the mass of the container and its contents from the digital display. Wait until the reading remains constant, then record the mass in your laboratory notebook, including all digits after the decimal point. For example, if the balance reads 3.610 g, do not record the mass as 3.61 g, because zeroes following the decimal point are significant. Then subtract the tare mass to obtain the mass of the solid.

To weigh a sample of a solid that is to be transferred to another container, such as a weighing dish or storage vial, place the container on the balance pan, press the tare button to zero the digital display, and transfer the solid to the container. With the draft shield in place, wait until the reading has stabilized and then read the mass of the sample directly from the digital display.

To measure out a specific quantity of a solid, such as a solid reactant, into a reaction flask or other container, first support the container on the balance pan and press the tare button to zero the digital display. Then use a spatula or Scoopula to add the solid in small portions until the desired mass appears on the digital display. With the draft shield in place, wait until the reading has stabilized and then read the mass of the sample directly from the digital display. Ordinarily you need not measure out the exact mass specified in a procedure, but try not to deviate from the specified mass by more than 2% or so, especially for a limiting reactant. Because the theoretical yield of a preparation is based on the actual mass of a starting material, always use your measured mass—not the mass specified in the procedure— for yield calculations.

For example, if a procedure requires 0.250 g of a limiting reactant, you should measure out between 0.245 g and 0.255 g of the reactant.

Weighing Liquids

Organic liquids should be weighed in screw-cap vials or other closed containers to prevent damage to the balance from accidental spillage and losses by evaporation. If liquid must be added to or removed from a weighed container, the container should be removed from the balance pan first. Any excess liquid should be placed in a waste container or otherwise disposed of, *not* returned to a stock bottle.

To measure the mass of a liquid sample in a tared or untared container, follow the directions for solids, but be sure to keep the container capped while it is on the balance pan. When using a tared container, subtract the tare mass to obtain the mass of the liquid.

To measure out a specific mass of a liquid from a reagent bottle, you should first measure the approximate quantity of the liquid by volume and then weigh that quantity accurately in an appropriate closed container. For example, if you need 3.71 g of 1-butanol ($d = 0.810$ g/mL) for a standard scale experiment, you can use a small graduated cylinder or a measuring pipet to measure [OP-5] about 4.6 mL (3.71 g $\div$ 0.810 g/mL) of the liquid into a tared container, then cap and weigh the container and liquid. (The balance can first be zeroed with the container and its cap on the balance pan—don't forget to include the cap!) If the measured mass is not close enough to 3.71 g, add or remove liquid with a clean Pasteur pipet or medicine dropper.

Take Care! Be careful not to spill liquids on the balance pan. If you do, clean up the spill immediately and inform your instructor.

For microscale work, it is best to weigh liquid reactants in the reaction vessel to avoid losses in transfer. A round-bottom flask or pear-shaped flask should be supported in a beaker or on a cork ring for weighing. Use an automatic pipet, a measuring pipet, a graduated syringe, or a bottle-top dispenser to measure the estimated volume of the liquid into the reaction vessel, and cap it before you weigh it.

OPERATION 5   Measuring Volume

Several different kinds of volume-measuring devices, described here, are used in the undergraduate organic chemistry laboratory. Although some kinds of measuring devices are used mostly for standard scale work, while others are used mostly for microscale work, all of the devices described here are applicable to both. Relatively large volumes of liquids are generally measured using graduated cylinders whose capacity may vary from 5 mL to 100 mL or more. Relatively small volumes of liquids can be measured using various kinds of pipets and syringes. Reagent bottles containing liquids may be provided with bottle-top dispensers that measure out a preset volume of the liquid. For a few experiments you may use a buret or a volumetric flask; their use is described in most general chemistry laboratory manuals.

Graduated Cylinders

Graduated cylinders are not highly accurate, but in standard scale work they are often used to measure specified quantities of solvents and wash liquids, or even some liquid reactants that are used in excess. In microscale work they are used mostly for measuring relatively large amounts of water or other solvents.

To use a graduated cylinder, transfer the liquid being measured to the cylinder by pouring it (for most standard scale work) or by using a Pasteur pipet (for microscale work) until the cylinder is filled to the graduation mark corresponding to the desired volume. Read the liquid volume from the bottom of the meniscus as shown in Figure A8. If necessary, add or remove liquid with a Pasteur pipet.

Bottle-Top Dispensers

A typical adjustable bottle-top dispenser (see Figure A9) has a moveable plunger that pumps liquid into a glass cylinder, from which it is dispensed through a discharge tube. The cylinder is usually surrounded by a protective sleeve that is raised to fill the cylinder and lowered to dispense the liquid. The dispenser is screwed onto a bottle containing the liquid and can be adjusted to dispense a specified volume of liquid, which is read from a scale on the sleeve or cylinder. Before its initial use, the dispenser must be *primed* by pumping it several times to fill the cylinder and discharge tube and to expel any air bubbles.

To use a bottle-top dispenser, first check to see that there are no air bubbles in the discharge tube (if there are, reprime the dispenser or inform your instructor). Then hold your container underneath the discharge tube outlet and raise the sleeve as high as it will go. Release the sleeve so that it drops by gravity, and then push it down gently until it moves no further. Touch the tip of the discharge tube to an inside wall of your container to remove the last drop of liquid. If the liquid is the limiting reactant for a preparation, you should weigh it accurately as described in OP-4.

Measuring Pipets and Volumetric Pipets

A *measuring pipet* has a graduated scale and is used to measure liquid volumes within a range of values; for example, a typical 1-mL measuring pipet can measure volumes up to 1.00 mL to the nearest 0.01 mL. A *volumetric pipet* measures only a single volume, but it is more accurate than a measuring pipet. Suction is required to draw the liquid into a measuring or volumetric pipet, but you should never use mouth suction because of the danger of ingesting toxic or corrosive liquids. A pipet pump is a simple and convenient suction device for filling such pipets. Other pipet fillers such as large rubber bulbs can also be used.

To use a measuring pipet with a pipet pump of the type shown in Figure A10, first see that the plunger is as far down as it will go. Then insert

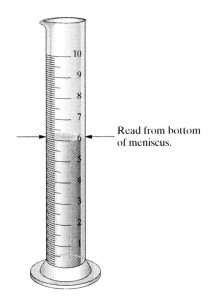

Read from bottom of meniscus.

Figure A8 Reading the volume contained in a graduated cylinder—in this case, 6.0 mL

A convenient "homemade" pipetting bulb is described in J. Chem. Educ. **1974**, *51,* 467.

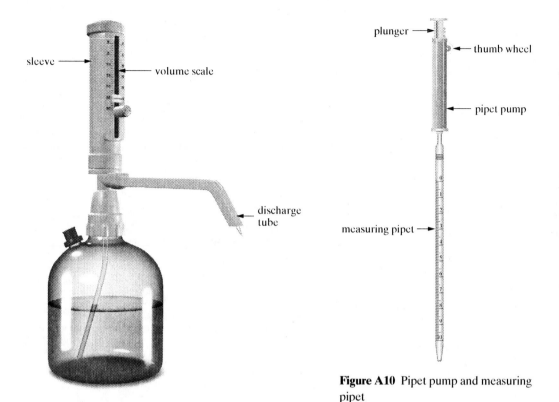

Figure A9 A bottle-top dispenser

Figure A10 Pipet pump and measuring pipet

Figure A11 Reading the volume delivered from a measuring pipet—in this case, 3.0 mL

the wide (untapered) end of the pipet firmly into the opening at the bottom of the pump. Place the tip of the pipet in the liquid and use your thumb to rotate the thumb wheel back toward you until the liquid meniscus rises a few millimeters above the zero graduation mark; be careful not to draw any liquid into the pump itself. Slowly rotate the thumb wheel away from you until the meniscus drops just to the zero mark. Measure the desired volume of liquid into a clean container by placing the pipet tip over the container and rotating the thumb wheel away from you until the meniscus drops to the graduation mark corresponding to the desired volume (Figure A11). Touch the tip of the pipet to the inside of the container to remove any adherent drop of liquid. If the pipet is one dedicated for use with a particular reagent bottle, the excess liquid can be drained into the bottle by depressing the pump's quick-release lever (if it has one) or by rotating the thumbwheel away from you as far as it will go. Otherwise, the liquid should be drained into another container or disposed of as directed by your instructor.

To use a volumetric pipet, obtain a bulb-type pipet filler or a pipet pump with a quick-release lever that allows the liquid to drain by gravity. Use the bulb or pump to fill the volumetric pipet to its calibration mark, hold the pipet tip over a receiving container, and use the quick-release lever or another device to let the liquid drain out until only a small amount of liquid is left in the tip. Touch the tip of the pipet to the inside of the container to remove any adherent drop of liquid, but do not expel the liquid in the pipet tip—its volume is accounted for when the pipet is calibrated.

Automatic Pipets

Automatic pipets (also called *pipetters*) provide a quick, convenient way of delivering a specified volume of liquid with a high degree of reproducibility (Figure A12). Most automatic pipets measure comparatively small volumes of liquids and are therefore most useful for microscale experiments. A variable-volume automatic pipet can be set to a specified volume within a certain range of volumes, such as 20 to 200 μL (0.020–0.200 mL). The volume is displayed on a digital display or an analog scale, usually in microliters (μL). To prevent contamination, liquid is drawn into a disposable tip and never inside the pipet itself. Whenever the pipet is used for a different liquid, the volume is reset (if necessary) and a new pipet tip is installed. The instructor or a lab assistant will ordinarily set the volume of an automatic pipet and designate it for a specific liquid. Do *not* try to reset the volume or use the pipet for a different liquid without explicit permission from your instructor.

To use an automatic pipet, depress the plunger to the first *detent* (stop) position, when you will feel resistance to further movement. (Using excessive force will move the plunger to its second detent position, causing an inaccurate measurement.) Insert the pipet tip into the liquid to a depth of about 1 cm or less; do not let it touch the bottom of the liquid's container, where impurities may be concentrated. Slowly release the plunger to draw liquid into the pipet tip. Place the tip inside the receiving container and depress the plunger to the first detent position, pause a second or two, then push the plunger down to the second detent position and touch the tip to the inner wall of the receiving container to expel the last droplet of liquid.

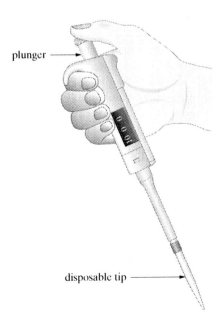

Figure A12 An automatic pipet

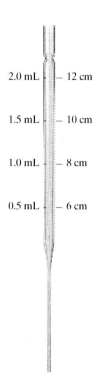

2.0 mL — 12 cm

1.5 mL — 10 cm

1.0 mL — 8 cm

0.5 mL — 6 cm

Figure A13 A calibrated Pasteur pipet

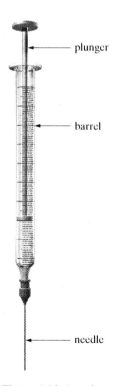

— plunger

— barrel

— needle

Figure A14 A syringe

Calibrated Pasteur Pipets

A calibrated Pasteur pipet can be used to measure approximate volumes of liquids such as washing and extraction solvents. For measuring volatile liquids, a filter-tip pipet (see OP-6) that has been calibrated should work better than an ordinary calibrated pipet.

  To calibrate a Pasteur pipet, first attach a latex rubber bulb to its wide end. Measure 0.50 mL of water into a conical vial or small test tube using a measuring pipet or another accurate measuring device. Carefully draw all the liquid into the Pasteur pipet so that there are no air bubbles in its tip (if necessary, squeeze the bulb *gently* to expel any air) and mark the position of the meniscus with an indelible glass-marking pen. Expel all the water and repeat this operation using 1.00 mL of water, and other volumes as desired. A quicker but less accurate way to calibrate a short $\left(5\frac{3}{4} \text{ inch}\right)$ Pasteur pipet is to use a ruler to mark lines at distances of 6 cm, 8 cm, 10 cm, and 12 cm from the narrow (capillary) tip of the pipet (see Figure A13). These lines mark volumes of approximately 0.5 mL, 1.0 mL, 1.5 mL, and 2.0 mL. Make sure that the capillary tip is intact; if part of it is broken off, this calibration method will not work.

  To use a calibrated pipet, hold the pipet (with its attached bulb) vertically over the liquid to be measured, squeeze the bulb to expel some of the air (ideally, an amount of air nearly equal to the volume of liquid required), and insert the tip in the liquid. With practice, you should learn how far to squeeze the bulb in order to draw in the desired amount of liquid. Then release the bulb until the liquid meniscus is at the level of the appropriate calibration mark. Without delay, raise the pipet tip out of the liquid, move it into position over the receiving container, and squeeze the bulb to expel all the liquid into the container. It takes practice to transfer the liquid without losing some in the process, so read OP-6 for additional tips about the use of Pasteur pipets.

Syringes

Syringes (Figure A14) are used to measure and deliver small volumes of liquid, often by inserting the needle through a *septum*—a rubber or plastic disk that can be penetrated by a needle but that remains more-or-less airtight after the needle is withdrawn. A syringe of appropriate size can be used to inject liquid samples into a gas chromatograph and (in some microscale experiments) to add liquid reagents to a reaction mixture during a reaction.

  To fill a syringe, hold it vertically with the needle pointing down, then place the needle tip in the liquid and slowly pull out the plunger until the barrel contains a little more than the required volume of liquid. (**Take Care!** Don't stick yourself with the needle!) If there are air bubbles in the liquid, remove them by holding the syringe vertically with the needle pointing up and tapping the barrel with your fingernail, or by expelling the liquid and filling the syringe again, more slowly. Hold the syringe with the needle pointing up and slowly push in the plunger to eject the excess liquid (collect it for disposal if requested) until the bottom of the liquid column is at the appropriate graduation mark. Wipe off the tip of the needle with a tissue, place the needle tip into the receiving vessel or through a septum, and expel the liquid by gently pushing the plunger in as far as it will go. Clean the syringe immediately after use by flushing it repeatedly with an appropriate solvent, such as acetone, or a soap solution. If you use soap for washing, rinse the syringe thoroughly with water afterward. Dry the syringe by pumping the plunger several times to expel excess solvent. Then remove the plunger to let the barrel dry. If the syringe is to be used again shortly, you can dry it by drawing air through the barrel with an aspirator or a vacuum line.

The plastic 1-mL syringes provided in some microscale lab kits are subject to contamination and are not compatible with certain organic solvents. For some volume-measuring applications, the barrel of such a syringe can be connected to the wide end of a clean Pasteur pipet with a small length of plastic or rubber tubing, so that the measured volume of liquid is drawn into the Pasteur pipet rather than the body of the syringe (see *J. Chem. Educ.* **1993**, *70*, A311).

Take Care! Don't bend the plunger.

Making Transfers

In many organic syntheses, losses during transfers constitute a substantial part of the total product loss, so they can have a major impact on your yield. Such losses can occur whenever you transfer a liquid or a solid from one container to another, whether the original container is a stock bottle, reaction flask, beaker, or funnel. Making complete transfers is particularly important in microscale work, where losing just a few crystals of a solid product or a drop of a liquid product may reduce your percent yield significantly.

 Transferring Solids

Bulk solids (such as those from a lab stock bottle) can be transferred from one container to another using spatulas of various shapes and sizes. For standard scale work, a Scoopula is preferred because it is curved to help keep the solid from sliding off. A flat-bladed spatula will also work, but unless you are careful some of the solid may spill over its sides. For microscale work, the U-shaped or V-shaped end of a Hayman-type microspatula (see Figure A15) can be used to transfer solids. Small amounts of solids can also be transferred with a plastic microscoop, made by cutting a 1-mL automatic pipet tip in half.

Solids can be conveniently transferred to small-mouthed containers such as test tubes and storage vials using a folded square of weighing paper or a square plastic weighing dish as a makeshift funnel. To transfer a solid from a plastic weighing dish, hold opposite corners of the dish between your thumb and middle finger and bend the dish to form a "spout" from which you can pour it or scrape it out. During such transfers, place the receiving container on a square of weighing paper or in a weighing dish (or clamp it above the paper or dish), so that any solid that misses the container can be recovered. If the solid you are transferring sticks to the sides of its container (such as a reaction vial or a Buchner funnel), use a flexible, flat-bladed spatula or microspatula to scrape as much as you can off the sides. If you need

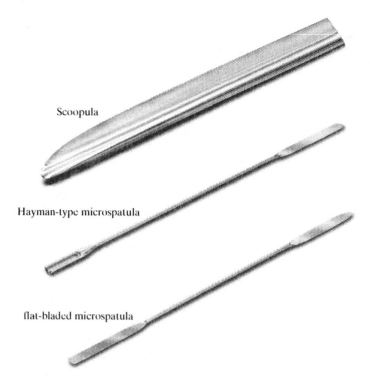

Scoopula

Hayman-type microspatula

flat-bladed microspatula

Figure A15 Spatulas for transferring solids

to transfer the last traces of a solid, you can dissolve the residual solid in a volatile solvent, make the transfer, and evaporate the solvent as described next for liquid transfers.

Transferring Liquids

In standard scale work, liquid transfers are usually accomplished by *decanting* (pouring) the liquid from its original container into another container. Various kinds of pipets and dispensers [OP-5] can also be used for standard scale liquid transfers.

In microscale work, most liquids (especially liquid reactants and products) should *not* be decanted because too much liquid will adhere to the inside of the original container and be lost. Instead, liquids are transferred using pipets or syringes.

Pipets and syringes are a common cause of contamination, so never allow a liquid to be sucked into a rubber bulb or pipet pump and clean [OP-1] all pipets and syringes thoroughly after use.

A Pasteur pipet fitted with a latex rubber bulb (Figure A16) can be used for most microscale transfers. Volatile liquids such as dichloromethane tend to partially vaporize during a transfer (especially on a warm day or when the pipet is warmed by your hand), causing some of the liquid to spurt out of the tip of the pipet. You may be able to avoid this problem by drawing in and expelling the liquid several times to fill the pipet with solvent vapors before you use it for the transfer. Alternatively, you can use a *filter-tip pipet.*

To make a filter-tip pipet, obtain a *very* small wisp of clean cotton, roll it into a loose ball, and use a straight length of thin (20 gauge) copper wire to push it past the narrow neck of a $5\frac{3}{4}$-inch Pasteur pipet into its capillary

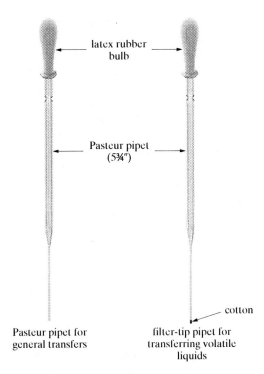

latex rubber
bulb

Pasteur pipet
(5¾")

cotton

Pasteur pipet for
general transfers

filter-tip pipet for
transferring volatile
liquids

Figure A16 Pipets for transferring liquids

end. Hold the pipet with the capillary tip pointed up as you use the wire to push the cotton as close to the tip of the pipet as you can (see Figure A16). If the cotton ball is too large, it will get hopelessly stuck in the capillary and you will have to start over with another pipet and cotton ball. You may have to poke the cotton plug repeatedly with the wire to get it in place; the capillary is very fragile, so be careful you don't break it. Mayo and Pike *et al.* [Bibliography, D4] recommend that the cotton plug in a filter-tip pipet be washed with 1 mL of methanol and 1 mL of hexanes (a mixture of C_6H_{14} isomers) and then dried before use, but this measure should be necessary only when the liquid transferred must be very pure and dry (when in doubt, follow your instructor's recommendation).

A filter-tip pipet is useful for transferring all types of liquids, not just volatile ones, because the cotton plug helps remove solid impurities from the liquid, if any are present. It also gives you better control over the transfer process, reducing the likelihood that some of your product will drip onto the bench top on the way to the collecting container. The main drawbacks of this kind of filter-tip pipet are that (1) it takes some time and practice to prepare one properly, (2) the cotton plug in an improperly prepared pipet may be so tight as to impede liquid flow or so loose that it will not stay in place, and (3) it is difficult to get the cotton plug out without breaking the fragile tip. A shortened filter-tip pipet has few of these disadvantages. You can prepare one by cutting off [OP-3] all but about 5 mm of the capillary tip of a $5\frac{3}{4}$-inch Pasteur pipet and pushing a wisp of cotton to the end of its shortened tip. A shortened filter-tip pipet is particularly useful for transferring hot recrystallization [OP-28] solutions.

A 1-mL (or smaller) syringe of the kind that comes with some microscale lab kits can also be used for microscale liquid transfers. It is particularly useful for transferring liquids to or from containers that are sealed with plastic or rubber septum caps to keep out air or moisture. To inject liquid through a septum with a syringe, fill the syringe with the desired volume of liquid (see OP-5), carefully push its needle through the septum while holding the needle to keep it from bending, and then gently push the plunger in as far as it will go. To withdraw liquid through a septum, push the plunger in as far as it will go, insert the syringe needle through the septum until its tip is below the liquid surface, and pull the plunger to withdraw the desired amount of liquid. See OP-5 for additional information about the use and care of syringes.

When you are transferring a liquid from one container to another, you can avoid losses by keeping the containers as close together as possible. For example, if you are transferring a liquid from a conical vial to a screw-cap storage vial, hold the vials together in the same hand with their mouths at the same level. Then use your free hand to make the transfer. This way, any dripping liquid should be caught by one or the other container rather than ending up on the bench top. Even after a careful transfer, an appreciable amount of liquid may remain behind in the original container. You can recover that liquid by adding a small amount of a volatile solvent to the original container, tilting and rotating the container to wash all the liquid off its sides, transferring the resulting solution to the receiving container, and evaporating [OP-19] the solvent with a stream of dry air or nitrogen. The volatile solvent must be one in which the liquid is appreciably soluble. Dichloromethane and diethyl ether are suitable solvents for most organic liquids.

B. Operations for Conducting Chemical Reactions

Conducting a chemical reaction is a little like baking a cake. You need to measure the ingredients accurately using basic operations discussed in the last section, combine them, heat the mixture (or cool it) while keeping its components well mixed, and occasionally mix in additional ingredients. In this section you will learn about a number of different methods for heating (OP-7) and cooling (OP-8) a reaction mixture, and how to monitor the temperature of the mixture (OP-9). You will also learn how to keep the reactants mixed (OP-10) and how to add reactants during the course of a reaction (OP-11). Certain reactions must be carried out in a dry atmosphere (OP-12) or an oxygen-free atmosphere (OP-13). Other reactions generate toxic or otherwise harmful gases and must be conducted using a gas trap (OP-14).

Beginning in Part B, certain operation descriptions, such as the one for heating under reflux (OP-7c), will contain a "troubleshooting" section titled *When Things Go Wrong*. Whenever you encounter problems while performing an operation, see this section for help and advice.

Heating   OPERATION **7**

a. Heat Sources

Chemical reactions are accelerated by heat, because heat (thermal energy) speeds up the reactant molecules so that they collide more often, and increases the energy of those collisions so that they are more likely to generate product molecules. Therefore, most organic reactions are performed using some kind of heat source so that they can be carried out in a reasonable period of time. Heating devices are used for other purposes as well, such as distilling liquid mixtures and evaporating volatile solvents.

A variety of heat sources are used in the organic chemistry laboratory. The choice of a heat source for a particular application depends on such factors as the temperature required, the flammability of a liquid being heated, the need for simultaneous stirring, and the cost and convenience of the heating device. Of the heat sources described here, heating mantles, steam baths, and oil baths are used mostly for standard scale work, while hot plates, heating blocks, sand baths, and hot water baths are used mostly for microscale work. Some heat sources, such as hot water baths, are useful for both standard scale and microscale work.

When a reaction mixture is being heated, there is always a chance that an exothermic reaction will "take off" (become too vigorous), in which case the reaction mixture may spew out of the reaction apparatus and create a potentially dangerous situation. For that reason it is always a good idea to have a container with cold water handy so that you can chill the reaction flask (after separating it from the heat source) to bring the reaction under control. If that can't be done safely and the reaction is being carried out under a fume hood, close the hood sash to isolate the reaction mixture and

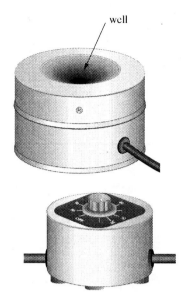

Figure B1 Heating mantle and heat control

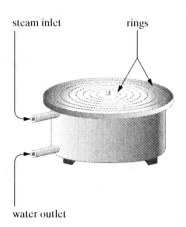

Figure B2 Steam bath

turn off the heat source by pulling the plug or by any other appropriate means. If a reaction becomes too vigorous while a reagent is being added [OP-11], the addition should be stopped or the rate of addition reduced. Some reactions mixtures will boil up vigorously if boiling chips (see part c of this Operation) have not been added or if the stirring rate [OP-10] is too slow. If you forget to use boiling chips, remember to cool the reaction mixture well below the boiling point before adding them.

Heating Mantles.   A heating mantle (Figure B1) is generally used to heat a round-bottom flask during a reaction or distillation. It is always used in conjunction with a voltage-regulating or time-cycling ("on–off") heat control to vary its heat output. A mantle can be used with a magnetic stirrer (see OP-10), and its heat output can be varied over a wide range, but its operating temperature cannot be monitored with a thermometer. Certain heating mantles, such as Thermowell ceramic flask heaters, are designed to heat round-bottom flasks over a range of sizes. For example, a 100-mL ceramic mantle can be used to heat 25-mL, 50-mL, and 100-mL round-bottom flasks efficiently. Most other mantles are designed for a specific flask size, so a 100-mL fiberglass heating mantle should only be used to heat a 100-mL round-bottom flask; the mantle will not heat efficiently and could even burn out if used with a flask of another size. Do not turn on an empty heating mantle or use it to heat an empty flask, because that might burn out its heating element. If you spill any chemicals into the well of a heating mantle, particularly if it is hot, unplug it and notify your instructor.

To operate a heating mantle, first support it on a ring support or a set of wood blocks so that it can be lowered and removed quickly if the rate of heating becomes too rapid. If one is available, you can use a *lab jack* instead. A lab jack is an adjustable platform that can be raised or lowered by rotating a knob. Clamp the flask in place so that it is in direct contact with the well of the heating mantle. If you are heating a small flask in a larger mantle, filling the well with glass wool up to the flask's liquid level may help distribute the heat more evenly (this is said to be unnecessary with a Thermowell mantle). See that the heat-control dial is set to zero, then plug the mantle into it—*never* directly into an electrical outlet—and adjust the heat-control dial until the desired rate of heating is attained. Note that the dial controls only the heating rate and cannot be set to a specific temperature. Because a heating mantle responds slowly to changes in the control setting, it is easy to exceed the desired heating rate by setting the dial too high at the start. If that happens, lower the mantle so that it is no longer in contact with the flask, reduce the dial setting, and allow sufficient time for the temperature to drop before raising the mantle again. Further adjustments may be needed to maintain heating at the desired rate. When you are done heating, lower the mantle, adjust the heat-control dial to its lowest or "off" setting, and let the mantle cool down before you attempt to remove it.

Steam Baths.   A steam bath (Figure B2) is a metal container having metal rings that can be removed or added to accommodate glassware of different sizes. It uses externally generated steam for heating, so it has only one operating temperature, 100°C. This limits its usefulness somewhat (for example, a steam bath can't be used to boil water or an aqueous solution), but the relatively low temperature helps prevent decomposition of heat-sensitive substances. Steam baths are often used to heat recrystallization mixtures, evaporate volatile solvents, and heat low-boiling liquids under reflux. Condensation of steam in the vicinity of a steam bath may be a nuisance, but it can be reduced by maintaining a slow rate of steam flow and by using enough rings to bridge any gaps between a flask (or another container) and the steam bath. Beyond a certain point, there is no advantage to increasing the steam flow rate, since the steam temperature is constant. If the flask and rings are positioned correctly, heating is quite uniform and efficient.

To use a steam bath, obtain two lengths of rubber tubing, attach one to the steam bath's *water outlet* tube and the other to the steam valve over the sink, and place the open ends of both rubber tubes in the sink. (If your steam bath has no water outlet tube, you will have to turn off the steam periodically to empty it of water.) Remove inner rings from the steam bath, leaving enough rings to safely support the container you wish to heat (unless it is supported by a clamp), but providing an opening large enough so that the steam will contact most of the container's bottom. If the container is a round-bottom flask that is clamped to a ring stand, remove enough rings so that the flask can be lowered through the rings to about its midpoint, leaving the smallest possible gap between the innermost ring and the flask. Directing the steam into the sink drain, open the steam valve fully and let it run until little or no water drips from the end of the rubber tube. Close the steam valve, connect it to the steam bath's *steam inlet* tube, then open it just enough to maintain the desired rate of heating with the container in place. You can adjust the heating rate somewhat by adding or removing rings, raising or lowering a clamped flask, or changing the steam flow rate. When you are done heating, turn off the steam valve completely and let the steam bath cool down. Then remove the rubber tubes, drain any water that remains in the steam bath, and put it and the rubber tubes back where you found them. (Don't leave rubber tubing in the sink!)

Oil Baths. An oil bath (Figure B3) provides very uniform heating and precise temperature control, reducing the likelihood of decomposition and side reactions caused by local overheating, and its operating temperature can be measured with a thermometer. But oil baths are seldom used in undergraduate organic chemistry labs because they are messy to work with, difficult to clean, and potentially hazardous. Hot oil can cause severe injury if accidentally spilled on the skin—the oil, which is difficult to remove and slow to cool, remains in contact with the skin long enough to produce deep, painful burns. An oil bath liquid can suddenly burst into flames if it is heated above its *flash point* when an ignition source, such as a spark or burner flame, is present. Hot oil can also catch fire if it splatters onto a hot surface, such as the top of a hot plate. Most oil fires can be extinguished by dry-chemical fire extinguishers or powdered sodium bicarbonate.

Water must be kept away from most oil baths because spilling water into a hot oil bath causes dangerous splattering. Oil that contains water should not be used until the water is removed, and a bath liquid that is dark and contains gummy residues should be replaced.

Mineral oil is probably the most commonly used oil bath liquid, but it presents a potential fire hazard and is hard to clean up. High-molecular-weight polyethylene glycols such as Carbowax 600 are water-soluble, which makes cleanup much easier, and they can be used at comparatively high temperatures

Take Care! Avoid contact with the steam, which can cause serious thermal burns.

The flash point of a liquid is the minimum temperature at which its vapors can be ignited by a small flame.

Oil bath liquids

Mineral oil
> Flash point ~190°C, but varies with composition
> Potential fire hazard

Glycerol
> Flash point 160°C
> Water-soluble, viscous

Dibutyl phthalate
> Flash point 171°C
> Viscous at low temperatures

Triethylene glycol
> Flash point 165°C
> Water-soluble

Polyethylene glycols (Carbowaxes)
> Flash point varies with molecular weight
> Water-soluble, some are solids at room temperature

Silicone oil, high temperature
> Flash point 315°C, usable range −40°C to 230°C
> Expensive; decomposition products are very toxic.

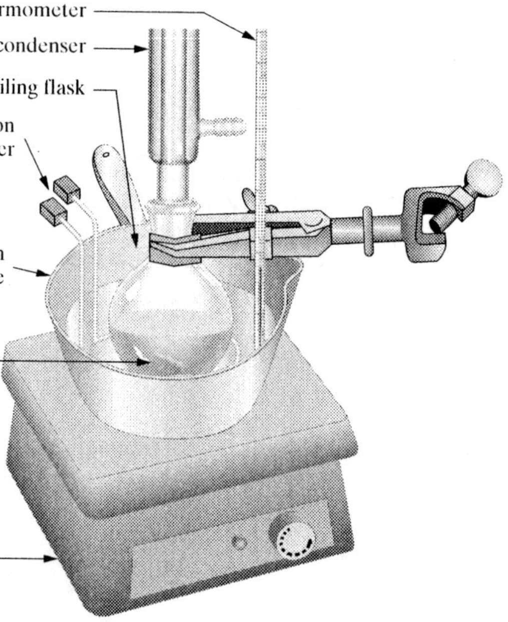

Figure B3 Oil bath assembly

Take Care! If the oil bath liquid starts smoking, discontinue heating and use fresh oil or an oil bath liquid with a higher flash point.

without appreciable decomposition. Some silicone oils can be used at even higher temperatures, but they are considerably more expensive than the other bath liquids. Flash points and other information about selected oil bath liquids are given in the margin. Note that flash points may vary with composition; check the label or ask your instructor if you are not sure about the flash point of a specific oil bath liquid.

An oil bath is ordinarily heated by a removable heating element, such as a coil of resistance wire or an immersion heater. (A hot plate can also be used, but it may cause a fire if any bath liquid spills onto the hot surface.) The output of the heating element is controlled by a variable transformer, and the temperature of the bath is measured with a thermometer suspended in the liquid. A large porcelain casserole (see Figure B3) makes a convenient bath container, since it is less easily broken than a glass container and has a handle for convenient placement and removal.

To use an oil bath, first place it on a lab jack, a set of wood blocks, or some other support that will allow it to be lowered quickly when necessary. Do not set it on a ring support because of the danger of spilling hot oil when the ring is raised or lowered. See that the apparatus containing the reaction flask or boiling flask is clamped securely to a ring stand. Then loosen (at the ring stand) the clamp holding the flask and lower the flask into the bath so that the liquid level inside it is 1 to 2 mm below the oil level. Clamp a thermometer so that its bulb is immersed in the oil but doesn't touch anything else in the oil bath. Drop in a stir bar, if desired, and use it to stir the bath gently; a smaller stir bar can be used to stir the flask contents. Switch on the variable transformer and adjust it until the desired temperature is obtained, then readjust it as needed to maintain that temperature. When you are done heating, turn off the heat and allow the oil bath to cool nearly to room temperature before you remove it. Transfer the oil to an appropriate container for reuse. Clean the bath container using a suitable solvent, such as petroleum ether for mineral oil, dichloromethane for silicone oil, or water for glycerol and polyethylene glycol.

🌡🔥 *Burners.* Bunsen-type burners are simple and convenient to operate, but they present a serious risk of fire in an organic chemistry lab, in which highly flammable solvents are often used. For that reason, burners should be used mainly for operations that cannot be conducted with flameless heat sources, such as bending and fire-polishing glass tubing.

Always check to see that there are no flammable liquids in the vicinity before you light a burner. Never use a burner near a heated oil bath or to heat a flammable liquid in an open container. Never leave a burner flame unattended; it may go out and cause an explosion due to escaping gas.

To operate a typical burner with a needle valve at the base, connect it to a gas outlet with a rubber hose and make sure that the valve at the gas outlet is turned off. Close the needle valve on the burner by rotating the knurled wheel clockwise until you feel resistance (don't close it tightly), then open it a turn or two. Open the gas valve and—without delay—ignite the burner with a burner lighter. If it doesn't light, rotate the barrel of the burner clockwise (or close a sleeve-type regulator) and try again. When the burner is lit, adjust the needle valve and rotate the barrel or sleeve to obtain a flame of the desired size and intensity. Rotating the barrel counterclockwise or opening the sleeve regulator to introduce more air produces a hotter, bluer flame. If you are using a burner to heat a nonflammable liquid in a beaker or other container, place the container on a ring support using a ceramic-centered wire gauze to spread out the flame and prevent superheating. The ring support should be positioned so that the bottom of the wire gauze is at the top of the inner blue cone of the flame, where it is hottest.

🌡🔥 *Hot Plates.* A hot plate (Figure B4) can be used to heat most liquids and solutions in flat-bottomed containers, such as Erlenmeyer flasks and beakers. It should *not* be used to heat low-boiling, flammable liquids that could splatter on the hot surface and ignite, or to heat round-bottom flasks directly. Hot plates can also be used to heat water baths, oil baths, sand baths, and heating blocks, which are in turn used to heat reaction mixtures and other liquids or solutions. Hot plates with built-in magnetic stirrers [OP-10], which can be used for operations that require

Take Care! Keep flames away from petroleum ether. Avoid contact with and inhalation of dichloromethane.

Safety Notes

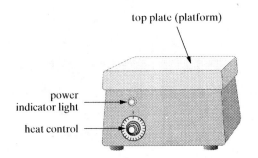

top plate (platform)

power indicator light

heat control

Figure B4 Hot plate

simultaneous stirring and heating, are particularly useful in the microscale laboratory.

To use a hot plate, plug it into an electrical outlet and adjust the heat-control dial to obtain the desired temperature or heating rate. If you are using a hot plate with a sand bath or an aluminum block, it's a good idea to prepare a calibration curve by measuring the equilibrium temperature at each dial setting (wait 10–15 minutes for the temperature to equilibrate at each setting) and plotting the temperature against the dial setting. Then you can adjust the heat control for the desired operating temperature when you use the same sand bath or aluminum block again. Note that the volume of sand in a sand bath must be kept constant in order for the calibration to be valid. Keep tongs, heat-resistant gloves, or other insulating materials handy so that you can quickly remove the container being heated when necessary. For example, you can fold a rectangle of paper toweling lengthwise several times and loop it around the neck of a hot Erlenmeyer flask to remove the flask from a hot plate.

Hot Water Baths.   Hot water baths are useful for heating low-boiling reaction mixtures, evaporating [OP-19] volatile solvents, and in other applications that require gentle heating. Although special metal water baths that resemble steam baths are available, a beaker can be used for most purposes. A typical hot water bath for microscale work is illustrated in Figure B5. Note that the water should fill the water-bath container about

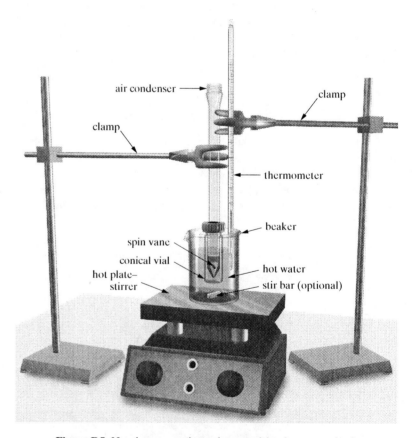

Figure B5 Heating a reaction mixture with a hot water bath

three-fourths full when the container being heated is immersed in it. The usual function of the air condenser shown is to return solvent vapors to a boiling reaction mixture (this process is described in OP-7c), but it also makes a convenient "handle" that can be used to lower or raise the container being heated and keeps it from tipping over in the water bath. So when you heat a jointed microscale container in a hot water bath, it's a good idea to attach a condenser whether or not the mixture will be boiling.

To prepare a hot water bath, measure an appropriate amount of water into the bath container and set it on a hot plate or hot plate–stirrer. Secure the container being heated inside the water bath so that the liquid level in the container is below the water level. Clamp a thermometer with its bulb beneath the water surface and at the same level as the mixture to be heated; its bulb shouldn't touch the bath container or the container being heated. If you are using a magnetic stirrer [OP-10], the stirring device inside the container should be close enough to the top and the center of the hot plate–stirrer to allow efficient stirring. If a specific bath temperature is required, adjust the heat setting and observe the thermometer reading until that temperature is reached. If the bath temperature rises above the specified value by 5°C or more, withdraw some of the bath water and replace it by an equal volume of cold water. A 10-mL (or larger) pipet equipped with a pipet pump can be used for this purpose. For most purposes, the bath temperature can be allowed to vary by ±5°C or so from the specified value. If a boiling water bath is required, add boiling chips or a stir bar to the water before you heat it to boiling. A stir bar is desirable even when the water isn't boiling, as it ensures more uniform heat distribution.

If precise temperature control isn't necessary, you can fill a beaker with preheated water from a hot water tap or another source and adjust the temperature by adding hot or cold water. As the bath cools, withdraw some of the bath water and replace it by fresh heated water.

Heating Blocks.   Aluminum heating blocks (see Figure B6) with holes or wells designed to accommodate small test tubes, round-bottom flasks, reaction vials, and similar containers can be used to conduct microscale reactions and for such operations as recrystallization [OP-28] and distillation [OP-30]. Copper heats and cools much faster than aluminum, so copper heat-transfer plates may also be used. These plates are not commercially available, but they can easily be fabricated [Bibliography, D4].

To use a heating block, set it on a hot plate or hot plate–stirrer and insert a thermometer (such as a nonmercury glass thermometer or a dial thermometer with a metal probe) to monitor its temperature, if desired. If you will be using a magnetic stirrer [OP-10], position the heating block so that the well holding the container being heated is close to the center of the hot plate–stirrer. A glass thermometer should be secured by a three-finger clamp on a ring stand and *carefully* lowered into a small hole (one drilled in the face of the block to accommodate it) until its bulb just rests on the bottom of the hole. Position a glass thermometer in the block *before* you begin heating, or the thermometer bulb may break. The metal probe of a dial thermometer can be inserted into a small hole drilled in one corner of a typical heating block. Note that it isn't always necessary to monitor the temperature of the block, which is invariably higher than the temperature inside the container it is heating; but by knowing the block temperature you can control

Take Care! If the bulb of a mercury thermometer breaks, especially in a heated block, it will release toxic mercury vapors into the atmosphere. Notify your instructor at once if this happens.

Figure B6 Heating a reaction mixture with a heating block

Take Care! A hot metal block looks just like a cold one, so never touch a heating block unless you are sure it is cold.

the heating rate more accurately and avoid overheating. Because a heating block takes some time to reach a desired temperature, it is a good idea to start heating it well before it will be needed, using a calibration curve (if you have prepared one) to select an appropriate heat-control setting.

Place the container to be heated in a well of the appropriate size. For the most uniform heating, the liquid level in the container should be just below the top of its well. If the liquid level is above the well top, you can insulate the container with glass wool held in place by aluminum foil. When very high temperatures are required, a conical vial can be provided with a commercially available split aluminum collar. Most glassware setups should be clamped to a ring stand, but a small test tube, a Craig tube, or a similar container can be supported adequately by the walls of its well. Adjust the heat control on the hot plate so that the temperature of the heating block is at least 20°C higher than the temperature you wish to attain inside the container you are heating. Then readjust the heat control, as necessary, until the desired heating rate is attained. For example, if you want to boil a reaction mixture in which water is the solvent, first raise the block temperature to 120°C. If the reaction mixture doesn't boil when the block is at that temperature, advance the heat control gradually until it does boil. You can control the heating rate to some extent by raising and lowering a container in its well without changing the heat setting. If you need to lower the temperature quickly, as when a recrystallization mixture threatens to boil over, raise the container out of its well first, *then* lower the heat setting or clamp the container at a higher level.

Sand Baths.   A typical sand bath (see Figure B7) consists of a flat-bottomed container, such as a cylindrical crystallization dish, that has been filled with fine sand to a depth of 10–15 mm. (Using a deeper sand layer may overload and damage a hot plate's heating element, especially at high temperatures.) Like a heating block, a sand bath is usually heated on a hot

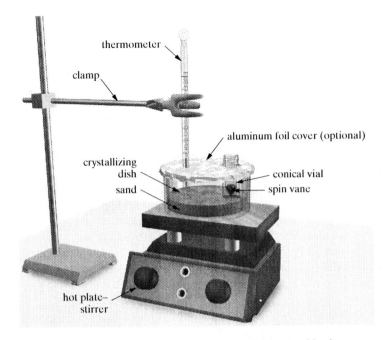

Figure B7 Heating a reaction mixture with a sand bath

plate or hot plate–stirrer, using a thermometer to monitor its temperature. For uniform heating, the container being heated should be pushed far enough into the sand so that the liquid level in the container is just below the top of the sand. When necessary, heating can be made more uniform and a higher temperature can be attained at a given heat setting by covering the bath container with aluminum foil. Holes must be cut in the foil to accommodate the thermometer and the container being heated. Although sand can be heated to a very high temperature, the glass container of a sand bath may break if it is heated much above 200°C. A convenient sand bath that is usable at high temperatures can be constructed by partly filling the ceramic well of a 100-mL Thermowell heating mantle with sand.

Sand baths heat and cool more slowly than aluminum blocks, but it is easier to observe changes in a reaction mixture and to swirl or shake a mixture in a sand bath. A sand bath also provides a temperature gradient, with lower temperatures near the top of the sand layer and higher temperatures near the bottom. Thus, it not necessary to control the measured temperature of a sand bath precisely, since you can vary the amount of heat applied to a container by varying its depth in the sand bath. For example, you can bring a reaction mixture to the boiling point quickly by immersing its container deep in the sand, then raise the container just enough to keep it boiling gently.

To use a sand bath, set it on a hot plate or hot plate–stirrer, then clamp a thermometer to a ring stand and lower it deeply enough into the sand so that its bulb is completely covered, but is not touching the bath container. Try to place the thermometer bulb at the same depth in the sand each time, because the temperature reading will vary with its depth. Begin heating the sand bath well before it will be needed, using a calibration curve (if you have prepared one) to select an appropriate heat setting. Adjust the heat

Take Care! Do not touch a sand bath or its container unless you are sure it is cold.

Take Care! If the bulb of a mercury thermometer breaks in a heated sand bath, it will release toxic mercury vapors into the atmosphere. Notify your instructor at once if this happens.

Figure B8 Heat lamp

control on the hot plate so that the temperature reading is at least 20°C higher than the temperature you wish to attain inside the container you are heating. Position the container, using clamps for support as necessary, with its bottom immersed in the sand. If you are using a magnetic stirrer, position the apparatus close to the center of the hot plate–stirrer. Raise or lower it in the sand, or readjust the heat control, until the desired heating rate is attained.

Other Heat Sources.   Heating devices such as infrared heat lamps (Figure B8) and electric forced-air heaters (heat guns) can be used in some heating applications. A heat lamp plugged into a variable transformer provides a safe and convenient way to heat comparatively low-boiling liquids. The boiling flask is usually fitted with an aluminum foil heat shield to concentrate the heat on the reaction mixture.

b. Smooth Boiling Devices

When a liquid is heated at its boiling point, it may erupt violently as large bubbles of superheated vapor are discharged from the liquid; this phenomenon is called *bumping*. A *boiling chip* prevents bumping by providing nucleating sites on which smaller bubbles can form. Boiling chips (also called boiling stones) are made from porous pieces of alumina, carbon, glass, marble (calcium carbonate), Teflon, and other materials. Alumina and calcium carbonate boiling chips may break down in strongly acidic or alkaline solutions, so boiling chips made of carbon, Teflon, or other chemically resistant materials should be used with such solutions. Wooden applicator sticks can be broken in two and the broken ends used to promote smooth boiling in nonreactive solvents; they should not be used in reaction mixtures because of the possibility of contamination. Boiling chips are not needed when a liquid being heated is stirred at a moderate rate with a magnetic stir bar or spin vane, because stirring [OP-10] causes turbulence that breaks up the large bubbles responsible for bumping.

Unless you are instructed differently, always add one or more boiling chips to any liquid or liquid mixture that will be boiled without stirring, such as a liquid to be distilled or a reaction mixture to be heated under reflux (see Section **c**). One or two small boiling chips are usually sufficient for microscale work; several boiling chips should be used for standard scale work. It is important to add boiling chips *before* heating begins, because the liquid may froth violently and boil over if you add them when it is hot. If you let a boiling liquid cool below its boiling point, add a fresh boiling chip if you reheat it later, because liquid fills the pores of a boiling chip and reduces its effectiveness when boiling stops.

c. Heating under Reflux

Most organic reactions are carried out by heating the reaction mixture to increase the reaction rate. The temperature of a reaction mixture can be controlled in several ways, the simplest and most convenient being to use a reaction solvent that has a boiling point within the desired temperature range for the reaction. Sometimes a liquid reactant itself may be used as the solvent. The reaction is conducted at the boiling point of the solvent, using a *condenser* to return solvent vapors to the reaction vessel so that no solvent is lost. This process of boiling a reaction mixture and condensing the solvent vapors back into the reaction vessel is known as *heating under reflux* (or more informally as "refluxing"), where the word reflux refers to the "flowing back" of the solvent. Usually a reaction time is specified for a reaction conducted under reflux. That interval should be measured from the time the reaction mixture begins to boil, *not* from the time heating is begun.

Round-bottom flasks are ordinarily used as the reaction vessels for a standard scale reaction. A typical standard scale lab kit contains round-bottom flasks with capacities of 25, 50, 100, 250, and 500 mL. A microscale lab kit may contain 3-mL and 5-mL conical vials, a 5-mL pear-shaped flask, and a 10-mL round-bottom flask, any of which can be used for carrying out a reaction. As a rule, the reaction vessel should be the smallest appropriate container that will be about half full or less when all of the reactants have been added. For example, if you will be dissolving 8 mL of reactant A in 20 mL of solvent before starting a reaction and adding 6 mL of reactant B during the reaction, the maximum volume of liquid in the reaction vessel will be approximately 34 mL. Since that volume would fill a 50-mL flask more than half full, a 100-mL round-bottom flask should be used.

Several different kinds of reflux condensers are available. A *water-cooled condenser* consists of two concentric tubes, with cold tap water circulating through the outer tube and solvent vapors from a boiling reaction mixture rising up the inner tube. The circulating water cools the walls of the inner tube, cooling the vapors and causing them to condense to liquid droplets that flow back into the reaction vessel. An *air condenser* is ordinarily a single tube whose walls transfer heat to the surrounding air, cooling and condensing the vapors of a boiling liquid.

For standard scale work, a water-cooled West condenser is generally used for reactions conducted under reflux (see Figure B9). A typical microscale lab kit contains both a water-cooled condenser and an air condenser. As a rule, air condensers should be used with nonaqueous solvents that boil at 150°C or above, or with small amounts of lower boiling solvents that are heated gently. Air condensers can also be used with aqueous solutions because any loss of water vapor into the laboratory air does not present a safety hazard. When in doubt, it is best to use a water-cooled condenser for its more efficient cooling action.

Figure B10a shows a microscale reflux apparatus with a round-bottom flask and a water-cooled condenser, and Figure B10b illustrates one with a conical vial and an air condenser. Either type of condenser can be used with either kind of reaction vessel, however.

A reflux apparatus consisting of a cold-finger condenser (a water-cooled tube) inserted through a notched rubber stopper into a test tube is convenient for some small-scale reactions, such as those used in qualitative analysis. If you have such a condenser, your instructor can show you how to use it.

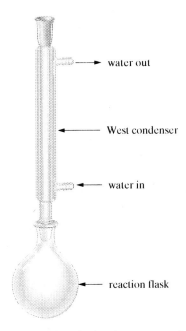

Figure B9 Standard scale apparatus for heating under reflux

Do not mistake a jacketed distilling column for a standard scale condenser; the condenser has a smaller diameter.

A West condenser (also called a Liebig–West condenser) is provided in most standard scale lab kits.

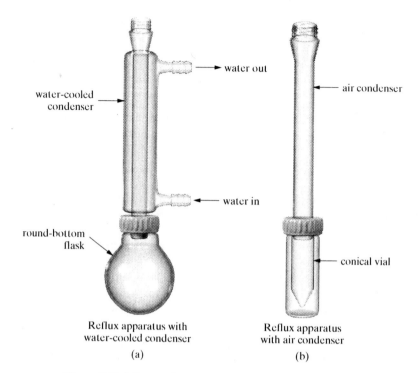

Figure B10 Microscale apparatus for heating under reflux

General Directions for Heating under Reflux

Safety Notes

> **Never heat the reaction flask before the condenser water is turned on; solvent vapors may escape and cause a fire or health hazard.**

 Standard Scale

Equipment and Supplies

heat source
round-bottom flask
West condenser
boiling chips or stir bar
two lengths of rubber tubing

Select an appropriate heat source, such as a heating mantle or steam bath. Position the heat source at the proper location on or near a ring stand so that it can be quickly removed if the flask should break or the reaction become too vigorous. Select a round-bottom flask of a size such that the reactants fill it about half full or less. Clamp this *reaction flask* securely to the ring stand at the proper location in relation to the heat source. Transfer [OP-6] the reactants and any specified solvent to the reaction flask.

Solids should be added through a powder funnel or with a weighing dish or a square of weighing paper, and liquids should be added through a stemmed funnel. Add a few boiling chips or a magnetic stir bar (see OP-10) and mix the reactants by swirling or stirring. Insert a West condenser into the flask, making sure that the joint is tight. Do *not* stopper the condenser because that will create a closed system, which may shatter violently when heated. Put a clamp near the top of the condenser to keep the apparatus from toppling over if jarred, but don't tighten its jaws completely.

Take Care! Never heat a closed system.

Connect the water inlet (the lower connector) on the condenser jacket to a cold water tap with a length of rubber tubing, and run another length of tubing from the water outlet (the upper connector) to a sink drain, making sure that it is long enough to prevent splashing when the water is turned on. If the rubber tubing slips off when pulled with moderate force, replace it by tubing of smaller diameter or secure it with wire or a tubing clamp. Turn on the water carefully so that the condenser jacket slowly fills with water from the bottom up, and adjust the water pressure so that a narrow stream flows from the outlet. The flow rate should be just great enough to (1) maintain a continuous flow of water in spite of pressure changes in the water line and (2) keep the condenser at the temperature of the tap water during the reaction. Excessively high water pressure may force the tubing off the condenser and spray water on you and your neighbors.

If you are using a stir bar, begin stirring at a moderate rate. Turn on the heat source and adjust it to keep the solvent boiling gently, measuring the reaction time from the time that boiling begins, when a continuous stream of bubbles will rise through the liquid. If the liquid bubbles violently or froths up, reduce the heating rate. Reflux has begun when liquid begins to drip into the flask from the condenser. The vapors passing into the condenser will then form a *reflux ring* of condensing vapors that should be clearly visible. Below this point, liquid will be seen flowing back into the flask; above it, the condenser should be dry. If the reflux ring is more than halfway up the condenser, reduce the heating rate or increase the water flow rate (for a water-cooled condenser) to prevent the escape of solvent vapors.

At the end of the reaction period, turn off the heat source and remove it from contact with the flask. Let the apparatus cool; then turn off the condenser water and pour the reaction mixture into a container suitable for the next operation. Clean [OP-1] the reaction flask as soon as possible so that residues do not dry on the glass.

Summary

1. Position heat source.
2. Clamp flask in or over heat source.
3. Add solvent and reactants.
4. Add boiling chips or a stir bar.
5. Insert condenser, clamp in place, and attach tubing.
6. Turn on condenser water and adjust flow rate.
7. Start stirrer (if used), adjust heat control so that reaction mixture boils gently.
8. Readjust water flow or heating rate as necessary, boil gently until end of reflux period.
9. Turn off and remove heat source, let flask cool, transfer reaction mixture.
10. Disassemble and clean apparatus.

Microscale

Equipment and Supplies

heat source
conical vial or round-bottom flask
air condenser or water-cooled condenser
boiling chip(s) or stirring device
two lengths of rubber tubing (for water-cooled condenser only)

As the *reaction vessel*, select a conical vial or round-bottom flask of a size such that the reactants fill it about half full or less. Transfer [OP-6] the reactants and any specified solvent to the reaction vessel. It is best to weigh limiting reactants directly into the reaction vessel. Add a boiling chip or a magnetic stirring device (see OP-10) and mix the reactants by swirling or stirring. Attach an appropriate condenser to the reaction vessel, making sure that the compression cap is screwed on securely (see OP-2). Do *not* stopper the condenser because that will create a closed system, which may shatter violently when heated. Clamp the apparatus to a ring stand and lower it into a heat source such as a heating block or sand bath.

Take Care! Never heat a closed system.

If you are using an air condenser, go to the next paragraph. If you are using a water-cooled condenser, connect the water inlet (the lower connector) on the jacket to a cold water tap with a length of rubber tubing and run another length of tubing from the water outlet (the upper connector) to a sink, making sure that it is long enough to prevent splashing when the water is turned on. If the rubber tubing slips off when pulled with moderate force, replace it with tubing of smaller diameter or secure it with wire or a tubing clamp. Turn on the water carefully so that the condenser jacket slowly fills with water from the bottom up and adjust the water pressure so that a narrow stream flows from the outlet. The flow rate should be just great enough to (1) maintain a continuous flow of water in spite of pressure changes in the water line and (2) keep the condenser at the temperature of the tap water during the reaction. Excessively high water pressure may force the tubing off the condenser and spray water on you and your neighbors.

If you are using a stirring device, begin stirring at a moderate rate. Adjust the heat setting or the depth of the reaction vessel in the heat source to keep the solvent boiling gently, measuring the reaction time from the time that boiling begins, when a continuous stream of bubbles will rise through the liquid. If the liquid bubbles violently or froths up, reduce the heat setting or reposition the reaction vessel in the heat source. If there is sufficient liquid in the reaction vessel, its vapors should form a *reflux ring* of condensing vapors in the condenser. Below this point, liquid will be seen flowing back into the reaction vessel; above it, the condenser should be dry. If the reflux ring is more than halfway up the condenser, reduce the heating rate or increase the water flow rate to prevent the escape of solvent vapors.

At the end of the reaction period, turn off the heat source and raise the apparatus on the ring stand, clamping it several inches above the heat source. When the reaction vessel has cooled nearly to room temperature, turn off the condenser water (if used). Unless the next operation will be carried out in the reaction vessel, transfer its contents to a container suitable for that operation. Clean [OP-1] the reaction vessel as soon as possible so that residues do not dry on the glass.

Summary

1. Transfer reactants and solvent to reaction vessel.
2. Add boiling chip or a stirring device.
3. Attach appropriate condenser.
4. Clamp apparatus over heat source.
   IF an air condenser is being used, GO TO 7.
5. Attach tubing to water-cooled condenser.
6. Turn on condenser water and adjust flow rate.
7. Start stirrer (if used), adjust heat so that reaction mixture boils gently.
8. Readjust water flow or heating rate as necessary, boil gently until end of reaction period.
9. Turn off heat source, let reaction vessel cool, transfer reaction mixture.
10. Disassemble and clean apparatus.

When Things Go Wrong

If, when you turn on the cooling water, there are air bubbles in the condenser jacket, you probably connected the rubber tubes to the wrong hose connectors. The tube from the water tap is attached to the condenser's bottom connector and the tube leading to a drain is attached to the top connector.

If, as you heat a reaction mixture, the liquid bumps or foams, you probably forgot to add boiling chips or to use a magnetic stirrer. Drop in a stirring device and start stirring, or let the apparatus cool for several minutes and then drop one or more boiling chips down the top of the condenser. If you were already using boiling chips or a stirrer, you are probably heating the reaction mixture too strongly. Reduce the heating rate by raising the apparatus or lowering the heat source a few millimeters (for rapid cooling) or by turning down the heat control (for slower cooling), or both. You can always return the apparatus to its original location once the heating rate has decreased.

If, as you heat a reaction mixture, the reflux ring of condensing vapors rises above the midpoint of the condenser or the liquid level in the reaction vessel goes down, either the cooling water isn't flowing fast enough (or at all), you are overheating the reaction mixture, or you are using an air condenser when you should be using a water-cooled condenser. If you are using a water-cooled condenser, it should be cool to the touch. If not, adjust (or turn on) the cooling water so that a steady stream flows through the condenser—not a strong stream that may force the hose off the condenser. If the reflux ring is still too high, reduce the heating rate as described previously until the reflux ring stays in the bottom half of the condenser. If you are using an air condenser, reduce the heating rate or replace the air condenser by a water-cooled condenser (let the reaction mixture cool down before switching condensers). If the liquid level in the reaction vessel has gone down, add more solvent to replace any that boiled away.

If, when you are taking the apparatus apart, you can't separate the condenser from the reaction flask or vial, the glass joints have probably frozen. See "Disassembling Glassware" in OP-2.

OPERATION 8

Cooling

Some reactions proceed too violently to be conducted safely at room temperature, or involve reactants or products that decompose at room temperature. In such cases, the reaction mixture is cooled with some kind of *cold bath*, which can be anything from a beaker filled with cold water to an electrically refrigerated device. Cold baths are also used to increase the yield of crystals from a reaction mixture or recrystallization mixture.

A setup like the one shown in Figure B5 (OP-7) for a hot water bath can also be used for a cold bath. A cold bath can be prepared using any suitable container, such as a beaker of suitable size, a crystallization dish, an evaporating dish, or a pair of nested Styrofoam cups. A beaker can be wrapped with glass wool and placed inside a larger beaker to keep it cold longer, if necessary.

A number of cooling media are used for cold baths. A mixture of ice (or snow) and water can be used for cooling in the 0 to 5°C range. The ice should be finely divided and enough water should be present to just cover the ice, since ice alone is not an efficient heat-transfer medium. An ice–salt bath consisting of three parts of finely crushed ice or snow to one part of sodium chloride can attain temperatures down to −20°C, and mixtures of $CaCl_2 \cdot H_2O$ containing up to 1.4 g of the calcium salt per gram of ice or snow can provide temperatures down to −55°C. In practice, these minimum values may be difficult to attain, as the actual temperature of an ice–salt bath depends on such factors as the fineness of the ice and salt and the insulating ability of the container. Temperatures down to −75°C can be attained by mixing small chunks of dry ice with acetone, ethanol, or another suitable solvent in a vacuum-jacketed container such as a Dewar flask.

Temperatures below −40°C can't be measured using a mercury thermometer because mercury freezes at that temperature.

Take Care! Never handle dry ice with bare hands.

General Directions for Cooling

 Standard Scale and Microscale

Equipment and Supplies

    cold bath container
    cooling medium
    thermometer
    air condenser (microscale, optional)

Obtain a suitable cold bath container and fill it with the cooling medium to a level depending on the size of the container to be cooled. When this container is immersed in the cold bath, the cooling medium should fill the cold bath container about three-fourths full. Clamp a thermometer [OP-9] so that its bulb is entirely immersed in the cooling medium but not touching either container. If you are using an ice–salt bath, mix in the appropriate salt in small portions, waiting for the temperature to equilibrate after each addition, until the desired temperature is attained. Lower the container to be cooled into the cooling bath so that the liquid level in that container is below the cooling fluid level. For microscale work, attach an air condenser to any threaded container being cooled (unless experimental conditions preclude this) and clamp the condenser to a ring stand. Otherwise, either clamp the neck of the container to a ring stand or hold the container in your hand so that it doesn't tip over. Keep the contents of the cold bath mixed by occasional stirring or swirling. The contents of the container being cooled can also be swirled or stirred for more efficient cooling. Add small portions of ice as needed to keep the temperature in the desired range, removing an equal amount of water (with a pipet, e.g.) to make room for it.

Summary

1. Fill container with cooling medium.
2. Insert thermometer in cooling medium, adjust temperature if necessary.
3. Insert container to be cooled.
4. Keep contents of container and cooling bath mixed, adjust temperature as needed.

Temperature Monitoring   OPERATION **9**

In the organic chemistry lab, thermometers are used to monitor the temperatures of heating devices, cooling baths, reaction mixtures, distillations, and for many other purposes. Such thermometers should have a range of at least $-10°C$ to $260°C$, and a wider range is desirable for some purposes. Most broad-range glass thermometers contain mercury, which is toxic and presents a safety hazard if a thermometer is broken, but broad-range nonmercury thermometers are also available. Short glass thermometers (about 15 cm long) are available for microscale work, and bimetallic thermometers with metal probes can be used to measure the temperatures of some heating devices. For the most accurate temperature readings, a thermometer should be *calibrated* and an *emergent stem correction* applied as described in OP-33, but this is generally not necessary for routine temperature monitoring.

To read a thermometer accurately, rotate it so that its graduation marks extend across the mercury column and view it with your line of sight perpendicular to the thermometer and level with the top of the mercury column. You should then be able to read the temperature accurately to at least the nearest half degree.

You can monitor the temperature of a liquid or a solid heating medium (such as sand) using a thermometer clamped with its bulb entirely immersed in the heating medium. It should be held in place by a three-fingered clamp or a special thermometer clamp, or inserted into a stopper that is held by a utility clamp. The thermometer should not touch the side or bottom of the container or anything inside the container.

To monitor the temperature of a reaction mixture that will be stirred [OP-10] in an open container such as an Erlenmeyer flask, clamp a thermometer so that its bulb is completely immersed in the mixture but doesn't contact the stirring device (a large stir bar could break the thermometer bulb). If a magnetic stirrer isn't available, you may have to hold a thermometer inside the reaction vessel as it is being swirled or shaken. Do this by holding the neck of the flask and nesting the thermometer stem in the "vee" between your thumb and index finger, so that the bulb of the thermometer is held securely inside the flask and continuously immersed in the liquid. With a little practice, you should be able to mix the contents of the flask quite vigorously without damage to the thermometer. If continuous mixing isn't necessary, you can insert the thermometer each time you stop shaking or swirling the reaction flask, read it when the temperature has stabilized, then remove it and resume mixing. Never use the thermometer itself for stirring because the bulb is fragile and breaks easily.

Using a Thermometer Adapter

You will ordinarily need a *thermometer adapter* to monitor the temperature of an operation (such as distillation) conducted in a jointed glassware setup (see Figure B11). Be certain that the thermometer adapter, when inserted in the apparatus, does not create a closed system. For example, never put a sealed thermometer adapter assembly on top of a reflux condenser that is attached directly to a reaction flask, because heating such a system may cause it to shatter.

To use a standard scale thermometer adapter, carefully insert the bulb of the thermometer through the rubber cap of the adapter using an appropriate lubricant (see OP-3). Then secure the adapter in the appropriate joint on the apparatus and carefully raise or lower the thermometer so that its bulb is positioned correctly. To prevent accidental breakage, remove the thermometer assembly from the apparatus before repositioning or removing the thermometer.

To use a microscale thermometer adapter, first slide a small rubber O-ring down the thermometer stem to near its center (the O-ring should come with the adapter). Secure the adapter, without its upper compression cap, in the appropriate joint on the apparatus. Insert the thermometer (bulb end down) through the adapter and move the O-ring up or down until its bulb is positioned correctly when the O-ring is resting on the adapter. Then slide the adapter's threaded cap down the thermometer stem and use it to secure the thermometer assembly to the apparatus.

Take Care! Never heat a closed system.

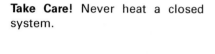
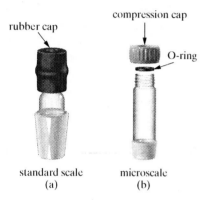

Figure B11 Thermometer adapters

standard scale (a)   microscale (b)

OPERATION 10

Mixing

Reaction mixtures are often stirred, shaken, or agitated in some other way to promote efficient heat transfer, prevent bumping, increase contact between the components of a heterogeneous mixture, or mix in a reactant as it is being added. Mechanical stirring, which requires a motor that turns a shaft connected to a stirring paddle, is seldom used in undergraduate organic chemistry labs and will not be described here.

Manual Mixing

If you are carrying out a reaction in an Erlenmeyer flask or a test tube, the reactants can be mixed by using a stirring rod or spatula, or by manual shaking or swirling. A motion combining shaking with swirling is more effective than swirling alone. For standard scale work, you can sometimes mix the reactants adequately by clamping the apparatus *securely* to a ring stand and carefully sliding the base of the ring stand back and forth. But when more efficient and convenient mixing is required, particularly over a long period of time, it is best to use a magnetic stirrer as described in the next section.

Manual mixing may be used at other times than during reactions. For example, liquids can be dried [OP-25] by swirling the liquid with a drying agent in an Erlenmeyer flask. This increases the amount of contact between the liquid and the particles of drying agent, increasing drying efficiency. Small quantities of liquids are often dried by stirring the liquid with the drying agent in a conical vial. You can do this by twirling the pointed end of a microspatula in the bottom of the vial. Similarly, you can twirl the rounded end of a flat-bladed microspatula inside a Craig tube or a small test tube to stir recrystallization solutions and other mixtures.

Magnetic Stirring

A magnetic stirrer (Figure B12) is an enclosed unit containing a motor that rotates a bar magnet underneath a metal or ceramic platform. As the bar magnet rotates, it in turn spins a Teflon-coated stirring device inside a container placed on or above the platform. The most common stirring device, called a *stir bar*, is an oblong (usually cylindrical) Teflon-coated magnet. Because no moving parts extend outside of the container in which stirring occurs, a reaction assembly that is to be stirred magnetically can be completely enclosed if necessary. The rate of stirring is controlled by a dial on the magnetic stirrer. For efficient stirring, the vessel (flask, beaker, conical vial, etc.) containing the stirring device should be positioned near the center of the stirring unit and as close to its platform as practicable.

Magnetic stirrers can be used in conjunction with heating mantles, oil baths, steam baths, and other heat sources that are constructed of nonferrous materials. A hot plate–stirrer has a heating device in the same unit that houses the magnetic stirrer, so it has two dials, one to control the heating unit and the other to control the stirrer. Hot plate–stirrers can be used to heat and stir flat-bottomed containers such as Erlenmeyer flasks directly, but they are also used in conjunction with heating devices that require an external heat source, such as aluminum blocks, hot water baths, and sand baths [OP-7].

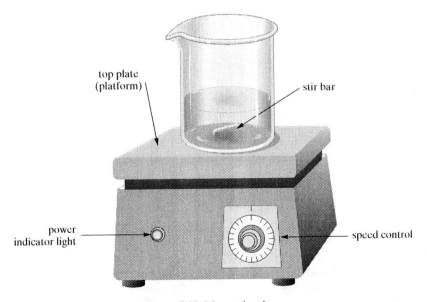

top plate (platform)

stir bar

power indicator light

speed control

Figure B12 Magnetic stirrer

Figure B5 (OP-7) shows a hot plate–stirrer being used to heat a hot water bath and stir both the bath liquid and a reaction mixture at the same time.

A hot plate–stirrer is nearly indispensable in the microscale laboratory. The stirring device can be either a *spin vane*, which is used in conical vials, or a small stir bar, which is used in round-bottom flasks, small Erlenmeyer flasks, beakers, and other vessels having flat or gently rounded bottoms. To visualize a spin vane, imagine a wedge-shaped piece of Teflon having a small bar magnet set crosswise on its short side (see Appendix I for an illustration). When the spin vane is inserted into a conical vial with its point down, the magnetic bar is quite distant from the platform, so the spin vane may not rotate properly unless it is centered correctly and the stirring rate is relatively low. If necessary, you can use a small stir bar to stir the contents of a conical vial. At low speeds, the stir bar wobbles around in the conical end of the vial and doesn't stir very efficiently, but at higher speeds, it should spin horizontally in the wider part of the vial and create a vortex that promotes efficient mixing.

General Directions for Stirring a Reaction Mixture Magnetically

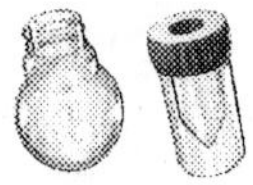

 Standard Scale and Microscale

Equipment and Supplies

    stirrer or hot plate–stirrer
    heating device
    reaction vessel (flask, conical vial, etc.)
    stir bar or spin vane

Set the heating device (heating mantle, aluminum block, sand bath, etc.), if any, directly on the platform of the magnetic stirrer or hot plate–stirrer. Position the reaction vessel in the heating device so that it is close to the center of the stirring unit's platform, secure it with a clamp if necessary, and drop in a stir bar or spin vane. (Don't add boiling chips; the stirring action prevents bumping.) If you are using a spin vane, position it with its sharply pointed end facing down. If you need to stir a heating bath as well, use a stir bar that is larger than the one in the reaction mixture. If necessary, attach a condenser or other device to the reaction vessel. Start circulating water through a water-cooled condenser, if you are using one. Start the magnetic stirrer and adjust the stirring rate dial carefully so that the stirring action is smooth but vigorous (you should see a vortex in the middle of the container). If the stirring rate is too high or the reaction vessel is not positioned correctly, the stirring device will flop around erratically rather than rotate smoothly. If that happens, reset the dial to a low value and increase it gradually until a suitable rate is attained, or reposition the reaction vessel to bring it closer to the top and center of the stirring unit's platform. High stirring rates may be needed for heterogeneous reaction mixtures, such as those involving two immiscible liquids, but in most cases a moderate stirring rate is suitable. If you need to heat the stirred reaction mixture, adjust the heating rate as described in OP-7 for the heating device you are using.

Summary

1. Place heating device on stirring unit.
2. Secure reaction vessel on or inside heating device.
3. Drop in stir bar or spin vane.
4. Adjust stirrer for appropriate stirring rate.

When Things Go Wrong

If, as you increase the spinning rate of a stirring device (stir bar or spin vane), the stirring device wobbles or otherwise moves erratically, try the following remedies in order. Make sure you're not using a ferrous container, such as a stainless steel beaker for an oil bath. Reduce the stirring rate and move the flask (or other vessel) that contains the stirring device closer to the center of the platform, then advance the speed control *slowly* until you attain a suitable stirring rate. If the stirring device still turns erratically, try lowering the container so that it is closer to the top of the magnetic stirrer's platform. If you are using a large beaker for a water bath, for example, replace it by a smaller, lower beaker or a porcelain dish. If that doesn't help, try a different magnetic stirrer or stirring device.

If the stir bar you are using for a reaction that requires thorough mixing (such as a heterogeneous reaction) isn't stirring the reaction mixture vigorously enough (it should produce a vortex in the liquid at high speed), switch to a larger stir bar.

Addition of Reactants

In many organic preparations, the reactants are not all combined at the start of the reaction. Instead, one or more of them is added during the course of the reaction. This is necessary when the reaction is strongly exothermic or when one of the reactants must be kept in excess to prevent side reactions. Solid reactants can be added slowly or at regular intervals from a plastic weighing dish that is bent to form a pouring spout. Solids can also be divided into small portions that are added at regular intervals. For standard scale work, liquids are added in portions or drop by drop using a separatory funnel or a special addition funnel. For microscale work, liquids are generally added using a syringe or a Pasteur pipet.

 ## Standard Scale Addition

An *addition funnel* has a cylindrical body, a drain tube controlled by a stopcock, and a pressure-equalizing tube to equalize the pressure in the reaction vessel and addition funnel, allowing its contents to flow freely into the reaction vessel. A *separatory funnel* usually has a pear-shaped body attached to a similar drain tube but lacks a pressure equalizing tube (see Figure B13). Since addition funnels are seldom available in undergraduate labs, a separatory funnel (which we will call a separatory–addition funnel here) is generally used for that purpose. An opening must be left at the top of the funnel for air to enter; otherwise the liquid outflow will create a vacuum and the flow will eventually stop. If a reaction is run in an open container, such as an Erlenmeyer flask, the separatory–addition funnel can

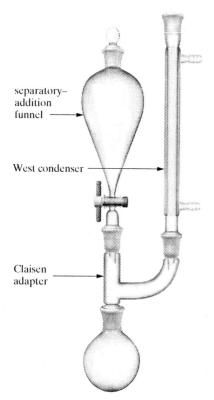

separatory–
addition
funnel

West condenser

Claisen
adapter

Figure B13 Apparatus for standard scale addition under reflux

simply be clamped to a ring stand above the flask, which is shaken and swirled during the addition. For a reaction conducted under reflux, see the general directions for standard scale addition that follow.

 ## Microscale Addition

Small amounts of reactants can be added directly to a reaction vessel using a short Pasteur pipet or through a reflux condenser into the reaction vessel using a long (9-inch) Pasteur pipet. When an addition must be made over an extended period, it is more convenient to use a syringe. The syringe is filled with a measured amount of the liquid to be added and then inserted through a septum so that its needle is directly over the reaction mixture. A rubber septum should remain airtight even after it has been punctured by a syringe needle. A Teflon-lined septum (the kind that fits inside a compression cap) may not remain airtight after being punctured, so if you are using such a septum, try to find an intact septum or one that has already been punctured but still fits tightly around the syringe needle. (Your instructor may want you to use a punctured one to keep the others intact.) Remember that a syringe needle is very sharp (unless it has been blunted to prevent injury) and may be contaminated with dangerous biological or chemical substances, so be careful not to stick yourself or anyone else with it.

General Directions for Addition under Reflux

 ## Standard Scale

Equipment and Supplies

> round-bottom flask
> Claisen adapter
> separatory–addition funnel
> West condenser
> stopper (can be omitted for most aqueous solutions or other nonvolatile liquids)

Measure the appropriate reactants into the flask and assemble the apparatus shown in Figure B13, placing the separatory–addition funnel on the straight arm of the Claisen adapter so that it is directly over the reaction flask. Clamp the flask and the Claisen adapter securely to a ring stand. Make sure the stopcock is closed; then place the liquid to be added in the separatory–addition funnel. Unless this liquid must be protected from atmospheric moisture, place a strip of filter paper between the stopper and the funnel's ground-glass joint. (If the liquid is moisture sensitive, insert a drying tube [OP-12a] filled with drying agent in the top of the funnel.) Add the liquid either in portions or continuously, as directed in the experiment. For portionwise addition, add small portions of the liquid at regular intervals by opening the stopcock momentarily while stirring magnetically or shaking to mix the reactants. For continuous addition, open the stopcock just far enough so that the liquid drips or drizzles slowly into the reaction flask and adjust the stopcock position to provide the desired rate of addition.

Continuous addition is usually carried out dropwise (drop by drop), with magnetic stirring or periodic shaking [OP-10] to keep the reactants mixed.

Some reactions conducted under reflux also require that the temperature of the reaction mixture be monitored [OP-9] or that the mixture be stirred with a mechanical stirrer (see OP-10), in which case the flask and Claisen adapter shown in Figure B13 are replaced by a three-necked flask (Figure B14). For temperature monitoring, the addition funnel should be inserted in the middle neck and the reflux condenser and thermometer in the outer necks. For mechanical stirring, the stirrer shaft is inserted in the middle neck and the addition funnel and reflux condenser in the outer necks.

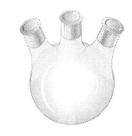

Figure B14 Three-necked flask

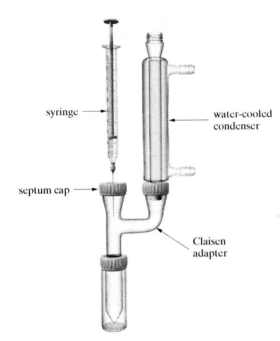

Figure B15 Apparatus for microscale addition under reflux

 Microscale

Equipment and Supplies

> reaction vessel (conical vial or round-bottom flask)
> Claisen adapter
> septum cap
> air condenser or water-cooled condenser
> syringe

Measure the appropriate reactants into the reaction vessel. Assemble an apparatus like that illustrated in Figure B15 by inserting a Claisen adapter in the reaction vessel and attaching an appropriate condenser to its long arm and a septum cap (a compression cap with a septum inside) to its short, straight arm. Clamp the apparatus securely to a ring stand. Fill the syringe with the desired volume of the liquid to be added (see OP-5). Hold the syringe vertically with its needle pointing down and carefully insert the needle through the septum. Add the liquid in small portions during the reaction, as directed in the experiment, by depressing the plunger. If necessary, remove the syringe to refill it, then immediately replace it on the apparatus. When the addition is complete, remove the syringe and clean it (see OP-5) without delay.

When Things Go Wrong

If, during a standard scale reaction, the liquid you are adding stops draining from a separatory–addition funnel, its stopper may be inserted too tightly. Remove the stopper and place a folded piece of paper between it and the joint. If the liquid still doesn't drain, close the stopcock and pour the liquid out the top of the funnel into another container. Filter [OP-15] the liquid if it contains solid impurities. Remove the stopcock, use a piece of wire (or a toothpick, etc.) to clean any debris out of the hole, and rinse the hole with distilled water or an appropriate organic solvent. Then insert the stopcock in the separatory-addition funnel, return the liquid to the funnel, and continue.

OPERATION 12   # Excluding Water from Reaction Mixtures

a. Using Drying Tubes

Some chemicals react with water vapor from the air, and some reactions are inhibited or prevented by traces of water. Water-sensitive reactions can be carried out and water-sensitive chemicals protected by attaching *drying tubes* (see Figures B16 and B17) wherever an apparatus is open to the atmosphere. The drying tube is filled with a suitable desiccant (drying agent) such as calcium chloride, calcium sulfate (Drierite), granular alumina, or silica gel. Calcium chloride is the least efficient of these, but it is adequate in many cases.

When a very dry atmosphere is required, the apparatus should be swept out by passing dry nitrogen or another dry gas through it as described in Operation 13.

Unless your instructor directs otherwise, put used desiccant in a designated container (*not* the container you got it from) when you are done. If you leave calcium chloride in a drying tube exposed to the atmosphere, its granules will eventually clump together in a solid mass that can only be removed by immersing the drying tube in water overnight or longer until the solid dissolves.

To prepare a standard scale drying tube, use a glass rod or applicator stick to push a small plug of dry cotton (or glass wool) into the drying tube until it covers the narrow opening and tamp it down gently to hold it in place. Add calcium chloride or another drying agent through the top of the drying tube until it is filled to within a few centimeters of the top; then insert another plug of cotton in the top to prevent spills. Push the connector at the bottom of the drying tube into a thermometer adapter and insert the adapter into the top of a reflux condenser (or into any other part of an apparatus that is open to the atmosphere) as shown in Figure B16. If the apparatus includes a vacuum adapter, the drying tube can be attached to its sidearm using a short length of rubber tubing.

To prepare a microscale drying tube (Figure B17), use an applicator stick or glass rod to push a plug of cotton or glass wool through its long end

Take Care! Never stopper the drying tube, as this will result in a closed system that might explode or fly apart when heated.

nearly to the bend in the tube. Holding the drying tube with its long arm upright (open end up), use a plastic weighing dish to add enough calcium chloride (~1.5 g) or another drying agent to form a layer 3–4 cm deep; then insert another plug of cotton or glass wool to keep the desiccant in place. Insert the drying tube into the top of a reflux condenser or any other part of an apparatus that is open to the atmosphere.

b. Water Separation

During some reactions that yield water as a by-product, it may be necessary to remove the water to prevent the decomposition of a water-sensitive product or to increase the yield. This is sometimes done by codistilling (see OP-20b) the water with an organic solvent that forms a low-boiling azeotrope (see OP-32) with water. This process is known as *azeotropic drying*. The solvent used must be less dense than water and immiscible in it. For example, water and toluene (bp 111°C, d = 0.866 g/mL) yield an azeotrope containing 13.5% water by mass that distills at 84°C, so toluene is often used to remove water from reaction mixtures.

The codistillation can be accomplished with an ordinary simple distillation apparatus [OP-30a], but for standard scale work it is more convenient to use a *water separator*, such as the Dean–Stark trap shown in Figure B18. The Dean–Stark trap is inserted in the reaction flask, filled to the level of its sidearm with the reaction solvent and fitted with a reflux condenser. As water forms during the reaction, its vapors and those of the reaction solvent condense inside the reflux condenser and drip down into the water separator. The organic solvent, which separates on top of the water layer, overflows through the sidearm and returns to the reaction flask, while the water stays in the bottom of the separator. The theoretical yield of water from the reaction can be calculated, so the volume of water in the separator is monitored to determine when the reaction is nearing completion.

If a Dean–Stark trap is unavailable, the parts from a standard scale organic lab kit can be assembled as shown in Figure B19 so that the 25-mL flask, still head, and vacuum adapter together function like a Dean–Stark trap. When you assemble this apparatus, clamp both flasks securely and close the vacuum adapter outlet with a rubber bulb from a medicine dropper. Use a funnel to fill the water separator with the organic solvent to a level just below the bottom of the sidearm. Heat the reaction mixture under gentle reflux, being careful that the adapter's drip tip doesn't become flooded with liquid (an adapter with a missing drip tip works better). When disassembling the apparatus after the reaction, leave the water separator clamped while the other components are removed, then tilt it carefully to pour liquid from its sidearm into a beaker.

For microscale work, a Hickman still can function as a water separator in an apparatus such as one pictured in Figure E11 [OP-30a]. The well of the Hickman still, which corresponds to the solvent reservoir of a Dean–Stark trap, can be previously filled with toluene or another suitable organic solvent. Alternatively, an amount of excess organic solvent equal to the well's capacity (1–2 mL) can be added to the reaction vessel, from which it distills into the well during the reaction. Any water that forms during the reaction distills into the Hickman still and collects in the bottom of its well, while the organic solvent that distills with the water overflows into the reaction flask.

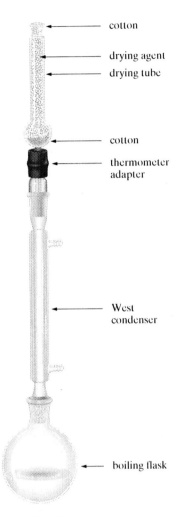

Figure B16 Standard-scale apparatus for heating under reflux in a dry atmosphere

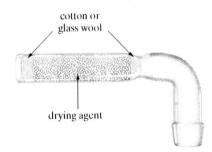

Figure B17 Microscale drying tube

Figure B18 Dean–Stark trap

A "homemade" Dean–Stark trap can be constructed as described in J. Chem. Educ, **1963,** *40,* 349.

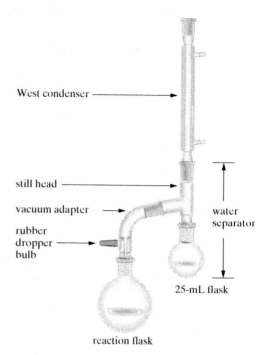

West condenser

still head

vacuum adapter

rubber dropper bulb

water separator

25-mL flask

reaction flask

Figure B19 Apparatus for standard scale water separation

OPERATION 13   Excluding Air from Reaction Mixtures

Some chemicals react readily with oxygen from the air, so reactions using such chemicals must be conducted in an inert (oxygen-free) atmosphere. The simplest and least expensive way to provide an inert atmosphere is to flush all parts of a reaction apparatus with nitrogen. This process also removes water vapor and thus provides a dry atmosphere as well.

To flush a standard-scale reflux assembly with nitrogen, first assemble the apparatus pictured in Figure B20a. Use glassware components that have been oven-dried and then cooled. Coat all the joints with a thin layer of joint grease [OP-1] and use Keck clips (see your instructor) to keep the joints from separating. Fold a rubber septum over the straight arm of the Claisen adapter and another rubber septum over the top of the condenser. Insert syringe needles (but not the syringes) through both septa. To make a gas bubbler that will monitor the nitrogen flow, attach a piece of glass tubing to one end of a length of rubber or plastic tubing and clamp it over a test tube partly filled with mineral oil, inserting its open end in the oil; then attach the other end of the tubing to the base of the syringe needle at the top of the condenser. Use another piece of tubing to attach the needle on the Claisen adapter to a nitrogen source. Open the nitrogen valve just far enough to produce a gentle stream of bubbles in the mineral oil. After a few minutes reduce the flow to maintain a small positive nitrogen pressure in the apparatus, and then carry out the reaction. If the reaction mixture is being heated, increase the flow rate when you turn off or remove the heat source to maintain a positive pressure.

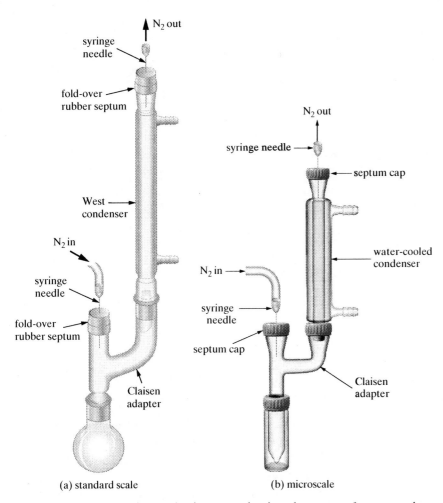

Figure B20 Apparatus for conducting a reaction in a dry, oxygen-free atmosphere

To flush a microscale reflux assembly with nitrogen, assemble an apparatus like that shown in Figure B20b. The glassware components should have been oven-dried and then cooled. Be sure all compression caps have undamaged O-rings and are screwed down tightly. Attach a septum cap to the top of the condenser and insert a syringe needle through it. To make a gas bubbler that will monitor the nitrogen flow, attach a piece of glass tubing to one end of a length of rubber or plastic tubing and clamp it over a test tube partly filled with mineral oil, inserting its open end in the oil; then attach the other end of the tubing to the syringe needle at the top of the condenser. Use another piece of tubing to attach the needle on the Claisen adapter to a nitrogen source. Turn on the nitrogen flow enough to produce a gentle stream of bubbles in the mineral oil. After a few minutes turn down the flow enough to maintain a small positive nitrogen pressure in the apparatus, and then carry out the reaction. If the reaction mixture is being heated, increase the flow rate when you turn off or remove the heat source to maintain a positive pressure while the reaction flask cools down.

If it will be necessary to add reactants [OP-11] during a microscale reaction, attach a balloon to the barrel of a plastic syringe (it may be necessary to cut off the top of the barrel) and fill it with nitrogen through a plastic tube secured over the syringe's needle. When the balloon is full, remove the tube, pinch the neck of the balloon to keep nitrogen from escaping, and immediately insert the needle through the condenser's septum. When the addition syringe is inserted through the adapter's septum cap, enough air should escape to allow the apparatus to fill with nitrogen.

OPERATION **14**   # Trapping Gases

The best way to keep toxic and smelly gases out of the laboratory air is to conduct all reactions under an efficient fume hood. If that is not possible, or if the hood isn't adequate, gases can be removed using either a gas trap or a water aspirator.

A gas trap containing a suitable gas-absorbing liquid or solid will remove most gases effectively. Water alone will dissolve some gases, but dilute aqueous sodium hydroxide (about 5% or 1 M) is generally used for acidic gases, such as HBr or SO_2. It converts them to salts that dissolve in the water.

$$HBr + NaOH \rightarrow NaBr + H_2O$$

$$SO_2 + NaOH \rightarrow NaHSO_3$$

Similarly, dilute aqueous HCl can be used to trap ammonia and other alkaline gases. Activated charcoal, in the form known as pelletized Norit, is quite effective for removing small quantities of gases during microscale reactions. It removes neutral gases as well as acidic and basic ones, so it can also be used to keep strong or offensive odors out of the lab. Glass wool moistened with water can remove some odors, but it is only effective for the odors of water-soluble organic compounds.

You can construct a simple gas trap for standard scale work by clamping an inverted narrow-stemmed funnel over a beaker containing a suitable gas-absorbing liquid and then lowering the funnel so that its rim just touches the surface of the liquid, as shown in Figure B21. Connect the gas trap to the reaction apparatus at any point that is open to the atmosphere (usually the top of a reflux condenser) using a short length of fire-polished glass tubing (see OP-3) inserted into a thermometer adapter or rubber stopper.

To remove water-soluble gases such as hydrogen halides with a water aspirator, attach a vacuum adapter to the top of your reflux condenser or any other part of your apparatus that is open to the atmosphere and use a length of heavy-walled rubber tubing to connect the sidearm of the adapter to a water aspirator. Turn on the aspirator while the gas is being generated. The gas should be drawn into the aspirator, dissolve in the water, and pass down the drain. Before using an aspirator for this purpose, make sure that it is legally permissible in your city or state to flush such gases into the wastewater system.

For small amounts of gases, a straight glass tube inserted into a test tube containing the gas-absorbing liquid may be adequate. Clamp the outlet of the glass tube about a millimeter *above* the surface of the liquid to keep it from backing up into the reaction apparatus when gas evolution ceases or heating is discontinued (see Figure B22). Connect it to the reaction apparatus as shown in Figure B21, using a thermometer adapter or rubber stopper.

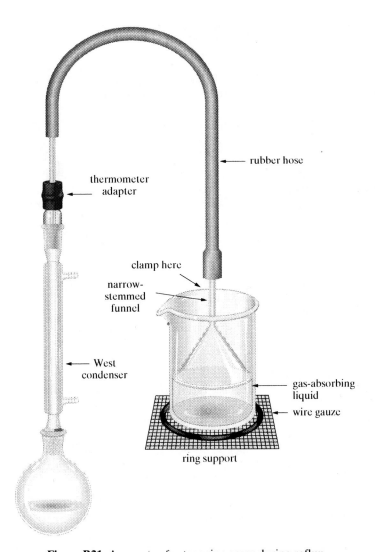

Figure B21 Apparatus for trapping gases during reflux

For microscale work you can use a Norit gas trap, which consists of a layer of pelletized Norit in a microscale drying tube (see Figure B17). To construct the gas trap, push a plug of cotton or glass wool nearly to the bend in the drying tube, then hold it with the long arm upright (open end up) and use a plastic weighing dish to add enough pelletized Norit (~0.5 g) to form a layer 3–4 cm deep. Insert another plug of cotton or glass wool to keep the Norit in place. Insert the drying tube in any part of the reaction apparatus that is exposed to the atmosphere, usually the top of a reflux condenser. If the reaction is to be conducted in a dry atmosphere, you can add a layer of calcium chloride on top of the Norit layer before inserting the second cotton plug.

A gas trap consisting of a plug of moistened glass wool in a microscale drying tube can effectively trap some water-soluble gases, but care must be taken to keep water from dripping into the reaction apparatus. Such a trap is prepared by filling most of the long arm of the drying tube with glass wool and using a Pasteur pipet to drop water onto it until it is moist but not dripping wet.

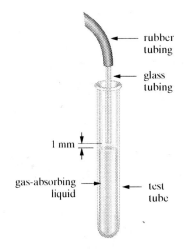

Figure B22 Small-scale gas trap

C. Separation Operations

After a chemical reaction has been carried out using operations from Section B, the product must be *separated* from the other components of the reaction mixture. Solid impurities can be removed from liquids by gravity filtration [OP-15], which is also used for the recrystallization operation [OP-28] described in part E. A solid product can be separated from the liquid part of a reaction mixture by vacuum filtration [OP-16] or, for some microscale experiments, by centrifugation [OP-17]. If the product is dissolved in an aqueous reaction mixture, it is often separated by extraction [OP-18] with a volatile organic solvent, which is then removed by evaporation [OP-19]. Some water-insoluble liquid products are separated from an aqueous reaction mixture by steam distillation [OP-20]. Column chromatography [OP-21] is a versatile separation method by which a liquid or solid product is separated from other components of a mixture as the mixture passes down a column of adsorbent material. Both thin-layer chromatography [OP-22] and paper chromatography [OP-23] can separate the components of very small amounts of mixtures; they are generally used to detect or identify components in a mixture rather than to separate reaction mixtures.

OPERATION **15**	Gravity Filtration

Filtration is used for two main purposes in organic chemistry:

- To remove solid impurities from a liquid or solution.
- To separate an organic solid from a reaction mixture or a crystallization solvent.

Gravity filtration is generally used for the first purpose and *vacuum filtration* [OP-16] for the second. *Centrifugation* [OP-17] can be used for either. In a gravity filtration, the liquid component of a liquid–solid mixture drains through a filtering medium (such as cotton or filter paper) by gravity alone, leaving the solid on the filtering medium. The filtered liquid, called the *filtrate*, is collected in a flask or another container. Gravity filtration is often used to remove drying agents (see OP-25) from dried organic liquids or solutions and solid impurities from hot recrystallization solutions.

    If the solid being removed is coarse and quite dense, it can sometimes be removed from a liquid by letting it settle to the bottom of the container (preferably an Erlenmeyer flask) and then slowly and carefully pouring the liquid into another container, leaving the solid behind. Some of the liquid may remain in the flask, but it can be transferred [OP-6] using a Pasteur pipet or a filter-tip pipet, if necessary. This process, called *decanting*, should not be used with finely divided solids, because some of the solid will inevitably be poured out with the liquid and contaminate it.

Standard Scale Gravity Filtration

Standard scale gravity filtration of organic liquids (see Figure C2) can be carried out using a funnel with a short, wide stem (such as a powder funnel) and a relatively fast, fluted filter paper. Circles of ordinary filter paper, such as 12.5 cm Whatman #1 (for fine particles) or Whatman #4 (for coarser particles), can be fluted (folded) as shown in Figure C1, but commercial fluted filter papers are available from chemical supply houses. Glass wool is sometimes used for very fast filtration of coarse solids. A thin layer of glass wool is placed inside the cone of a short-stemmed funnel, covering the outlet hole, and the mixture to be filtered is poured directly onto the glass wool. Because fine particles will pass through the glass wool fibers, this method is most often used for prefiltration of mixtures that will be filtered again.

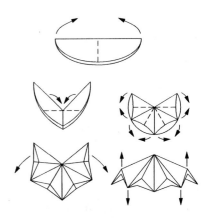

Figure C1 Making a fluted filter paper

Microscale Gravity Filtration

Small amounts of solid–liquid mixtures (~10 mL or less) can be filtered using a *filtering pipet*—a Pasteur pipet containing a small plug of cotton or glass wool (see Figure C3). A filtering pipet is usually made with a standard $5\frac{3}{4}$-inch Pasteur pipet. For filtering hot recrystallization [OP-28] solutions, it is best to use a shortened filtering pipet made by cutting off [OP-3] most of the capillary tip to leave a 5-mm stub; otherwise, crystals may form in the narrow capillary and block the liquid flow. A glass-wool plug can be used when large particles are being removed or when the mixture being filtered

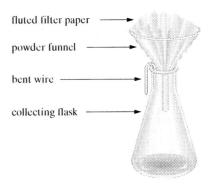

Figure C2 Apparatus for standard scale gravity filtration

fluted filter paper

powder funnel

bent wire

collecting flask

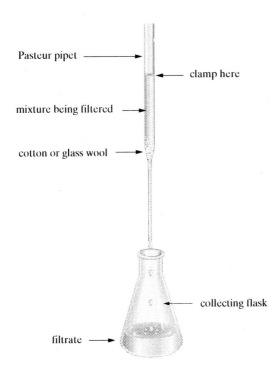

Pasteur pipet

clamp here

mixture being filtered

cotton or glass wool

collecting flask

filtrate

Figure C3 Filtration with a filtering pipet

contains an acid or another substance that may react with cotton. Because glass wool doesn't filter out fine particles, cotton is preferred for most applications. Very fine particles may pass through cotton also; a filtering pipet containing a 2-mm layer of chromatography-grade alumina or silica gel on top of a cotton plug can be used to remove such particles from a mixture.

A microscale gravity filter with a larger capacity can be constructed by cutting the rounded top off the bulb of a plastic Beral-type pipet and packing some cotton or glass wool in its neck (see *J. Chem. Educ.* **1993**, *70,* A204). The plastic may not be compatible with some organic solvents, however.

A filter-tip pipet, prepared as described in OP-6, can also be used to filter small amounts of liquids. This method is not really gravity filtration but it accomplishes the same purpose; removing a solid from the liquid being filtered. It has the disadvantage that some solid may adhere to the cotton and be transferred to the collecting container. When the solid particles are fairly coarse, like the sodium sulfate granules used for drying liquids [OP-25], you may be able to use a Pasteur pipet without a filter tip. Use the pipet tip to push aside any solid that is in the way, then hold it flat against the bottom of the container as you draw liquid into it.

General Directions for Gravity Filtration

 Standard Scale

Equipment and Supplies

> powder funnel
> fluted filter paper
> bent wire or paper clip
> Erlenmeyer flask or other collecting container

If you are filtering the solid–liquid mixture into a narrow-necked container such as an Erlenmeyer flask, support the funnel on the neck of the flask, placing a bent wire or paper clip between them to leave a gap for pressure equalization (Figure C2). Alternatively, you can set the funnel in a ring support or other funnel support positioned directly over the collecting container. Open the fluted filter paper to form a cone and insert it snugly into the funnel, trying not to flatten out any of its folds. If the solid isn't finely divided or if there isn't much of it, decant (pour) the mixture directly into the filter paper cone. If the mixture contains a considerable amount of finely divided solid, let the solid settle first, then carefully decant the liquid into the filter paper cone so that most of the solid remains behind until the end of the filtration (this helps keep the fine solid from clogging the pores of the filter paper and slowing the filtration). In either case, add the mixture fast enough to keep the filter paper cone about two-thirds full, without allowing any liquid to rise above the top of the filter paper, until it has all been added. Swirl a small amount of an appropriate pure solvent (usually the solvent present in the mixture being filtered) in the decanting container to wash any residual solid, then pour it into the filter paper cone and let it drain into the receiving container. Wash the solid on the filter paper with this solvent by stirring gently as the solvent drains, but be careful not to tear

Unless you need to save the solid as well as the filtrate, there is no need to transfer all of the solid to the filter paper.

the filter paper with your stirring rod. This washing step should reduce losses due to adsorption of dissolved organic materials on the solid.

Summary

1. Support funnel over collecting container.
2. Insert fluted filter paper.
3. Pour mixture being filtered into filter cone, let drain.
4. Wash solid in decanting vessel and on filter paper.
5. Clean up.

 Microscale

Equipment and Supplies

two $5\frac{3}{4}$-inch Pasteur pipets
rubber bulb (2-mL capacity)
cotton (or glass wool)
applicator stick or stirring rod
collecting containers

Roll a small amount of cotton (or glass wool) between your fingers to form a loose ball and insert it in the top of a $5\frac{3}{4}$-inch Pasteur pipet. (Alternatively, you can use a Beral-type pipet with the top cut off.) Then use a small stirring rod or a wooden applicator stick to push it down the body of the pipet, forming a cotton plug that ends about where the capillary section of the pipet begins, as shown in Figure C3. Don't pack it too tightly because that will slow down the filtration. Clamp the resulting filtering pipet vertically over a small beaker. Rinse the plug with a suitable wash solvent (usually the solvent present in the mixture being filtered) by using a second Pasteur pipet to transfer about 0.5 mL of the solvent to the top of the filtering pipet, letting it drain, and using a rubber bulb to force any remaining solvent through. Replace the beaker by another collecting container, such as a conical vial or a small flask, and use the second Pasteur pipet to transfer the mixture being filtered, in several portions if necessary, to the top of the filtering pipet. Let the liquid drain by gravity into the collecting container. If the filtration rate is very slow, you can use a pipet pump (see OP-5) to apply a gentle, constant pressure to the top of the filtering pipet, depressing the quick release lever (if it has one) when you detach it from the pipet. (Don't use excessive pressure because that may force particles into the filtrate.) Wash the solid and the plug with a small amount of wash solvent as described previously, collecting the solvent in the container that holds the filtrate. Use a rubber bulb or a pipet pump to force out the last few drops of liquid. Let the plug dry, then remove it either by pulling it out with a wooden applicator stick (twirl it to snag the fibers) or snagging it with a copper wire bent to form a "J" at one end.

Summary

1. Prepare filtering pipet by inserting cotton or glass-wool plug in Pasteur pipet.
2. Clamp filtering pipet over beaker, rinse with solvent.
3. Replace beaker with collecting container.

4. Transfer mixture being filtered to filtering pipet, let drain.
5. Wash filtering pipet with solvent.
6. Remove plug, clean up.

When Things Go Wrong

If a liquid is filtering very slowly, the filter paper or other filtering medium may be too retentive or the particles may be so fine that they plug the pores in the filter paper. Try using a coarser filtering medium, such as a coarse prefolded filter paper rather than one folded from Whatman #1 paper (the most widely used grade of filter paper). For a microscale filtration, try using glass wool rather than cotton in a filtering pipet; if necessary you can *gently* force solvent through a filtering pipet with a rubber bulb or a pipet pump.

If fine particles pass into the filtrate, refilter using a more retentive filtering medium.

 ## General Directions for Filtration with a Filter-Tip Pipet

Equipment and Supplies

$5\frac{3}{4}$-inch Pasteur pipet
rubber bulb (2-mL capacity)
cotton
collecting containers

Construct a filter-tip pipet as described in OP-6 and rinse it with a suitable wash solvent (usually the solvent present in the mixture being filtered) by drawing in about 0.5 mL of the solvent and slowly ejecting it into a waste container. Use the pipet and attached bulb to draw up liquid from the mixture being filtered and transfer it to a suitable collecting container, leaving the solid behind in the original container, until all of the liquid has been transferred. To avoid transferring adherent solid along with the liquid, try to keep the pipet's tip away from the solid when you draw liquid into it. To transfer the last few drops of liquid, push the solid aside with the tip of the pipet and hold it flat against the bottom of the original container as you withdraw the liquid. Stir a little wash solvent into the solid, draw it into the filter-tip pipet, and combine it with the rest of the filtrate.

Summary

1. Construct filter-tip pipet, rinse with wash solvent.
2. Draw liquid being filtered into pipet.
3. Transfer liquid to collecting container.
4. Wash solid, transfer wash solvent to collecting container.
5. Clean up.

OPERATION 16 Vacuum Filtration

Vacuum filtration (also called suction filtration) provides a fast, convenient method for isolating a solid from a liquid–solid mixture and for removing solid impurities from a relatively large quantity of liquid. In a typical standard

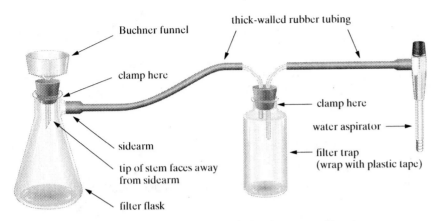

Figure C4 Apparatus for standard scale vacuum filtration

scale vacuum filtration, a circle of filter paper is laid flat on a perforated plate inside a porcelain *Buchner funnel*, which is attached by an airtight connector to a heavy-walled *filter flask* having a sidearm on its neck (see Figure C4). The filter flask's sidearm is connected to the inlet of a *water aspirator* or to a vacuum line, often by way of a trap (described below), with a short length of heavy-walled rubber tubing. In the water aspirator, a rapid stream of water passes by a small hole at the inlet, creating a vacuum there and in the attached filter flask, and exits into a sink. An aspirator should always be run "full blast" because its efficiency decreases and the likelihood of water back-up increases at lower flow rates. When the mixture being filtered is poured into the Buchner funnel, the liquid is forced through the paper by the unbalanced external pressure and collects in the filter flask, while the solid remains on the filter paper as a compact *filter cake*.

Small quantities of solids can be filtered by essentially the same method using a *Hirsch funnel* attached to a small filter flask or a sidearm test tube (see Figure C5). This kind of apparatus is suitable for many microscale experiments and can be used for some standard scale experiments as well. A porcelain

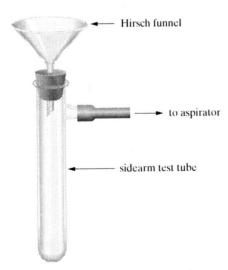

Figure C5 Apparatus for small-scale vacuum filtration with sidearm test tube

Hirsch funnel contains an integral perforated plate about 1–2 cm in diameter; plastic Hirsch-type funnels with separate fritted disks are also available. Hirsch funnels use very small filter paper circles. These are available commercially, but they can also be cut from ordinary filter paper using a sharp cork borer on a flat cutting surface, such as the bottom of a large cork. For a porcelain Hirsch funnel, the filter paper should be about equal in size to the perforated plate or slightly smaller, but large enough to completely cover all of its holes.

Experimental Considerations

Filter Traps and Cold Traps. You should ordinarily interpose a *filter trap* between the filter flask (or sidearm test tube) and a water aspirator to keep water from backing up into the flask when the water pressure changes, as it often does. Using a filter trap is most important when the filtrate is to be saved, because any backed-up water will contaminate the filtrate, but it is a good idea to use the trap for all vacuum filtrations using an aspirator. A suitable filter trap can be constructed by wrapping a thick-walled Pyrex jar with transparent plastic tape (to reduce the chance of injury in case of implosion) and inserting a rubber stopper fitted with two L-shaped connecting tubes, constructed as described in Minilab 1. Note that the connecting tubes closest to the aspirator should be longer than the other (see Figure C4). Another kind of filter trap with a pressure-release valve is illustrated in Figure E15 of OP-31.

Take Care! Never use a thin-walled container, such as an Erlenmeyer flask, as a trap; it may shatter under vacuum.

If you are using a vacuum line connected to a central mechanical vacuum pump, you may need to use a *cold trap* to protect the pump from solvent vapors that might damage it (your instructor will inform you if this is necessary). A filter trap can function as a cold trap if it is immersed in an appropriate cold bath [OP-8].

Filtering Media. Because of the external pressure on the mixture being filtered, solid particles are more likely to pass through the filter paper than with gravity filtration, so a slower (finer grained) grade of filter paper should be used. An all-purpose filter paper, such as Whatman #1, is adequate for filtration of most solids.

When filtering a finely divided solid impurity from a liquid, you may need to use a *filtering aid* (such as Celite) to keep the solid from plugging the pores in the filter paper. In a stoppered flask, shake the filtering aid vigorously with a suitable solvent to form a slurry (a thick suspension). Without delay, pour the slurry onto the filter paper under vacuum until a bed about 2–3 mm thick has been deposited. Remove the solvent from the filter flask before continuing with the filtration. This method cannot be used when the solid is to be saved, since it would be contaminated with the filtering aid.

Washing. Unless otherwise instructed, you should always *wash* the solid on the filter paper with an appropriate solvent, usually the same solvent as the one from which it was filtered. To reduce losses, the wash solvent should be cooled in ice water. For example, if you filter a solid from an aqueous solution, use cold distilled water as the wash solvent. If you filter a solid from a mixture of solvents, as in a mixed-solvent recrystallization [OP-28b], you should ordinarily use the solvent in which the solid is least soluble. For more information about washing solids see OP-26a.

Drying. A solid that has been collected by vacuum filtration is usually *air-dried* after the last washing by leaving it on the filter for a few minutes with the vacuum turned on. The vacuum draws air through the solid, which increases the drying rate. If the solid is still quite wet, you can place a *rubber dam* (a thin, flexible rubber sheet) or a sheet of plastic wrap over the mouth of the funnel. The vacuum should cause the sheet to flatten out on top of the filter cake, forcing water out of it. Unless a solid is filtered from a very low-boiling solvent, it ordinarily requires further drying by one of the methods described in OP-26b.

 ## General Directions for Vacuum Filtration

Equipment and Supplies

Buchner funnel or Hirsch funnel
filter flask (or sidearm test tube)
1-hole rubber stopper or neoprene adapter
filter trap
filter paper
thick-walled rubber tubing
flat-bottomed stirring rod (optional)
flat-bladed spatula
wash solvent

Clamp the filter flask and filter trap (if you are using one) securely to a ring stand and connect them to an aspirator or a vacuum line as shown in Figure C4. Use thick-walled rubber tubing that will not collapse under vacuum for all connections. If you are using a water aspirator, connect the longer glass tube on the filter trap (the tube that extends farther into the trap) to the aspirator and the shorter glass tube to the filter flask. If you are using a vacuum line that must be protected by a cold trap, connect the trap to the vacuum line and filter flask and secure it inside a cooling bath as directed by your instructor. Insert a Buchner or Hirsch funnel into the filter flask using a neoprene filter flask adapter or a snug-fitting rubber stopper to provide a tight seal. Obtain a circle of filter paper of the correct diameter and place it inside the funnel so that it covers all of the holes in the perforated plate but does not extend up the sides of the funnel.

   Moisten the filter paper with a few drops of wash solvent — a solvent that is present in the mixture being filtered or one that is miscible with it. Open the aspirator tap or vacuum-line valve as far as it will go. Direct the water stream from an aspirator into a large beaker or another container to prevent splashing (do not attach a rubber tube to the aspirator outlet to reduce splashing because it will also reduce the aspirator's effectiveness). If the solid is finely divided, let it settle before you decant the liquid into the funnel, and transfer the bulk of the solid near the end of the filtration. Otherwise, stir or swirl the mixture just before decanting to transfer more of the solid and leave less behind in the decanting vessel. If the volume of the filtration mixture is greater than the capacity of the funnel, add the mixture

Replace the term "filter flask" by "sidearm test tube" if you are using the latter.

rapidly enough to keep the funnel about two-thirds full throughout the filtration, until it has all been added. Transfer any remaining solid to the filter paper with a flat-bladed spatula, using a small amount of the filtrate or some cold wash solvent to facilitate the transfer. Leave the vacuum on until only an occasional drop of liquid emerges from the stem of the funnel. If there is a possibility that water collecting in a trap will back up into the filter flask, or if you are using an aspirator with no filter trap, break the vacuum using a pressure-release valve or disconnect the rubber tubing at the vacuum source before you turn off the vacuum.

Rule of Thumb: Use about 1–2 mL of wash solvent per gram of solid unless directed otherwise.

With the vacuum off, add enough previously chilled wash solvent to cover the solid. Being careful not to disturb the filter paper, stir the mixture *gently* with a spatula or a flat-bottomed stirring rod until the solid is suspended in the liquid. (For microscale work, stirring may be omitted because of the likelihood of product loss.) Without delay, turn on the vacuum to drain the wash liquid. For standard scale work, the washing step can be repeated with another portion of chilled wash solvent; one portion is usually adequate for microscale work. Work quickly to avoid dissolving an appreciable amount of solid in the wash solvent. After the last washing, leave the vacuum on for 3–5 minutes to air-dry the solid on the filter and make it easier to handle. Run the tip of a small flat-bladed spatula around the circumference of the filter paper to dislodge the filter cake, then invert the funnel carefully over a square of glazed paper, a watch glass, a weighing dish, or another suitable container to remove the filter cake and filter paper. Use your spatula to scrape any remaining particles onto the paper or into the container. Dispose of the filtrate as directed by the experimental procedure or your instructor and dry the solid by one of the methods described in OP-26b. To reduce losses, dry the filter paper along with the filter cake and scrape off any additional solid after it is dry, being careful not to scrape any filter paper fibers into your product.

Summary

1. Assemble apparatus for vacuum filtration.
2. Position and moisten filter paper, turn on vacuum.
3. Add mixture being filtered to funnel.
4. Transfer any remaining solid to funnel.
5. Wash solid on filter with chilled wash solvent.
6. Air-dry solid on filter paper.
7. Transfer solid to container, remove filtrate from filter flask.
8. Disassemble and clean apparatus.

When Things Go Wrong

If a liquid is filtering very slowly, the filter paper may be too retentive or its pores may be plugged with fine particles. Replace the filter paper with a new one (keep any solid from the used one) and let the solid settle to the bottom of the mixture you are filtering, then carefully decant the liquid into the filtering funnel and wait until the end of the filtration process to transfer the solid with any remaining liquid. If one is available, you can use a larger filtering device, such as a small Buchner funnel rather than a Hirsch funnel. You can also try using a coarser filtering medium, such as a circle of Whatman #4 paper rather than Whatman #1, but you may have to refilter the filtrate if fine particles pass through the paper. If you don't need to save the solid, you can use a filtering aid such as Celite. (See *Filtering Media.*)

If water backs up into the filter flask when you turn off an aspirator, you probably didn't use a filter trap. Remove the water, then connect a trap between the filter flask and the aspirator as shown in Figure C4. Alternatively, pull off the tubing at the aspirator before you turn it off. (See *Filter Traps and Cold Traps.*)

If an appreciable amount of solid product disappears when you wash it, the wash solvent may not have been adequately chilled, you may have used too much of it, or it may be inappropriate for washing your product. If the filtrate contains a relatively low boiling solvent, you should be able to recover the solid by evaporating [OP-19] the solvent, but it should then be purified and washed more carefully. Alternatively, some solid may have passed under the filter paper and into the filtrate. Try re-filtering the filtrate, being careful not to displace the filter paper when you wash the recovered solid. (See *Washing.*)

Centrifugation

OPERATION 17

Centrifugation is used to separate different phases from one another by centrifugal force. When a mixture in a centrifuge tube is placed inside a *centrifuge* (Figure C6) and whirled around a circular path at high speed, the denser phase (often a solid) is forced to the bottom of the tube, leaving the other phase on top. In the microscale laboratory, centrifugation is often used to collect solids

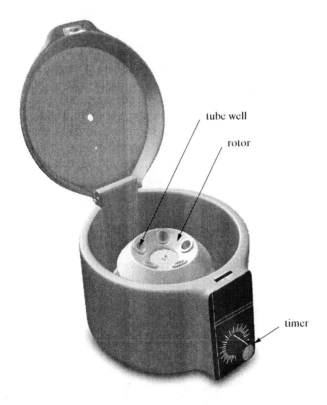

tube well

rotor

timer

Figure C6 A benchtop centrifuge

that have crystallized from solution in a *Craig tube.* Centrifugation may also be used to separate a finely divided solid from a liquid, or to separate two immiscible liquids sharply after an extraction [OP-18].

 General Directions for Centrifugation

Equipment and Supplies

benchtop centrifuge
2 centrifuge tubes
Pasteur pipet (optional)
flat-bladed spatula (optional)

Transfer the mixture to be centrifuged to a conical centrifuge tube with a capacity of ~15 mL, or another centrifuge tube specified by your instructor or the experimental procedure. (If you are using a Craig tube, see OP-28.) Obtain an empty matched centrifuge tube and add enough water to it so that the two tubes, with their contents, have approximately equal masses. Usually you can estimate the amount of water by volume, but you may have to weigh the tubes to ensure proper balance. (Alternatively, find another student who is doing the same operation and use his or her centrifuge tube to balance your own. Label the tubes so that you don't mix them up.) Place the two centrifuge tubes directly opposite one another in the centrifuge's rotor; they fit into tube wells, which may be cushioned to help prevent breakage. Close and lock the centrifuge lid, set the centrifuge's timer (if it has one) to 3–5 minutes, and start the centrifuge. When the time is up (or when you switch off the centrifuge), the rotor will slowly come to a stop. Wait until its whirring sound has stopped; then open the lid and remove the centrifuge tubes.

If you are centrifuging a liquid–solid mixture and the solid is not firmly packed in the bottom of the tube, continue centrifuging until the solid is compacted and there are no floating particles. Then carefully decant the liquid or remove it with a Pasteur pipet or a filter-tip pipet, leaving the solid behind. Use the pointed end of a flat-bladed microspatula to remove the solid. If you are centrifuging a mixture of two immiscible liquids, separate the liquids after centrifugation by removing the lower layer with a Pasteur pipet as described in OP-18. Use a tapered centrifuge brush, if one is available, to clean the centrifuge tube.

Summary

1. Transfer mixture to centrifuge tube.
2. Place centrifuge tube and a second tube of comparable mass in opposite tube wells.
3. Run centrifuge for 3–5 minutes.
4. Let centrifuge stop, remove tubes.
5. Separate liquid layer from solid or second liquid layer.

When Things Go Wrong

If the centrifuge vibrates badly, rattles loudly, or stops before the time is up, the centrifuge tubes are not properly balanced. Remove and balance them and then resume centrifugation.

If a centrifuge tube breaks while the centrifuge is operating, the broken tube may have been faulty, it may have been the wrong kind for the centrifuge you are using, it may not have been balanced properly, or the tube wells may not be adequately cushioned. Clean up the mess and try to identify the problem so that it doesn't happen again. If there is little or no cushioning in the wells you may be able to push in some Styrofoam that has been cut to fit. See your instructor first, since the padding may raise the tubes so high that they contact the lid of the centrifuge when it spins.

Extraction   OPERATION **18**

If you shake a bromine/water solution with some dichloromethane, the red-brown color of the bromine fades from the water layer and appears in the dichloromethane layer as you shake. These color changes show that the bromine has been transferred from one solvent (water) to another (dichloromethane). The process of transferring a substance from a liquid or solid mixture to a solvent is called *extraction*, and the solvent is called the *extraction solvent.* An extraction solvent is usually a low-boiling organic solvent that can be evaporated [OP-19] after extraction to isolate the desired substance.

Extraction is used for the following purposes in organic chemistry:

- To separate a desired organic substance from a reaction mixture or some other mixture.
- To remove impurities from a desired organic substance, which is usually dissolved in an organic solvent.

The second process is described in OP-24, "Washing Liquids."

a. Liquid–Liquid Extraction

Principles and Applications

Liquid–liquid extraction is based on the principle that if a substance is soluble to some extent in two immiscible liquids, most of it can be transferred from one liquid to the other by a process that involves thorough mixing of the liquids. For example, acetanilide is partly soluble in both water and dichloromethane. If a solution of acetanilide in water is shaken with a portion of dichloromethane, some of the acetanilide will be transferred to the organic (dichloromethane) layer (see Figure C7). The organic layer, being denser than water, separates below the water layer and can be removed and replaced with another portion of dichloromethane. When the fresh dichloromethane is shaken with the aqueous solution, more acetanilide passes into the new organic layer. This new layer can then be removed and combined with the first. By repeating this process enough times, virtually all of the acetanilide can be transferred from the water to the dichloromethane.

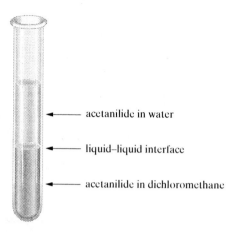

Figure C7 Distribution of a solute between two liquids

The ability of an extraction solvent (S2) to remove a solute (A) from another solvent (S1) depends on the partition coefficient (K) of solute A in the two solvents, as defined in Equation **1**:

$$K = \frac{\text{concentration of A in S}_2}{\text{concentration of A in S}_1} \qquad \textbf{(1)}$$

In the example of acetanilide in water and dichloromethane, the partition coefficient is given by

$$K = \frac{[\text{acetanilide}]_{\text{dichl.}}}{[\text{acetanilide}]_{\text{water}}}$$

The larger the value of K, the more solute will be transferred to the organic layer with each extraction, and the fewer portions of dichloromethane will be required for essentially complete removal of the solute. A rough estimate of K can be obtained by using the ratio of the solubilities of the solute in the two solvents—that is,

$$K \sim \frac{\text{solubility of A in S}_2}{\text{solubility of A in S}_1}$$

This approximate relationship can be helpful in choosing a suitable extraction solvent.

Extraction Solvents

Most extraction solvents are organic liquids that are used to extract nonpolar and moderately polar solutes from aqueous solutions. A good organic extraction solvent should be immiscible with water, dissolve a wide range of organic substances, and have a low boiling point so that it can be removed by evaporation after the extraction. The substance being extracted should be more soluble in the extraction solvent than in water; otherwise, too many steps will be required to extract it.

Table C1 Properties of commonly used extraction solvents

Solvent	bp, °C	d, g/mL	Comments
water	100	1.00	for extracting polar compounds, generally using a reactive solute such as NaOH or HCl
diethyl ether	35	0.71	good general solvent; absorbs some water; very flammable
dichloromethane	40	1.33	good general solvent; suspected carcinogen
toluene	111	0.87	for extracting aromatic and nonpolar compounds; difficult to remove
petroleum ether	~35–60	~0.64	for extracting nonpolar compounds; very flammable
*hexane	69	0.66	for extracting nonpolar compounds; flammable

*The mixture of C_6H_{14} isomers called hexanes is cheaper than pure hexane, and is often used in place of it.

Diethyl ether and dichloromethane (methylene chloride) are the most commonly used organic extraction solvents. Diethyl ether (also called ether or ethyl ether) has a very low boiling point (35°C) and can dissolve both polar and nonpolar organic compounds, but it is extremely flammable and tends to form explosive peroxides on standing. Dichloromethane is denser than water, which usually simplifies the extraction process, and it isn't flammable. Dichloromethane has a tendency to form emulsions, which can make it difficult to separate cleanly, and it must be handled with caution because it is a suspected carcinogen. These and other extraction solvents and their properties are listed in Table C1. Organic extraction solvents that are less dense than water ($d = 1.00$ g/mL) will separate as the top layer during the extraction of an aqueous solution; extraction solvents that are more dense than water will ordinarily separate as the bottom layer.

Just as organic solvents are used to extract substances from aqueous solutions, water and aqueous solutions can be used to extract certain polar substances from organic solutions. An aqueous solution may function as a *chemically active* extraction solvent if it contains a solute that reacts with the substance to be extracted, changing its distribution between the aqueous and organic layers. For example, dilute aqueous sodium hydroxide can be used to extract carboxylic acids from organic solvents by first converting them to carboxylate salts, which are much more soluble in water and less soluble in organic solvents than are the original carboxylic acids.

$$RCOOH + NaOH \rightarrow RCOO^- Na^+ + H_2O$$

If the carboxylic acid is sufficiently insoluble in water, it can be recovered by acidifying the aqueous extract to precipitate the acid, which is then collected by vacuum filtration [OP-16]. Similarly, dilute hydrochloric acid is used to extract basic solutes such as amines from organic solvents by converting the amines to ammonium salts, which are much more soluble in water and less soluble in organic solvents than are amines.

$$RNH_2 + HCl \rightarrow RNH_3^+Cl^-$$

Potential hazards should be considered when selecting and using an extraction solvent. For example, solvents such as benzene, trichloromethane (chloroform), and tetrachloromethane (carbon tetrachloride) should not be used as extraction solvents in an undergraduate laboratory because of their toxicity and carcinogenic potential. Precautions must be taken with all organic solvents to minimize skin and eye contact and inhalation of vapors. Flames must not be allowed in the laboratory when highly flammable solvents, such as diethyl ether and petroleum ether, are in use.

Experimental Considerations

Extraction Methods. A standard scale liquid–liquid extraction is ordinarily carried out in a *separatory funnel*, which has a stopcock in its stem (see Figure C9). This makes it possible to drain the lower layer into a separate container, leaving the upper layer behind in the separatory funnel. Separatory funnels are expensive and break easily. Never prop a separatory funnel on its base; set it in a ring support or some other stable support. If your separatory funnel has a glass stopcock, lubricate it by applying thin bands of stopcock grease on both sides, leaving the center (where the drain hole is located) free of grease to prevent contamination (see Figure C8). A glass stopcock is secured to the separatory funnel by a compression clip or a rubber ring, which should be tight enough to keep it from leaking. A Teflon stopcock (which is *not* lubricated) is secured by a threaded Teflon nut. The stopcock is inserted into the bore hole at the base of the separatory funnel and a Teflon washer and rubber ring are placed over its threaded end, in that order. The Teflon nut is then screwed on tightly enough to prevent leakage, but not so tightly as to prevent smooth rotation of the stopcock.

In the microscale laboratory, an extraction is usually performed by shaking the liquids in a conical vial or conical centrifuge tube and separating them with a Pasteur pipet. A 5-mL conical vial can be used with liquid volumes up to ~4 mL, and a 15-mL centrifuge tube can be used with liquid volumes up to ~12 mL. A screw-cap centrifuge tube is preferable to one having a snap-on cap, since it is less likely to leak. A conical vial is usually capped with a compression cap having a Teflon-faced silicone liner; the liner should be inserted in the cap so that its Teflon-coated (white) side faces down when the vial is capped. A conical vial may leak if its rim is chipped or if the liner is damaged, so shake some water in the capped vial to check for leaks before using the vial for an extraction. Do this for a centrifuge tube also, and replace the cap if it leaks.

Liquid layers are ordinarily separated by removing the *lower* layer with a Pasteur pipet and transferring it to another container. This way, the interface between the layers is at the narrowest part of the container when the last of the lower layer is removed, making a sharp separation possible. It takes some practice and a steady hand to remove all of the bottom layer without including any of the top layer, but it is important that you learn how to do so. Otherwise, you will lose part of your product or the product will be contaminated

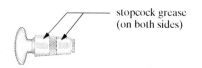

stopcock grease (on both sides)

Figure C8 Lubricating a glass stopcock

with material from the layer being extracted. Accurate separation is most important for the last extraction step because any extraction solvent that isn't recovered during earlier steps can be recovered in the last one. Most extraction solvents have a high vapor pressure, which may cause them to spurt out of the tip of a Pasteur pipet during transfer. Product loss due to spurting can be reduced or prevented by one or more of the following measures:

- Make sure that the extraction solvent, the liquid being extracted, and the Pasteur pipet are at room temperature or below.
- Rinse the Pasteur pipet with the extraction solvent two or three times to fill it with solvent vapors just before use.
- Use a filter-tip pipet.

Volume of Extraction Solvent. The volume of extraction solvent and the number of extraction steps are sometimes specified in an experimental procedure. If they are not, use a volume of extraction solvent about equal to the volume of liquid being extracted, divided into at least two portions. For example, you can extract 12 mL of an aqueous solution with two successive 6-mL (or three 4-mL) portions of extraction solvent. Note that it is more efficient to use several small portions of extraction solvent rather than one large portion of the same total volume.

Rule of Thumb: Total volume of extraction solvent ≈ volume of liquid being extracted.

Getting Good Separation. Under some conditions, the liquid layers do not separate sharply, either because an *emulsion* forms at the interface between the two liquids or because droplets of one liquid remain in the other liquid layer. Emulsions can often be broken up by using a wooden applicator stick to stir the liquids gently at the interface. If that doesn't work, mix in some saturated aqueous sodium chloride solution (or enough solid NaCl to saturate the aqueous layer) and allow the extraction container to stand open and undisturbed for a time. To consolidate the liquid layers, use an applicator stick to rub or stir any liquid droplets that form on the sides or bottom of the extraction container. You can also use an applicator stick to remove small amounts of insoluble "gunk" that sometimes form near the interface. Larger amounts of insoluble material can be removed by filtering [OP-15] the mixture through a loose pad of glass wool in a powder funnel or filtering pipet.

An emulsion usually contains microscopic droplets of one liquid suspended in another.

Saving the Right Layer. *Always keep both layers until you are certain which layer contains the desired product!* All too often, a student will unthinkingly discard the extraction layer that contains the product and will have to repeat an experiment from the beginning. The safest practice is to keep both layers until you have actually isolated the product from one of them, but you can usually determine which is the right layer before then. In most extractions, the product is extracted from an aqueous solution into an organic solvent such as dichloromethane or diethyl ether. If you make careful observations when you add the extraction solvent, you can usually tell whether it floats on top of the aqueous layer or sinks below it. Since diethyl ether is less dense than water, it will form the upper layer when it is used to extract an aqueous solution. Dichloromethane is denser than water, so it will ordinarily form the lower layer with an aqueous solution (but see *When Things Go Wrong*). If you still aren't sure which is the right layer, add a drop or two of water to a drop or two of each layer separately. The one in which the water dissolves is the aqueous layer.

General Directions for Extraction

 Standard Scale

Equipment and Supplies

> separatory funnel with stopper
> ring stand
> support for separatory funnel
> stemmed funnel
> extraction solvent
> graduated cylinder
> wooden applicator stick
> 2 flasks

If you use a metal ring support, cushion it with three short lengths of split rubber tubing to prevent damage to the separatory funnel.

Support a separatory funnel on a ring support of suitable diameter or another appropriate support. Close the stopcock by turning its handle to a horizontal position and pour the liquid to be extracted into the separatory funnel, preferably using a stemmed funnel to avoid getting liquid on the glass joint. The liquid should be at room temperature (or below) to prevent vaporization of a volatile extraction solvent. Measure the required volume of extraction solvent using a graduated cylinder (the exact volume is not crucial) and pour it through the stemmed funnel into the separatory funnel. The total volume of both liquids should not exceed three-quarters of the funnel's capacity. If it does, use a larger separatory funnel or carry out the extraction in two or more steps, using a fraction of the liquid to be extracted in each step.

Take Care! Wear gloves during an extraction to protect your hands in case of leakage.

Moisten the stopper with water and insert it firmly with a twisting motion. Then pick up the funnel in both hands and partly invert it, with your right hand holding the stopcock (or your left hand, if you're a southpaw) and the first two fingers of your left hand holding the stopper in place (see Figure C9). Holding the separatory funnel with its outlet above the liquid level and its stem pointed away from you and your neighbors, *vent* it by slowly opening the stopcock to release any pressure buildup. Close the stopcock, shake the separatory funnel gently for a few seconds (still keeping its stem end higher than its stoppered end), and vent it as before. Then shake the funnel more vigorously, with occasional venting, for 2–3 minutes. As you shake the funnel, rotate the wrist holding its stem end so that its contents are swirled as well as shaken; this motion is more efficient than shaking alone. Avoid overly vigorous mixing if the solvent tends to form emulsions.

Venting should not be necessary after there is no longer an audible hiss of escaping vapors when the stopcock is opened.

Replace the funnel on its support, remove the stopper, and allow the funnel to stand until there is a sharp dividing line between the two layers. If the layers do not separate cleanly, take the appropriate measures described in *Getting Good Separation*. Begin to drain the bottom layer into a labeled Erlenmeyer flask (flask **A**) by opening the stopcock fully. As the interface approaches the bottom of the funnel, partly close the stopcock to slow the drainage rate. Close it completely, separating the layers cleanly, just as the interface reaches the stopcock. Follow method **1** if the extraction solvent is *more* dense than the liquid being extracted (forming the

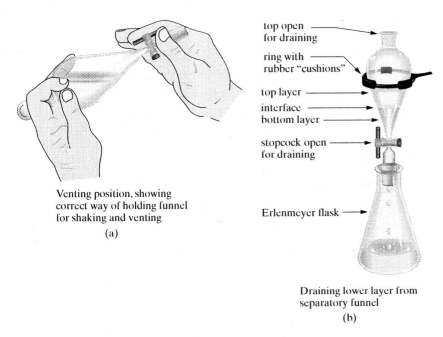

Venting position, showing
correct way of holding funnel
for shaking and venting
(a)

top open
for draining

ring with
rubber "cushions"

top layer
interface
bottom layer

stopcock open
for draining

Erlenmeyer flask

Draining lower layer from
separatory funnel
(b)

Figure C9 Standard-scale extraction techniques

lower layer) and method **2** if it is *less* dense than the liquid being extracted
(forming the upper layer).

1. Extract the liquid that remains in the separatory funnel (the original top
   layer) with a fresh portion of the same extraction solvent; then drain the
   bottom layer into container **A** as before, combining the extracts. Repeat
   the process, as necessary, with fresh extraction solvent. After the bottom
   layer has been removed following the last extraction, pour the remaining liquid out of the *top* of the separatory funnel into a separate labeled
   container (**B**) and retain it for later disposal.
2. Pour the liquid that remains in the separatory funnel (the original top
   layer) out of the *top* of the separatory funnel into a separate labeled container (flask **B**). Then return the liquid in **A** to the separatory funnel and
   extract it with a fresh portion of extraction solvent. Again drain the bottom layer into **A** and pour the top layer into **B**. Repeat the process, as
   necessary, with fresh extraction solvent, combining all extracts in **B**.
   Retain the liquid in flask **A** for later disposal.

Summary

1. Add liquid to be extracted to separatory funnel.
2. Add extraction solvent, stopper funnel, invert, and vent.
3. Shake and swirl funnel, with venting, to extract solute into extraction
   solvent.
4. Remove stopper, let layers separate.
5. Drain lower layer into flask **A**.
   IF extraction solvent forms lower layer, GO TO 6.
   IF extraction solvent forms upper layer, GO TO 7.

6. Stopper flask **A**.
   IF another extraction step is needed, GO TO 2.
   IF extraction is complete, empty and clean separatory funnel; STOP.
7. Pour upper layer into flask **B** and stopper it.
   IF another extraction step is needed, return contents of **A** to separatory funnel, GO TO 2.
   IF extraction is complete, clean separatory funnel; STOP.

Microscale

Equipment and Supplies

> conical vial or centrifuge tube with cap
> support for extraction container
> 2 Pasteur pipets with bulbs, one calibrated
> extraction solvent
> wooden applicator stick
> 1–2 containers (conical vials, screw-cap vials, test tubes, etc.)

Depending on the amount of liquid to be extracted, obtain a 3- or 5-mL conical vial with a compression cap and unperforated liner or a 15-mL conical centrifuge tube with a screw cap. Check the extraction container for leaks. To keep a conical vial from tipping over, place it in a small beaker; set a centrifuge tube in a test-tube rack or other suitable support. Add the liquid to be extracted, which should be at room temperature or below. If its volume is less than 1 mL, it is advisable to add enough of a suitable solvent (usually water) to give it a total volume of at least 1 mL. Use a calibrated Pasteur pipet (or other measuring device) to add a measured portion of extraction solvent and cap the container tightly. Shake the extraction container gently and unscrew the cap slightly after 5–10 shakes to release any pressure inside the container. Tighten the cap and shake the mixture vigorously at least 100 times, with occasional venting. Shake less vigorously but for a longer time if the extraction solvent is dichloromethane, which tends to form emulsions. (Alternatively, you can use a spin vane to stir [OP-10] the contents of a conical vial vigorously for at least 1 minute, or use a vortex mixer as directed by your instructor.) Loosen the cap and let the mixture stand until there is a sharp interface between the layers. Use a wooden applicator stick to help consolidate the layers, if necessary. If you are using a centrifuge tube, it can be spun in a centrifuge [OP-17] to facilitate layer separation. If the layers do not separate cleanly, take the appropriate measures described in *Getting Good Separation*. Follow Method **1** (illustrated in Figure C10) if the extraction solvent is *more* dense than the liquid being extracted (forming the lower layer) and Method **2a** (illustrated in Figure C11) or **2b** if it is *less* dense than the liquid being extracted (forming the upper layer). Method **2b** can be used only if both liquid layers will fit into the Pasteur pipet. Although this method requires fewer transfers than Method **2a**, it is more difficult to perform proficiently.

1. Squeeze the bulb of a Pasteur pipet to expel air and insert the pipet vertically so that its tip just touches the bottom of the "vee" in the conical extraction container. Slowly withdraw the *bottom* (organic) layer,

Take Care! Wear gloves during an extraction to protect your hands in case of leakage.

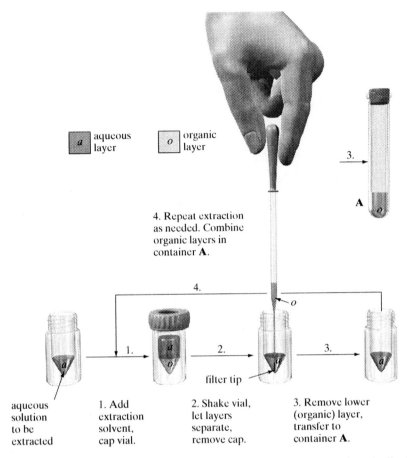

	aqueous layer			organic layer

4. Repeat extraction as needed. Combine organic layers in container **A**.

aqueous solution to be extracted

1. Add extraction solvent, cap vial.

filter tip

2. Shake vial, let layers separate, remove cap.

3. Remove lower (organic) layer, transfer to container **A**.

Figure C10 Microscale extraction using an extraction solvent denser than the liquid being extracted

taking care not to mix the layers, and transfer it to a suitable labeled container (container **A**; see Figure C10). Add another measured portion of pure extraction solvent to the liquid remaining in the extraction container and shake to extract as before, transferring the extract (the bottom layer) to **A**. If another extraction is necessary repeat the process, combining all of the extracts in **A**. Retain the liquid in the extraction container for later disposal.

*The kind of container you use for **A** depends on the next operation (washing, drying, etc.) that you will use.*

2a. Squeeze the bulb of a Pasteur pipet to expel air and insert the pipet vertically so that its tip just touches the bottom of the "vee" in the conical extraction container. Slowly withdraw the *bottom* (aqueous) layer, taking care not to mix the layers, and transfer it to a labeled test tube or another small container (container **A**; see Figure C11). Transfer the contents of the extraction container to a different labeled container (**B**) and return the contents of **A** to the extraction container. Add another measured portion of pure extraction solvent and shake to extract as before, then transfer the lower layer to **A** and combine the upper layer with the extract in **B**. If another extraction is necessary, return the contents of **A** to the extraction container and repeat the process. Retain the liquid in **A** for later disposal.

*The kind of container you use for **B** depends on the next operation (washing, drying, etc.) that you will use.*

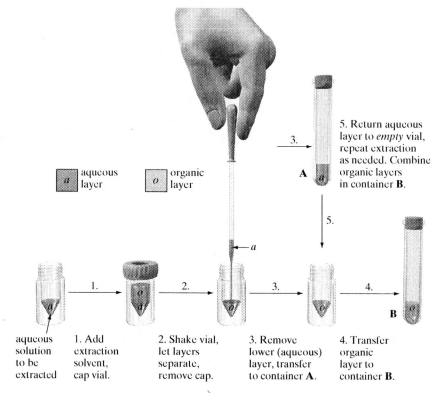

Figure C11 Microscale extraction using an extraction solvent less dense than the liquid being extracted

2b. (For small amounts of liquid.) Slowly draw *both* liquid layers into the Pasteur pipet (avoid drawing in much air) and wait until a sharp interface forms between the layers. Then carefully return the bottom layer to the empty extraction container and transfer the top layer to a labeled container (**A**). Add another measured portion of pure extraction solvent to the liquid in the extraction container and shake to extract as before. Again draw both layers into the Pasteur pipet, let them separate, and then return the bottom layer to the empty extraction container and transfer the top layer to **A**. Repeat this process for any subsequent extractions, combining all of the extracts in **A**. Retain the liquid in the extraction container for later disposal.

Summary (Methods **1** and **2a**)

1. Add liquid to be extracted to extraction container.
2. Add extraction solvent to extraction container, cap tightly.
3. Shake gently, vent, and continue shaking to extract solute into extraction solvent.
4. Loosen cap, let layers separate.
5. Uncap extraction container, transfer lower layer to container **A**.
   IF extraction solvent forms lower layer, GO TO 6.
   IF extraction solvent forms upper layer, GO TO 7.

6. Cap container **A**.
   IF another extraction step is needed, GO TO 2.
   IF extraction is complete, clean up, STOP.
7. Transfer upper layer to container **B** and cap it.
   IF another extraction step is needed, transfer contents of container **A** to extraction container, GO TO 2.
   IF extraction is complete, clean up, STOP.

When Things Go Wrong

If there is foaming when you mix a chemically active extraction solvent with the liquid to be extracted, the extraction solvent is reacting with a component of the liquid to yield a gas, usually carbon dioxide. Don't shake the mixture yet; stir it until the reaction has subsided, then begin shaking *gently*, with frequent venting.

If you're not sure which layer is the organic one, test the one you think is organic by adding a drop or two of water to a drop or two of that layer. If the water doesn't mix to give a homogeneous liquid, you've selected the organic layer; if the water mixes completely, you've got the aqueous layer. It's a good idea to label the containers for the extraction layers so you won't lose track of what's in them. (See *Saving the Right Layer.*)

If the organic and aqueous layers don't separate cleanly, see *Getting Good Separation.*

If you're draining the bottom layer from a separatory funnel and it stops draining, you forgot to take out the stopper. Close the stopcock and remove it, then continue draining.

Suppose you're trying to extract a substance from an aqueous solution with two portions of diethyl ether. You carried out the extraction with the first portion of ether and transferred the bottom layer to a different container, leaving the top layer in a separatory funnel (or conical vial, etc.). Then you poured the second portion of ether into the funnel and no layer separated. This is one of the most common extraction errors—adding the second portion of extraction solvent to the first portion of the *same* solvent rather than to the aqueous solution being extracted. Keep in mind that the combined ether layers still contain all of the product that was extracted by the first portion of ether. All you have to do is transfer the combined layers to another container, return the former bottom layer to the separatory funnel, extract it with a third portion of ether, and combine that portion with the rest of the ether. If you then evaporate [OP-19] the solvent it will take a little longer to remove the extra ether, but that's better than starting over.

Suppose you saturated an aqueous reaction mixture with potassium carbonate to salt out [OP-18b] the organic product and then extracted the mixture with dichloromethane. Since water ($d = 1.00$ g/ml) is less dense than dichloromethane ($d = 1.33$ g/ml), you saved the bottom layer, only to discover that the solvent in that layer wouldn't evaporate [OP-19] under the usual conditions. Although pure water is less dense than dichloromethane, adding a solute increases its density, and saturated aqueous potassium carbonate actually has a density of 1.56 g/mL. You could still salvage the situation by evaporating the solvent from the top (dichloromethane) layer—unless you discarded it, in which case you're out of luck.

In the James Bond movies, Bond always insists that his martinis be "shaken, not stirred." He's just being contrary; according to tradition, a proper martini should be stirred, not shaken.

b. Salting Out

Adding an inorganic salt (such as sodium chloride or potassium carbonate) to an aqueous solution containing an organic solute usually reduces the solubility of the organic compound in the water and thus promotes its separation. This *salting-out* technique is often used to separate a sparingly soluble organic liquid from its aqueous solution. For example, when an ester is hydrolyzed, the product mixture may be distilled to remove the alcohol, which can then be recovered by saturating the aqueous distillate with potassium carbonate and extracting it with diethyl ether. Salting out can also be used during an extraction to increase the amount of organic solute transferred from the aqueous to the organic layer and to remove excess water from the organic layer.

To salt out an organic liquid from an aqueous solution containing the liquid, add enough of the salt to saturate the aqueous solution, then stir or shake it to dissolve the salt. You can estimate the amount of salt needed by using its solubility or data from the Table "Saturated Solutions" in *The Merck Index*. If you don't know how much salt to use, add it in small portions until a portion no longer dissolves completely. If any undissolved salt remains, filter the mixture through a plug of glass wool. Transfer the mixture to a separatory funnel, a conical vial, or another suitable extraction vessel for separation or extraction.

To improve separation during an extraction, add enough of the salt to the extraction vessel to saturate the aqueous layer and shake to dissolve the salt. If necessary, filter the contents of the extraction container through a plug of glass wool to remove any undissolved salt. Then proceed as for a normal extraction. You can also add a saturated solution of the salt, rather than the pure salt, to the extraction mixture, but the salting-out process is less complete because the resulting salt solution is more dilute.

c. Liquid–Solid Extraction

Liquid–solid extraction involves the removal of one or more components of a solid by mixing the solid with an extraction solvent and separating the resulting solution from the solid residue. This method is often used to separate substances from natural products and other solid mixtures. For example, the red-orange pigment *lycopene* can be extracted from tomato products by petroleum ether and other solvents, as described in Experiment 9.

For a standard scale liquid–solid extraction, mix the solid intimately with the extraction solvent in a beaker or another suitable container, using a flat-bottomed stirring rod or a flexible flat-bladed spatula. Press or crush the solid against the bottom and sides of the container with the flat end of the stirring rod or spatula to extract as much of the solute as you can. Filter the mixture by gravity filtration [OP-15] or vacuum filtration [OP-16] and return any solid that collects on the filter to the beaker for the next extraction. Repeat the extraction as many times as necessary and combine the liquid extracts in a single collecting container.

For a microscale liquid–solid extraction, shake the solid vigorously with the extraction solvent in a capped centrifuge tube or another suitable container, then use a flat-bladed microspatula to crush and rub the solid against the sides of the tube. Repeat the shaking and crushing sequence several times. Centrifuge [OP-17] the mixture and transfer the extract to a suitable container, leaving the residue in the centrifuge tube. Alternatively, use a filter-tip pipet to separate the extract from the solid residue. Repeat the extraction as many times as necessary and combine the liquid extracts in a single collecting container.

Evaporation

Evaporation is the conversion of a liquid to vapor at or below the boiling point of the liquid. Evaporation can be used to remove a volatile solvent, such as diethyl ether or dichloromethane, from a comparatively involatile liquid or solid. Complete solvent removal is used to isolate an organic solute after such operations as extraction [OP-18] or column chromatography [OP-21]. Partial solvent removal, or *concentration*, can be used to bring a recrystallization solution to its saturation point (see OP-28).

Experimental Considerations

Because of possible health and fire hazards, you should never evaporate an organic solvent by heating an open container outside a fume hood. Even when using the methods described here, you should know and allow for the hazards associated with each solvent. In standard scale work, solvents are generally removed by distillation or under vacuum, while in microscale work, they are more often evaporated using a stream of nitrogen or dry air, but there are exceptions to these rules. For example, the standard scale apparatus shown in Figure C12c can be adapted for microscale use by using a 10-mL round-bottom flask and a microscale thermometer adapter.

Evaporation under a Fume Hood.   The easiest (but slowest) way to evaporate small quantities of a volatile solvent from a solution is to put the solution in a tared wide-mouth container, such as a large vial (set inside a beaker for stability) or a small beaker, and to leave the container under a fume hood for several hours. Make sure that the hood is turned on. Large quantities of solution may need to be left under the hood overnight or for several days. The process can be accelerated by clamping an inverted stemmed funnel over the container and drawing air through it with an aspirator or vacuum pump.

Distillation.   High-boiling solvents and relatively large quantities of low-boiling solvents can be removed by simple distillation [OP-30] or vacuum distillation [OP-31]. This procedure is often used to concentrate a solution that is then evaporated further by one of the methods described later. In microscale work, for example, a solution can be concentrated to a volume of 0.5–1.0 mL by distillation into a Hickman still and the remaining solvent can be evaporated under a stream of nitrogen or dry air, as described below.

Evaporation under Vacuum.   Solvents can be evaporated under vacuum using one of the setups pictured in Figure C12. The test tube or flask containing the liquid to be evaporated is heated [OP-7a] gently with a hot water bath or steam bath, and an aspirator or vacuum line is used to reduce the pressure inside the apparatus, increasing the evaporation rate.

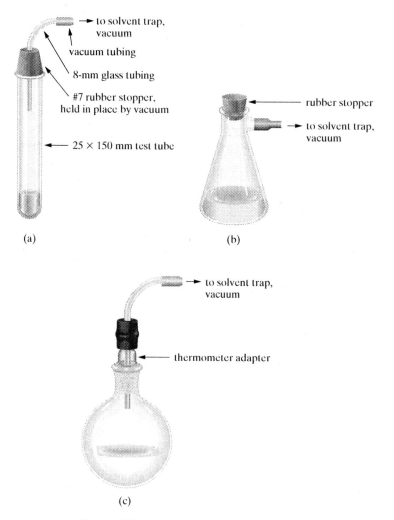

to solvent trap, vacuum

vacuum tubing

8-mm glass tubing

#7 rubber stopper, held in place by vacuum

25 × 150 mm test tube

(a)

rubber stopper

to solvent trap, vacuum

(b)

to solvent trap, vacuum

thermometer adapter

(c)

Figure C12 Apparatus for evaporation under vacuum

Swirling or stirring the solution continuously during evaporation speeds up the process and reduces foaming and bumping. Magnetic stir bars, because they tend to retain some of the residue, are not recommended for microscale evaporation unless the residue will later be dissolved in another solvent or undergo an operation requiring magnetic stirring.

Evaporation under vacuum requires constant attention because excessive heat or a sudden pressure decrease may cause liquid to foam up and out of the container. One way to control the vacuum and reduce the likelihood of boilover is to replace the stopper shown in Figure C12b with a Hirsch funnel assembly. If you hold your thumb over the holes in the porcelain plate of the Hirsch funnel, you can decrease the internal pressure by pressing down with your thumb or increase the pressure by raising it.

A trap similar to the one pictured in Figure C4 of OP-16 should ordinarily be interposed between the evaporation container and the aspirator to collect the evaporated solvent, which should then be returned to a solvent recovery container. To recover a low-boiling solvent such as diethyl ether, the solvent trap should be immersed in an appropriate cold bath [OP-8].

Commercial *flash evaporators* are used to evaporate solvents rapidly under reduced pressure, but they are seldom available in undergraduate organic chemistry labs because of their high cost.

Evaporation under Nitrogen or Dry Air. Small quantities of a volatile solvent can be evaporated by passing a slow stream of nitrogen or

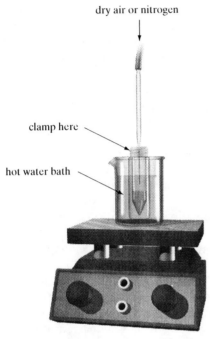

(a) Evaporation under a stream of dry air or nitrogen

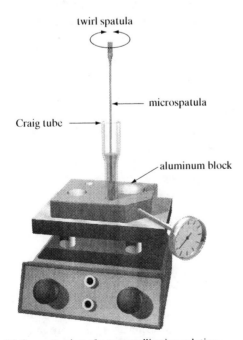

(b) Concentration of a recrystallization solution

Figure C13 Apparatus for microscale evaporation

dry air over the solution. Nitrogen is preferred because the oxygen in air may react with easily oxidized solutes, but clean, dry air is suitable for most purposes. The gas stream sweeps solvent molecules away from the surface of the liquid, accelerating the evaporation rate. Evaporation cools the remaining liquid, however, so heating is needed to maintain a rapid evaporation rate and to prevent condensation of water vapor in the product. This operation must be carried out under a fume hood to keep solvent vapors out of the laboratory. Your laboratory may have a hood with an "evaporation station" where a number of Pasteur pipets are attached to a source of nitrogen or dry air and supported above a large hot plate. The solution to be evaporated is placed in a conical vial (or another suitable container), which is heated gently in a hot water bath or warm sand bath while a stream of the dry gas is directed over the surface of the liquid (see Figure C13a). The bath temperature should be about 5–10°C below the boiling point of the solvent so that the solvent evaporates rapidly without boiling.

It is possible to use an aluminum block or hot plate as the heat source, but you must be very careful to avoid overheating, which may cause the residue to decompose. The evaporation container should not be placed directly on the hot surface; instead, clamp it above the hot surface at a level that will allow evaporation without boiling the solvent.

An aluminum block or similar heat source can be used safely to concentrate a solution by boiling, because decomposition is unlikely while some of the solvent remains. Twirling the end of a microspatula in the solution helps prevent bumping and boilover.

To avoid losses in transfer in microscale work, it is best to carry out an evaporation in the container that will be used in the next step, whenever possible. For example, if the residue left after evaporation will be the final product from a preparation, you can carry out the evaporation in a tared storage vial. If the residue will later be purified by distillation, evaporate the solvent in the conical vial or other container from which the residue will be distilled. If that vial isn't large enough to hold all of the liquid, you can carry out the evaporation in stages by adding one portion of the liquid, evaporating most of it, adding the next portion, and so on.

General Directions for Evaporation under Vacuum

Equipment and Supplies

    evaporation container (Figure C12)
    rubber tubing
    solvent trap
    aspirator
    heat source

Assemble one of the setups pictured in Figure C12 using heavy-walled rubber tubing that will not collapse under vacuum. Be sure to check all glassware for cracks, star fractures, and other imperfections that might cause them to implode under vacuum. Add the solution to be evaporated, stopper the evaporation container, and connect the apparatus to a solvent trap and the trap to a vacuum source. Turn on the vacuum and heat the evaporation container gently over a steam bath or in a hot water bath [OP-7a], swirling it throughout the evaporation to minimize foaming and bumping. (In some cases, the liquid can be stirred magnetically [OP-10].) Adjust the steam flow rate or water bath temperature to attain a satisfactory rate of evaporation; the liquid may boil gently but should not foam up. Be ready to remove the evaporation container from the heat source immediately if it starts to foam up; otherwise, your product may be carried over to the trap.

If you are using a steam bath, wear gloves or use a towel to protect your hands.

Continue evaporating until the residue, if it is a liquid, no longer appears to decrease in volume with time, or if it is a solid, appears dry. At this time, boiling should have stopped and the odor of the solvent should be gone. When evaporation appears complete, discontinue heating and remove the evaporation container from the heat source. Break the vacuum by detaching the vacuum hose, opening a pressure-release valve on the trap (if it has one) or—for the apparatus in Figure C12a—sliding the stopper off the mouth of the test tube. Then turn off the vacuum source. If you are not certain that all of the solvent has evaporated, dry the outside of the evaporation container to remove moisture from the heating bath, weigh it when it has cooled, and resume evaporation for a few minutes; then dry and weigh it again. Repeat this process as necessary until the mass no longer decreases significantly between weighings.

If you only need to concentrate the solution, stop the evaporation when sufficient solvent has been removed.

If you need to transfer [OP-6] the residue to another container, let the evaporation container cool down first. You can transfer the last traces of residue by rinsing the container with a small amount of a volatile solvent (such as diethyl ether or dichloromethane) and allowing the solvent to evaporate under a hood or in a stream of nitrogen or dry air. Place the solvent from the trap in a solvent recovery container.

Summary

1. Assemble apparatus for evaporation under vacuum.
2. Add liquid, connect evaporation container to trap and vacuum source.
3. Turn on vacuum.
4. Apply heat with swirling or stirring until evaporation is complete.
5. Discontinue heating, turn off vacuum.
6. Transfer residue and recover solvent.
7. Disassemble and clean apparatus.

When Things Go Wrong

If you are evaporating a liquid under vacuum and it keeps foaming up, see the second paragraph in *Evaporation Under Vacuum.*

If you are evaporating a liquid under vacuum and some of the liquid foams out of the evaporating flask, you can return it directly to the flask from the trap. If the trap isn't clean, you should purify the residue after the solvent has evaporated (you may be doing this anyway).

Suppose you are trying to isolate a solid by heating a solution under vacuum to evaporate the solvent, but no solid appears; you obtain only a liquid that doesn't decrease in volume with further heating. Remove the flask from the heat source and cool it under cold tap water to see if the liquid solidifies. If it doesn't, try to induce crystallization by rubbing the tip of a glass stirring rod against the inside of the container as directed under *Inducing Crystallization* [OP-28], or use another appropriate technique described there. If that doesn't work, your solution may contain a high-boiling solvent along with the more volatile solvent. You may be able to evaporate it with stronger heating under vacuum (using a hot plate rather than a steam bath, for example), or by distillation [OP-30], but be careful that you don't evaporate or decompose the desired product.

General Directions for Evaporation with Nitrogen or Dry Air

Equipment and Supplies

> evaporation container (conical vial, etc.)
> Pasteur pipet, rubber tubing
> drying tube with drying agent (optional)
> hot water bath, sand bath, or other heating device
> hot plate

The liquid to be evaporated should be in a tared conical vial or another container suitable for evaporation. If an evaporation station is not available, prepare a suitable gas-drying tube [OP-12], and then connect one end to an air line and the other end to a Pasteur pipet (the gas delivery pipet) using short lengths of rubber tubing. If dry nitrogen is available, omit the drying tube. Clamp the gas delivery pipet (your own or one from an evaporation station) vertically above the heat source. Heat the hot water bath or sand bath to a temperature 10°C or so below the solvent's boiling point and immerse the evaporation vial in it. Holding the vial in your hand provides more control over the evaporation rate, but clamping or otherwise securing it in the bath can help prevent accidental spillage. Position the gas delivery pipet so that its tip is 1–2 cm above the surface of the liquid in the vial and *slowly* increase the flow rate so that the gas distorts the surface of the liquid

To avoid the likelihood of product loss, you can adjust the gas flow rate using a vial of water or another solvent rather than the solution being evaporated.

but causes no splashing. As necessary, adjust the heating rate, the position of the vial, or the position of the gas delivery pipet so that the solvent evaporates quite rapidly but the liquid doesn't boil or foam up. If you are holding the vial in your hand, swirl it gently to increase the evaporation rate and help prevent boilover.

If you only need to concentrate the solution, stop the evaporation when sufficient solvent has been removed.

Continue evaporating until the residue, if it is a liquid, no longer appears to decrease in volume with time, or if it is a solid, appears dry. Remove the vial from the heat source. If you are not certain that all of the solvent has evaporated, clean or dry the outside of the vial, and then weigh it when it has cooled. Hold it under the gas stream for another minute or so and reweigh it. If the masses are essentially identical, evaporation is complete; if not, continue evaporating until the mass no longer decreases significantly between weighings.

If you need to transfer [OP-6] the residue to another container, let the evaporation container cool down first. You can transfer the last traces of residue by rinsing the container with a small amount of a volatile solvent (such as diethyl ether or dichloromethane) and allowing the solvent to evaporate under a hood or in a stream of nitrogen or dry air.

Summary

1. Transfer liquid being evaporated to evaporation vial.
2. Place vial in heated water bath or sand bath.
3. Position gas delivery pipet over liquid and adjust gas flow.
4. Apply gentle heat until evaporation is complete.
5. Transfer residue, if necessary.
6. Clean apparatus.

Steam Distillation

Distillation of a mixture of two (or more) immiscible liquids is called *codistillation*. When one of the liquids is water, the process is usually called *steam distillation. External steam distillation* is carried out by passing externally generated steam (usually from a steam line) into a boiling flask containing the organic material (see Figure C14). The vaporized organic liquid is carried over into a receiver along with the condensed steam. *Internal steam distillation* can be carried out by boiling a mixture of water and an organic material in a distillation setup such as the standard scale apparatus in Figure C15, causing vaporized water (steam) and organic liquid to distill into a receiver.

Both kinds of steam distillation are used to separate organic liquids from reaction mixtures and natural products, leaving behind high-boiling residues such as tars, inorganic salts, and other relatively involatile components. Steam distillation is particularly useful for isolating the volatile oils of plants from various parts of the plant. Steam distillation is not useful for the final purification of a liquid, however, because it cannot effectively separate components having similar boiling points.

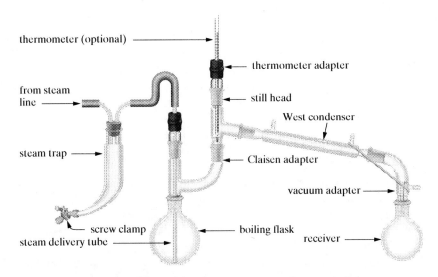

thermometer (optional)

thermometer adapter

from steam line

still head

West condenser

steam trap

Claisen adapter

vacuum adapter

screw clamp

boiling flask

receiver

steam delivery tube

Figure C14 Apparatus for standard-scale external steam distillation

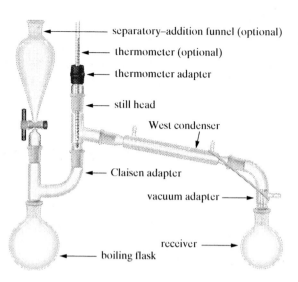

separatory–addition funnel (optional)

thermometer (optional)

thermometer adapter

still head

West condenser

Claisen adapter

vacuum adapter

receiver

boiling flask

Figure C15 Apparatus for standard-scale internal steam distillation

Principles and Applications

If you are not familiar with the principles of distillation, see OP-30.

When a *homogeneous mixture* of two liquids is distilled, the vapor pressure of each liquid is lowered by an amount proportional to the mole fraction of the other liquid present. This usually results in a solution boiling point that is somewhere between the boiling points of the separate components. For example, a solution containing equal masses of cyclohexane (bp = 81°C) and toluene (bp = 111°C) boils at 90°C.

When a *heterogeneous mixture* of two immiscible liquids, A and B, is distilled, each liquid exerts its vapor pressure more or less independently of the other. The total vapor pressure over the mixture (P) is thus approximately equal to the sum of the vapor pressures that would be exerted by the separate pure liquids (P_A^{o} and P_B^{o}) at the same temperature.

$$P \approx P_A^{o} + P_B^{o}$$

This has several important consequences. First, the vapor pressure of a mixture of immiscible components will be *higher* than the vapor pressure of its most volatile component. Because raising the vapor pressure of a liquid or liquid mixture lowers its boiling point, the boiling point of the mixture will be *lower* than that of its most volatile (lowest boiling) component. Because the vapor pressure of a pure liquid is constant at a constant temperature, the vapor pressure of the mixture of liquids will be constant as well. Thus, the boiling point of the mixture will remain constant throughout its distillation as long as each component is present in significant quantity.

For example, suppose that you are distilling a mixture of the immiscible liquids toluene and water at standard atmospheric pressure (760 torr, 101.3 kPa). The mixture will start to boil when the sum of the vapor pressures of the two liquids is equal to the external pressure, 760 torr. This occurs at 85°C, where the vapor pressure of water is 434 torr and that of toluene is 326 torr. Because the vapor pressures of the two components are additive, the mixture distills well below the normal boiling point of either toluene (bp = 111°C) or water. According to Avogadro's law, the number of moles of a component in a mixture of ideal gases is proportional to its partial pressure in the mixture, so the mole fraction of toluene in the vapor should be about 0.43 (326/760) and that of water should be about 0.57 (434/760). (These calculations are approximate because the vapors are not ideal gases.) In other words, about 43% of the molecules in the vapor are toluene molecules. Because toluene (M.W. = 92) molecules are heavier than water (M.W. = 18) molecules, they make a greater contribution to the total mass of the vapor. For example, in 1.00 mol of vapor there will be 0.43 mol of toluene and 0.57 mol of water, so the mass of toluene in the vapor will be about 40 g (0.43 mol × 92 g/mol) and the mass of water will be about 10 g (0.57 mol × 18 g/mol). The mass of one mole of the vapor is thus about 50 g, of which toluene makes up 40 g or 80%. The liquid that collects in the receiver during a distillation—the *distillate*—is merely condensed vapor, so distilling a mixture of toluene and water will yield a distillate that contains about 80% toluene, by mass.

Because of its comparatively low molecular weight and its immiscibility with many organic compounds, water is nearly always one of the liquids used in a codistillation involving an organic liquid. The organic liquid must be insoluble enough in water to form a separate phase, and it must not react with hot water or steam. As shown in Table C2, the higher the boiling point of the organic liquid, the lower will be its proportion in the distillate, and the closer the mixture boiling point will be to 100°C.

Because the distillation boiling point is never higher than 100°C at 1 atm—well below the normal boiling points of most water-immiscible organic liquids—thermal decomposition of the organic component is minimized.

Table C2 Boiling points and compositions of heterogeneous mixtures of various organic compounds, with water as Component B

Component A	bp of A (°C)	bp of A/B mixture (°C)	mass % of A in distillate
toluene	111°	85°	80%
chlorobenzene	132°	90°	71%
bromobenzene	156°	95°	62%
iodobenzene	188°	98°	43%
quinoline	237°	99.6°	10%

a. External Steam Distillation

Externally generated steam is preferred for most standard scale steam distillations, especially those involving solids or high-boiling liquids, because external steam produces a rapid distillation rate and helps prevent bumping caused by solids and tars. The steam is usually obtained from a steam line; if another kind of steam generator is to be used, your instructor will show you how to use it. A *steam trap* is used to remove condensed water and foreign matter, such as grease or rust, from externally generated steam. A steam trap that includes a valve for draining off excess water, such as the one illustrated in Figure C14, works best. With other kinds of traps, distillation may have to be interrupted periodically to drain the trap.

The boiling flask should be large enough so that the liquid will not fill it much more than half full throughout the distillation. Some steam will condense during the distillation, raising the water level in the boiling flask; during an extended distillation, excessive water can be removed by external heating, if necessary. A Claisen adapter is used to help prevent mechanical transfer of liquids or particles from the boiling flask to the receiver. If the organic distillate is fairly volatile, a thermometer can be used to indicate when the end of the distillation is near. For example, with a toluene–water mixture, the temperature will rise rather rapidly from 85°C to about 100°C when the toluene is nearly gone. With liquids having boiling points of 200°C or higher a thermometer is of little use because the vapor temperature will be close to 100°C throughout the distillation. The distillation should be carried out rapidly to minimize condensation in the boiling flask and to compensate for the large volume of water-laden distillate that may have to be collected to yield much of the organic component. Owing to the rapid distillation rate and the high heat content of steam, efficient condensing is essential. The vacuum adapter should be cool to the touch throughout the distillation and no steam should escape from its outlet.

General Directions for External Steam Distillation

 Standard Scale

Equipment and Supplies

heat source
ring stand, ring supports, clamps
rubber tubing
steam delivery tube
large round-bottom flask
Claisen adapter
still head (connecting adapter)
thermometer and thermometer adapter (optional) or stopper
West condenser
vacuum adapter
receiving flask
steam trap (bent adapter, two-hole rubber stopper, bent glass tubes, rubber tubing, screw clamp)

Assemble the apparatus pictured in Figure C14 using a large round-bottom boiling flask and, as the steam delivery tube, a 6-mm o.d. (outer diameter) glass tube extending to within about 0.5 cm of the bottom of the flask (it must not touch the bottom). If the component being separated boils above ~200°C, use a stopper in place of the thermometer adapter assembly.

Position the boiling flask high enough so that external heat can be applied, if necessary. Connect the steam delivery tube to a bent glass tube on the steam trap, which should be clamped to a ring stand over a beaker, and see that the screw clamp on the steam trap is closed. Add the organic mixture and a small amount of water (unless the mixture already contains water) to the boiling flask, which should be no more than one-third full at the start. Be sure that the condenser hoses fit tightly before you turn on the condenser water; the water should flow at a comparatively rapid rate. Connect a rubber hose to the steam valve over the sink and turn on the steam, directing it into the sink, until only a little water drips from the end of the hose. Turn off the steam and connect the hose to the other bent tube on the steam trap.

Take Care! Do not burn yourself with the steam.

Open the steam valve cautiously so that the liquid in the flask is agitated, but not too violently. The liquid should soon begin to boil, after which distillate will begin to pass over into the receiver. Adjust the steam flow to maintain a rapid rate of distillation without causing liquid in the flask to splash up into the condenser. Check the vacuum adapter periodically; if it becomes warm, and especially if vapor begins to escape from its outlet, you should increase the cooling water flow rate, cool the receiver in an ice/water bath, or reduce the steam flow rate. Check the connection between the condenser and still head frequently to make sure that no vapor is escaping; this joint sometimes separates because of the violent action of the steam (you can use a strong rubber band or a joint clip to prevent this). Drain the trap periodically to remove condensed water. If you must interrupt the distillation for any reason, open the steam-trap valve (or raise the steam delivery tube out of the liquid) *before* you turn off the steam; otherwise, liquid in the boiling flask may back up into the steam trap.

When the distillate appears clear *and* the temperature is near 100°C (if you used a thermometer), collect and examine a few drops of fresh distillate on a watch glass. Continue distilling if the fresh distillate is cloudy, contains oily droplets, or has a pronounced odor, and collect and examine more distillate at 5- or 10-minute intervals. When the distillate is water clear and distillation appears complete, open the steam-trap valve fully (or raise the steam delivery tube out of the liquid), and then turn off the steam.

The organic liquid can be separated from the distillate using a separatory funnel or by extraction [OP-18] with diethyl ether or another suitable solvent. Extraction is necessary if the volume of the organic liquid is small compared to that of the water. If the aqueous layer is cloudy, you can saturate it with sodium chloride or another salt to salt out [OP-18b] the organic liquid.

Summary

1. Assemble apparatus for external steam distillation.
2. Add organic mixture and water (if necessary) to boiling flask.
3. Turn on condenser water.
4. Purge steam line, connect to steam trap, turn on steam.
5. Distill rapidly until distillate is clear; drain trap periodically.
6. Open steam-trap valve, turn off steam.
7. Separate organic liquid from distillate.
8. Disassemble and clean apparatus.

When Things Go Wrong

Most of the things that go wrong during a simple distillation can also go wrong during an external steam distillation, so you can refer to *When Things Go Wrong* in OP-30a for help. The following cases apply only to external steam distillation.

If during a steam distillation the water level in the boiling flask rises well above its midpoint, check to make sure that the water level in the steam trap is well below the tube leading to the distillation apparatus. If not, drain the steam trap. If it is, heat the boiling flask externally with a steam bath or heating mantle to reduce condensation.

If during a steam distillation vapor escapes from the vacuum adapter outlet, first check to see that the condenser is cool to the touch and if not, increase the cooling water's flow rate. If that doesn't help, cool the receiver in an ice/water bath. It may also be necessary to reduce the steam flow rate.

b. Internal Steam Distillation

An organic liquid can sometimes be separated from a reaction mixture or another mixture by internal steam distillation (codistillation with water). The procedure is essentially the same as that for simple distillation [OP-30a], except that additional water may need to be added during the distillation. If water is to be added during a standard scale steam distillation, the apparatus illustrated in Figure C15 should be used; otherwise, a standard scale simple distillation apparatus (see Figure E7, OP-30) is suitable. If the organic distillate is fairly volatile, a thermometer can be used to indicate when the end of the distillation is near.

General Directions for Internal Steam Distillation

See OP-30 for more detailed directions for conducting a distillation.

 ### Standard Scale

Equipment and Supplies

 heat source
 ring stand, ring supports, clamps
 condenser tubing
 round-bottom flask
 boiling chips or stir bar
 still head (connecting adapter)
 thermometer and thermometer adapter (optional) or stopper
 West condenser
 vacuum adapter
 receiver
 Claisen adapter (optional)
 separatory–addition funnel (optional)

If it will be necessary to add more water during the distillation, assemble the apparatus shown in Figure C15; otherwise, assemble the apparatus pictured in Figure E7 of OP-30. If the component being separated boils above ~200°C, replace the thermometer adapter and thermometer by a stopper. Add the mixture to be steam distilled, boiling chips or a stir bar, and enough water to fill the boiling flask about one-third to one-half full (unless enough water is already present). Turn on the condenser water and the stirrer (if you are using one) and heat the flask with an appropriate heat source [OP-7a] to maintain a rapid rate of distillation. If necessary, add water to replace that lost during the distillation. When the distillate appears clear *and* the temperature is near 100°C (if you used a thermometer), collect and examine a few drops of fresh distillate on a watch glass. Continue distilling if the fresh distillate is cloudy, contains oily droplets, or has a pronounced odor, and collect and examine more distillate at 5- or 10-minute intervals. When the distillate is water clear and distillation appears complete, discontinue heating. Separate the organic liquid from the distillate as described for external steam distillation.

See the Summary following the Microscale directions.

Microscale

Equipment and Supplies

    heat source (aluminum block or sand bath)
    round-bottom flask
    Hickman still
    water-cooled condenser
    stir bar (optional)
    condenser tubing
    ring stand, clamp
    thermometer (optional)
    $5\frac{3}{4}$-inch Pasteur pipet
    9-inch Pasteur pipet (optional)
    collecting container

If it will be necessary to add more water during the distillation, measure out the approximate amount of water that will be needed. Assemble an apparatus similar to the one pictured in Figure E11 of OP-30, but attach the Hickman still to a round-bottom flask and use a water-cooled condenser. If the component being separated boils above ~200°C, omit the thermometer. If there is any likelihood that the mixture being distilled will foam up (as when certain plant products, such as ground cloves, are subjected to steam distillation), use the largest available round-bottom flask for boiling. Add the mixture to be steam distilled, a stir bar, and enough water to fill the boiling flask about one-half full (unless enough water is already present). In some cases, a stir bar may cause excessive foaming and should be omitted. Turn on the stirrer, if you are using one, and heat the flask with an appropriate heat source to maintain a rapid rate of distillation without allowing any boilover into the Hickman still. As the well of the Hickman still fills with liquid, use a short Pasteur pipet to transfer the distillate to an appropriate collecting container.

As necessary, use a long Pasteur pipet to add water through the condenser to replace that lost during distillation. Discontinue heating when the fresh distillate appears to contain no more of the organic component *or* after a designated volume of liquid has been collected. When distillation is complete, the distillate should no longer be cloudy or contain droplets of organic liquid. If droplets of the organic component have collected on the inner walls of the Hickman still, use a Pasteur pipet to rinse the walls into the well with some of the distillate or an appropriate solvent, and transfer the contents of the well to the collecting container. Separate the organic liquid from the distillate by microscale extraction [OP-18] with a suitable solvent.

Summary

1. Assemble apparatus for internal steam distillation.
2. Add organic mixture and water to boiling flask.
3. Turn on stirrer (if used) and condenser water.
4. Distill until distillate is clear, adding water as necessary.
5. Separate organic liquid from distillate.
6. Disassemble and clean apparatus.

When Things Go Wrong

Most of the things that go wrong during a simple distillation can also go wrong during an internal steam distillation, so you can refer to *When Things Go Wrong* in OP-30a for help.

OPERATION 21   # Column Chromatography

If you touch the tip of a felt-tip pen to a piece of absorbent paper, such as a coffee filter, and then slowly drip isopropyl rubbing alcohol onto the spot with a medicine dropper, the spot will spread and separate into rings of different color—the dyes of which the ink is composed. This is a simple example of *chromatography,* the separation of a mixture by distributing its components between two phases. The *stationary phase* (the coffee filter in this example) remains fixed in place, while the *mobile phase* (the rubbing alcohol) flows through it, carrying components of the mixture along with it. The stationary phase acts as a "brake" on most components of a mixture, holding them back so that they move along more slowly than the mobile phase itself. Because of differences in such factors as the solubility of the components in the mobile phase and the strength of their interactions with the stationary phase, some components move faster than others, and the components therefore become separated from one another.

Different types of chromatography can be classified according to the physical states of the mobile and stationary phases. In *liquid–solid* chromatography, which is applied in column chromatography [OP-21] and thin-layer chromatography [OP-22], a liquid mobile phase filters down or creeps up through the solid phase, which may be cellulose, silica gel, alumina, or some other *adsorbent.* The adsorbent is a finely divided solid that attracts solute molecules from the mobile phase onto its surface. In *liquid–liquid chromatography,* which is used in high-performance liquid chromatography [OP-38], the mobile phase is usually an organic solvent and the stationary phase can be a high-boiling liquid that is adsorbed by or chemically bonded to a solid *support*. In *gas–liquid chromatography*—the most common type of gas chromatography [OP-37]—the mobile phase is a gas that passes through a hollow or packed column containing a high-boiling liquid on a solid support. Gas chromatagraphy (GC) and high-performance liquid chromatography (HPLC) are instrumental methods that are used primarily for analyzing mixtures, so they will be discussed in the *Instrumental Analysis* section of the operations.

a. Liquid–Solid Column Chromatography

Principles and Applications

The usual stationary phase for liquid–solid column chromatography is a finely divided solid adsorbent, which is packed into a glass tube called the *column*. The mixture to be separated (the *sample*) is placed on top of the column and *eluted*—washed down the column—by the mobile phase, which is a liquid solvent or solvent mixture. Different components of the sample are attracted to the surface of the adsorbent more or less strongly depending on their polarity and other structural features. The more strongly a component is attracted to the adsorbent, the more slowly it will move down the column. So, as the mobile phase—also called the *eluant*—filters down through the adsorbent, the components of the sample spread out to form separate *bands* of solute, some passing down the column rapidly and others lagging behind.

For example, consider a separation of carvone and limonene on a silica gel adsorbent using hexane as the eluant. At any given time, a molecule of one component will either be adsorbed on the silica gel stationary phase or dissolved in the mobile phase. While it is adsorbed, the molecule will stay put; while it is dissolved, it will move down the column with the eluant. Molecules with polar functional groups are attracted to polar adsorbents such as silica gel and are relatively insoluble in nonpolar solvents such as hexane. So a molecule of carvone, with its polar carbonyl group, tends to spend more time adsorbed on the silica than dissolved in the hexane. It will therefore pass down the column very slowly with this solvent. On the other hand, a nonpolar molecule of limonene is quite soluble in hexane and only weakly attracted to silica gel, so it will spend less time sitting still and more time moving than will a carvone molecule. As a result, limonene molecules pass down the column rapidly and are soon separated from the slow-moving carvone molecules.

Carvone and limonene are major constituents of spearmint oil.

limonene   carvone

The separation attained by column chromatography depends on a number of factors, including the nature of the components in the mixture, the quantity and kind of adsorbent used, and the polarity of the mobile phase. The lists that follow show how strongly different functional groups are attracted to polar adsorbents and how strongly different adsorbents attract polar molecules.

Experimental Considerations

Adsorbents. A number of different adsorbents are used for column chromatography, but alumina and silica gel are the most popular. Adsorbents are available in a wide variety of activity grades and particle size ranges;

Table C3 Alumina activity grades (Brockmann scale)

Grade	Mass % water
I	0
II	3
III	6
IV	10
V	15

alumina can be obtained in acidic, basic, and neutral forms as well. The *activity* of an adsorbent is a measure of its attraction for solute molecules, the most active grade of a given adsorbent being one from which all water has been removed. The most active grade may not be the best for a given application, since too active an adsorbent may catalyze a reaction or cause bands to move down the column too slowly. Less active grades of alumina, for example, are prepared by adding different amounts of water to the most active grade (see Table C3).

Approximate strength of adsorption of different functional groups on polar adsorbents

COOH	strongest
OH	
NH$_2$	
SH	
CHO	
C=O	
COOR	
OR	
C=C	
Cl, Br, I	weakest

Common chromatography adsorbents in approximate order of adsorbent strength

Alumina (Al$_2$O$_3$)	strongest
Activated carbon (polar) (C)	
Silica gel (SiO$_2$)	
Magnesia (MgO)	weakest

Note: Adsorbent strength varies with grade, particle size, and other factors.

Some samples should not be separated on certain kinds of adsorbents. For example, basic alumina would be a poor choice to separate a mixture containing aldehydes or ketones, which might undergo aldol reactions on the column. Neutral alumina of activity II or III is a suitable adsorbent for most purposes. Silica gel, which is less active than alumina, is a good all-purpose adsorbent that can be used with most organic compounds.

Since all polar adsorbents are deactivated by water, it is important to keep their containers tightly closed and to minimize their exposure to atmospheric moisture.

The particle size of the adsorbent is also important, since the flow rate will be too slow if the particles are too small and the separation will be poor if the particles are too large. Particle size is often indicated by a range of mesh numbers, where a *mesh number* is the number of openings per linear inch in the finest screen that will allow the particles to pass through. Thus a higher mesh number indicates smaller particles. Typical particle sizes for column chromatography are 70–230 mesh for silica gel and ~150 mesh for alumina.

The amount of adsorbent required for a given application depends on the sample size and the difficulty of the separation. If the components of a mixture differ greatly in polarity, a long column of adsorbent should not be necessary, since the separation will be easy. The more difficult the separation, the more adsorbent will be needed. About 20 to 50 g of adsorbent per gram of sample is recommended for most separations, but easy separations may require less adsorbent and difficult separations may require more.

Eluants. In a column chromatography separation, the eluant acts primarily as a solvent to differentially remove molecules of solute from the surface of the adsorbent. In some cases, polar solvent molecules will also *displace* solute molecules from the adsorbent by becoming adsorbed themselves. If the solvent is too strongly adsorbed, the components of a mixture will spend most of their time in the mobile phase and will not separate efficiently. For this reason, it is usually best to start with a solvent of low polarity, and then (if necessary) increase the polarity gradually to elute the more strongly adsorbed components. Table C4 lists a series of common chromatographic solvents in order of increasing eluting power from alumina and silica gel. Such a listing is called an *eluotropic series*.

Table C4 Eluotropic series for alumina and silica gel

Alumina	Silica gel
pentane	cyclohexane
petroleum ether	petroleum ether
hexane	pentane
cyclohexane	trichloromethane
diethyl ether	diethyl ether
trichloromethane	ethyl acetate
dichloromethane	ethanol
ethyl acetate	water
2-propanol	acetone
ethanol	acetic acid
methanol	methanol
acetic acid	

Elution Techniques. Many chromatographic separations cannot be performed efficiently with a single solvent, so several solvents or solvent mixtures are used in sequence, starting with the weaker eluants—those near the top of the eluotropic series for the adsorbent being used. Such eluants will wash down only the most weakly adsorbed components, while strongly adsorbed solutes remain near the top of the column. The remaining solute bands can then be washed off the column by more powerful eluants.

In practice, it is best to change eluants gradually by using solvent mixtures of varying composition, rather than to change directly from one solvent to another. In *stepwise elution*, the strength of the eluting solvent is changed in stages by adding varying amounts of a stronger eluant to a weaker one. The proportion of the stronger eluant is increased more or less exponentially. For example, 5% dichloromethane in hexane may be followed by 15% and 50% mixtures of these solvents. According to one rule of thumb, the eluant composition should be changed after about three column volumes of the previous eluant have passed through. For example, if the packed volume of the adsorbent is 15 mL, the eluant composition should be changed with every 45 mL or so of eluant.

Columns. There are many different kinds of chromatography columns, ranging from a simple glass tube with a constriction at one end to an elaborate column with a detachable base and a porous plate to support the adsorbent. A buret, preferably one with a Teflon stopcock, is adequate for standard scale and some microscale separations, but the lack of a detachable base makes it difficult to remove the adsorbent afterward. If a column doesn't have a stopcock, the tip can be closed with a piece of flexible tubing equipped with a screw clamp. Unless the tubing is resistant to the eluants (polyethylene and Teflon will not contaminate most solvents), it should be removed before elution begins.

In selecting a column for a standard scale chromatographic separation, first estimate the amount of adsorbent needed for the sample you will be separating. Then choose a column of such a size that, after it has been packed with adsorbent, the adsorbent's surface will be about 10 cm or more below the top of the column and the height of the adsorbent column will be at least 10 times its diameter. If the column contains a porous plate to support the packing, no additional support is necessary; otherwise, the column packing should be supported on a layer of glass wool and clean sand.

A $5\frac{3}{4}$-inch Pasteur pipet packed with a suitable adsorbent makes an adequate chromatography column for simple microscale separations. Such a column provides the best separation when used with no more than 100 mg of sample, but it can be used to remove small amounts of impurities from larger quantities of sample. A microburet or graduated pipet may also be used to construct a microscale column. A microburet with an integral funnel top works well because the funnel top functions as a solvent reservoir, so that fresh eluant can be added less frequently. A funnel top may also make it easier to pack the column.

Figures C16 and C18 in the "General Directions" sections illustrate packed chromatography columns for standard scale and microscale use.

Flow Rate. The rate of eluant flow through the column should be slow enough so that the solute can attain equilibrium, but not so slow that the solute bands will broaden appreciably by diffusion. For most purposes, a flow rate of between 5 and 50 drops per minute should be suitable. Microscale columns have lower flow rates than standard scale columns (usually 5–10 drops/minute), and difficult separations require the slowest rates. The flow rate can be reduced by partly closing the stopcock or pinch clamp on the column (if it has either) or by reducing the *solvent head*—the depth of the eluant layer above the adsorbent. The flow rate can be increased by opening the stopcock or pinch clamp fully and by maintaining a high solvent head.

For a microscale column, the flow rate can be increased using a pipet pump with a quick release lever (see OP-21b). This method can decrease the separation efficiency, however, and removing the pump during elution without proper pressure release may damage the column, so it should be used only when absolutely necessary.

Packing a Column. To achieve good separation with a chromatographic column, it must be packed properly. The packing must be uniform, without air bubbles or channels, and its surface must be even and horizontal. Standard scale columns using alumina can be packed by pouring the dry adsorbent through a layer of solvent; microscale columns are often packed without solvent. Columns using silica gel are usually packed with a slurry containing the adsorbent suspended in a solvent. Pasteur-pipet columns cannot easily be slurry-packed, so they are most often used with alumina as the adsorbent. Once a column's adsorbent is moistened with solvent, it must be kept covered with the solvent at all times; allowing it to dry out creates channels that lead to uneven bands and poor separation. The following general directions give detailed instructions for packing standard scale and microscale columns. The directions for standard scale chromatography can also be used for microscale separations using a microburet and 5–20 g of adsorbent.

General Directions for Column Chromatography

Standard Scale

Equipment and Supplies

> chromatography column
> buret funnel (or small powder funnel)
> column-packing solvent
> glass wool
> tapper (pencil and one-hole stopper)
> clean sand
> adsorbent
> collectors (flasks, test tubes, vials, etc.)
> Pasteur pipet with bulb
> eluant(s)
> separatory–addition funnel (optional)

Packing the Column. Obtain an appropriate column and clamp it securely to a ring stand so that it is as nearly vertical as possible. Be sure that the stopcock or screw clamp at the outlet of the column is closed. Construct a "tapper" by, for example, inserting one end of a pencil into a small one-hole rubber stopper. Measure out the amount of adsorbent you will need to prepare the column and keep it in a tightly closed container. Number and weigh the collectors you will be using to collect the eluant fractions. Then pack the column with adsorbent by one of the following methods (omit the glass wool and sand support if it has a porous plate). Method **1** is generally used for silica gel and Method **2** for alumina.

1. *Slurry packing the column.* Fill the column about half full with the least polar eluting solvent to be used in the separation, or a solvent recommended in the experimental procedure. Use a long glass rod to push a plug of glass wool to the bottom of the column, and tamp the glass wool

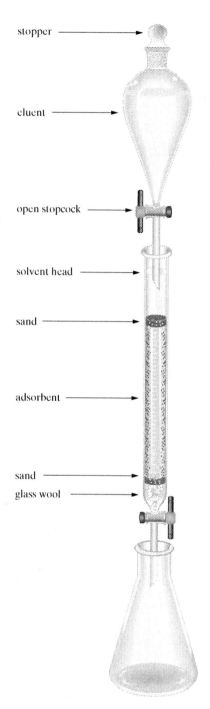

stopper

eluent

open stopcock

solvent head

sand

adsorbent

sand

glass wool

Figure C16 Packed column with continuous-feed reservoir

down gently to form a level surface and press out any air bubbles. Using a *dry* funnel, slowly pour in enough clean sand to form a 1-cm layer at the bottom of the column. As the sand filters down through the solvent, tap the column gently and continuously with the tapper so that the sand layer is uniform and level. The column should be tapped near its center, where it is clamped, to avoid displacing it from the vertical.

Mix the measured amount of silica gel (or another suitable adsorbent) thoroughly with enough of the column-packing solvent to make a fairly thick, but pourable, slurry. With the column outlet closed, slowly pour part of this slurry through a funnel, with tapping, until the adsorbent forms a layer about 2 cm thick at the bottom of the column. With a flask or beaker under the column, open the column outlet so that the solvent drains slowly as you add the rest of the slurry, tapping constantly to help settle and pack the adsorbent. If the slurry becomes too thick to pour, add more solvent to it. There should be enough solvent in the column so that the solvent level is well above the adsorbent level at all times; if necessary, add more solvent.

When all of the adsorbent has been added, close the outlet. The surface of the adsorbent should be as even and horizontal as possible, so continue tapping until it has settled. Gently stirring the top of the solvent layer as the adsorbent is settling can also help form a level surface. Use a Pasteur pipet containing the solvent to rinse down any adsorbent that adheres to the sides of the column. Add enough clean sand to form a protective layer about 0.5 cm thick on top of the adsorbent; the sand surface should also be level. Open the outlet until the solvent surface drops to within 1–2 cm of the sand surface, then close the outlet and stopper the column tightly. Keep the adsorbent covered with solvent at all times; allowing it to dry out creates channels that lead to uneven bands and poor separation.

2. *Packing the column with dry adsorbent.* Fill the column about two-thirds full with the least polar eluting solvent to be used in the separation, or a solvent recommended in the experimental procedure. Use a long glass rod to push a plug of glass wool to the bottom of the column and tamp the glass wool down gently to form a level surface and press out any air bubbles. Using a *dry* funnel, slowly pour in enough clean sand to form a 1-cm layer at the bottom of the column. As the sand filters down through the solvent, tap the column gently and continuously with the tapper so that the sand layer is uniform and level. The column should be tapped near its center, where it is clamped, to avoid displacing it from the vertical.

With the column outlet closed, slowly pour enough alumina (or another suitable adsorbent) through a dry funnel, while tapping, until the adsorbent forms a layer about 2 cm thick at the bottom of the column. With a flask or beaker under the column, open the column outlet and add the rest of the dry alumina as the solvent drains, tapping constantly to help settle and pack the adsorbent. There should be enough solvent in the column so that the solvent level is well above the adsorbent level at all times; if necessary, add more solvent.

When all of the adsorbent has been added, close the outlet. The surface of the adsorbent should be as even and horizontal as possible, so continue tapping until it has settled. Gently stirring the top of the solvent layer as the adsorbent is settling can also help form a level surface. Use a Pasteur pipet containing the solvent to rinse down any adsorbent that adheres to the sides of the column. Add enough clean sand to form a protective layer about 0.5 cm thick on top of the adsorbent; the sand surface should also be level. Open the outlet until the solvent surface drops to within 1–2 cm of the sand surface, then close the outlet and stopper the column tightly. Keep the adsorbent covered with solvent at all times.

Separating the Sample. If the sample is a solid, dissolve it in a minimum amount of a suitable nonpolar solvent; use liquid samples without dilution. Open the column outlet until the solvent surface comes down *just* to the top of the sand layer, and then close it. Use a Pasteur pipet to apply the sample around the circumference of the sand so that it spreads evenly over the surface. Open the outlet until the sample's surface comes down to the top of the sand layer, and then close it. Pipet a small amount of the initial eluant around the inside of the column to rinse down any adherent sample. Open the outlet again until the eluant's surface comes down to the top of the sand layer, and then close it.

Clamp a solvent reservoir such as a separatory/addition funnel over the column and measure the initial eluant into it. (Alternatively, add the eluants through an ordinary funnel or use a continuous-feed reservoir as shown in Figure C16, moistening the stopper with solvent to provide an airtight seal.) Add enough eluant to nearly fill the column. Place a tared collector at the column outlet, open the outlet, and continue adding eluant to keep the liquid level near the top of the column throughout the elution. If you need to change eluants during the elution, let the previous eluant drain to the top of the sand layer before adding the next one.

If the components are colored or can be observed on the column by some visualization method (such as irradiation with ultraviolet light), change collectors each time a new band of solute begins to come off the column *and* when it has almost disappeared from the column. If two or more bands overlap, collect the overlapping regions in separate collectors to avoid contaminating the purer fractions. Unless directed otherwise, evaporate [OP-19] the solvent from any fractions that contain the desired component(s).

If the components are not visible and the procedure doesn't specify the fraction volumes, collect equal-volume fractions (about 25 mL each) in tared collectors. Evaporate the solvent from each fraction, weigh the collectors and their contents, and plot the mass of each residue versus the fraction number to obtain an elution curve such as the one illustrated in Figure C17. From the elution curve, you should be able to identify separate components and decide which fractions can be combined before evaporation.

To remove the contents of the column, let it dry completely, then invert it over a beaker and tap it as needed to dislodge the adsorbent. If necessary, attach the column outlet to an air line, hold its other end over the beaker, and *slowly* open the air valve to blow out the glass-wool plug and any remaining adsorbent. Dispose of any unused or recovered solvents as directed by your instructor.

Summary

1. Pack column using appropriate adsorbent, solvent, and packing method.
2. Drain column to top of sand layer, add sample, drain, rinse, drain again.
3. Add eluant, put collector in place, open column outlet.
4. Elute sample, keeping eluant level nearly constant.

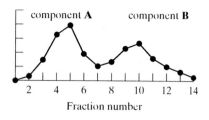

Figure C17 Elution curve

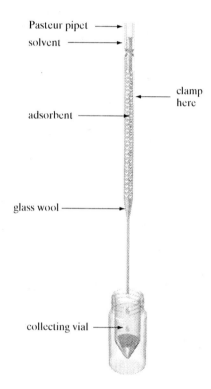

Figure C18 Apparatus for microscale column chromatography

5. To change eluants, drain current eluant to top of sand, add next eluant. IF components are visible, GO TO 6.
IF components are not visible, GO TO 7.

6. Change collectors when new band starts or ends and where bands overlap. GO TO 8.

7. Change collectors after a predetermined volume has been collected.

8. Stop elution after last fraction has been collected.

9. Evaporate and weigh appropriate fractions.

10. Disassemble and clean apparatus, dispose of solvents as directed.

Microscale

Equipment and Supplies

Two $5\frac{3}{4}$-inch Pasteur pipets, one with rubber bulb

column-packing solvent

plastic weighing dish

glass wool

pencil

clean sand

adsorbent

collectors (vials, test tubes, etc.)

eluant(s)

Packing the Column.   Obtain a clean, dry $5\frac{3}{4}$-inch Pasteur pipet and use a wooden applicator stick or glass stirring rod to push a small plug of glass wool to the point at which the pipet narrows (see Figure C18). Be careful not to tamp it down so tightly as to restrict the solvent flow. Clamp the pipet vertically to a ring stand. If desired, add about 50 mg of fine sand while gently tapping the column with the eraser end of a pencil to form a layer on top of the cotton (this isn't essential, but it keeps fine alumina particles from passing through the glass wool). The top of the sand layer should be even and horizontal. Measure the approximate amount of adsorbent you will need (usually 1.5–2.0 g) and keep it in a tightly closed container. Number and weigh the collectors you will be using to collect the eluant fractions, then pack the column by one of the methods that follow. Method **2** gives a more evenly packed column when alumina is used as the adsorbent; it is not satisfactory for silica gel. If you use Method **2**, you should be prepared to add your sample immediately after filling the column.

1. *Packing the column with no solvent.* Put the adsorbent in a small, square plastic weighing dish. (Alternatively, you can use a microscoop made by cutting a 1-mL automatic pipet tip in half.) Bend the dish to form a spout (see OP-6) and add the adsorbent *slowly* to the top of the column, while continuously tapping the column near its center with the eraser end of a pencil, until the specified amount of adsorbent has been added *or* the adsorbent level is about 1 cm below the upper constriction in the column. The surface of the adsorbent should be as even and horizontal as possible; if necessary, tap gently near the surface to make it

so. If desired, you can sprinkle a thin, even layer of clean sand on top to protect the adsorbent layer.

2. *Packing the column using a solvent.* This method requires constant attention because the adsorbent layer must be completely covered with solvent at all times. Put the adsorbent in a square plastic weighing dish. Put a small flask or beaker under the column and use a Pasteur pipet to fill it about half full with the least polar eluant to be used in the chromatography (or another suitable solvent). Immediately bend the dish to form a spout (see OP-6) and add the adsorbent *slowly* to the top of the column as the solvent drains, while continuously tapping the column near its center with the eraser end of a pencil. If the solvent is draining too fast, stop the flow temporarily by holding your fingertip on top of the column and allow the adsorbent to settle before adding the next portion of adsorbent. Continue adding adsorbent until the specified amount has been added *or* until the adsorbent level is about 1 cm below the upper constriction in the column. The surface of the adsorbent should be as even and horizontal as possible; if necessary, tap gently near the surface to make it so. If desired, you can sprinkle a thin, even layer of clean sand on top to protect the adsorbent layer. Be ready to add your sample *immediately* when the solvent surface drops to the top of the adsorbent layer. (If you are not ready, stopper the column with a size 000 cork to stop the solvent flow.)

*A different method for filling Pasteur pipet chromatography columns is described in J. Chem. Educ. **2006**, 83, 1200.*

Separating the Sample. Once it is moistened with solvent, the column of adsorbent must not be allowed to dry out until the separation is complete. If the sample is a solid, dissolve it in a minimum amount of a nonpolar solvent; use liquid samples without dilution. Be sure that there is a suitable collector under the column. If the column was packed without solvent, use a Pasteur pipet to add enough of the initial eluant to moisten all of the adsorbent and cover the adsorbent (or sand) layer to a depth of about 1 cm. When the solvent surface *just* drops to the top of the adsorbent (or sand) layer, add the sample with a Pasteur pipet. Quickly rinse this pipet with a very small amount of the initial eluant and use it to rinse the sides of the column. When the liquid surface has fallen to the top of the adsorbent (or sand) layer, use a Pasteur pipet to add enough of the initial eluant to nearly fill the column. Be careful not to disturb the surface of the adsorbent as you add it. Continue adding eluant as necessary to keep the solvent level near the top of the column throughout the elution. If you need to change eluants, allow the previous eluant to drain to the level of the adsorbent (or sand) before adding the next one. Collect the fractions as described previously for standard scale column chromatography. Stop the elution when a designated volume of eluant has been added or when the desired sample bands have been collected. Unless directed otherwise, evaporate [OP-19] the solvent from any fractions that contain the desired component(s).

To remove the contents of the column, let it dry completely, and then invert it over a beaker and tap it with the eraser end of a pencil. If that doesn't work, attach a rubber hose from an air line to the column at its narrow end (be careful not to break the tip!); then hold the open end over a beaker and *slowly* open the air valve until the air blows out the adsorbent. To remove the glass wool plug, which should now be near the wide end of the column, twirl the end of an applicator stick in the glass wool until it catches the fibers, and then pull it out. Dispose of any unused or recovered solvents as directed by your instructor.

Summary

1. Clamp column vertically.
2. Pack column with adsorbent.
   IF column was packed using a solvent, GO TO 4.
3. Cover adsorbent with eluant.
4. Drain column to top of adsorbent, add sample, rinse, drain again.
5. Add eluant to elute sample, keeping eluant level nearly constant.
6. To change eluants, drain current eluant to top of adsorbent, add next eluant.
   IF components are visible, GO TO 7.
   IF components are not visible, GO TO 8.
7. Change collectors when new band starts or ends and where bands overlap. GO TO 9.
8. Change collectors after a predetermined volume has been collected.
9. Stop elution after last fraction has been collected.
10. Evaporate and weigh appropriate fractions.
11. Disassemble and clean apparatus, dispose of solvents as directed.

When Things Go Wrong

Suppose you are separating the components of a reaction mixture by column chromatography but analysis of the product shows that it is still impure. To find out what went wrong, ask yourself the following questions. If the answer to a question is yes, consult the paragraph indicated, reread the General Directions more carefully, and repeat the chromatographic separation. (1) Was the amount of sample too large for the amount of adsorbent used? (See *Adsorbents.*) (2) Was the flow rate too slow, causing broad or diffuse bands? (See *Flow Rate.*) (3) Was the column unevenly packed or was its surface disturbed, causing channeling, uneven bands, or nonhorizontal bands? (See *Packing a Column.*) (4) Did you change the eluant polarity too rapidly, resulting in channeling or cracks in the column? (See *Elution Techniques.*)

b. Flash Chromatography

Flash chromatography is a variation of liquid–solid column chromatography that uses a single elution solvent and takes less time than the standard method. Pressurized nitrogen, air, or another gas is applied to the top of the column to force eluant through the adsorbent, which is usually finely divided (~230–400 mesh) silica gel. Since only one eluant is used, it must be selected carefully to ensure good separation. Mixtures of ethyl acetate or dichloromethane and low-boiling petroleum ether work for many separations. A prospective solvent system can be tested by spotting a silica-gel TLC plate [OP-22] with the sample and developing it with the solvent system; a suitable solvent system should move the desired component at least a third of the way from the starting line to the solvent front, and it should give good separation between the desired component and any impurities. Separation is poor if the sample size is too large; the following procedure works best with 0.25 g of sample or less. Larger samples usually require more eluant and a column of greater diameter.

Commercial flash chromatography systems may require expensive columns, pumps, and flow controllers, so they are seldom available for use in undergraduate laboratories. An inexpensive flash chromatography system can be constructed using the "homemade" flow controller shown in Figure C19. The flow controller is assembled by inserting the bottom of a small plastic T-tube into a one-hole rubber stopper that fits into the top of the column, and then attaching two lengths of $\frac{3}{8}$-inch i.d. (inner diameter) Tygon tubing to the straight ends of the T and securing them with copper wires. One tube is attached to the gas inlet on a Bunsen burner base (the barrel can be removed) and the other to a source of clean, dry compressed air. (The pressure can also be controlled, but less precisely, with the user's thumb rather than the burner base.) The column can be a 50-mL buret packed to a depth of 15 cm or so with adsorbent, but a 12 × 1.5-cm polypropylene column packed with 5-6 g of adsorbent has also been used successfully and is less likely to break under pressure.

This flash chromatography apparatus is described in J. Chem. Educ. **1992**, *69*, 939.

You should have read OP-21a before attempting to carry out a flash chromatography separation. Pack an appropriate column no more than half full with finely divided silica gel while tapping the column gently; then add a uniform layer of clean sand. Fill the column with eluant, open the column outlet (with a receiving flask in place), and apply pressure with dry air or nitrogen to saturate the adsorbent with solvent. Save the excess eluant, which can be reused. Drain the solvent to the top of the sand layer and immediately introduce the sample, as a 25% solution in the eluant, onto the sand layer with a long pipet. When all of the sample is within the sand layer (pressure can be applied to force it down), close the column outlet and carefully fill the column to within 1–2 cm of the top with eluant. Insert the stopper of the flow controller into the top of the column with a firm twist to keep it from popping out (it should stay in place at the desired pressure). Open the needle valve on the burner base, then open the column's stopcock with a collector in place. Carefully turn on the air valve to pressurize the system and adjust the needle valve so that the eluant level decreases at a rate of ~5 cm per minute. Collect and evaporate the fractions as described in OP-21a. If the separation

See J. Chem. Educ. **2000**, *77*, 263.

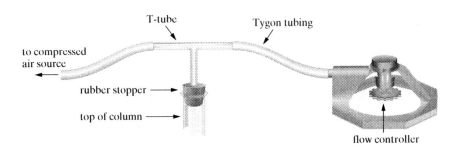

Figure C19 Flow controller for flash chromatography

is unsatisfactory, try using a longer column and a larger volume of eluant, or change to a more suitable eluant.

It is difficult to do a true flash chromatography separation using a microscale setup like that in Figure C18, but you can speed up a microscale separation by preparing the column and adding the sample as described in the General Directions, then adding the eluant and applying pressure to the top of the Pasteur-pipet column with a pipet pump (Figure A10, OP-5) that has a pressure-release valve. After the eluant has been added, secure the pipet pump over the mouth of the column and rotate its thumbwheel slowly to apply a constant, even pressure, so that the eluant drips from the tip at a more-or-less constant rate. When the solvent surface reaches the surface of the adsorbent, depress the pressure release valve and *carefully* remove the pipet pump, then immediately add more eluant and continue to apply pressure. Removing the pipet pump too rapidly or failing to release the pressure may damage the column of adsorbent and spoil the separation. Repeat the process as needed until the desired volume of eluant has been added. This method is not suitable for difficult separations, because speeding up the elution rate reduces the separation efficiency.

c. Reversed-Phase Column Chromatography

Adsorbents are polar solids that attract polar compounds more strongly than nonpolar ones, so nonpolar solutes are eluted from adsorbent-packed columns more rapidly than polar ones. Another kind of stationary phase can be prepared by coating particles of silica gel with a high-boiling nonpolar liquid. With such a stationary phase, components of the sample are partitioned between the liquid mobile phase and the liquid layer of the stationary phase, where to *partition* a solute means to distribute it between two phases. If the mobile phase is more polar than the stationary phase, the usual order of elution will be reversed; that is, polar compounds will be eluted before nonpolar ones. This general method is called *reversed-phase chromatography*. The mobile phases for reversed-phase column chromatography are usually polar solvents such as water, methanol, and acetonitrile, or mixtures of such solvents. Stationary phases can be prepared by coating a specially treated (silanized) silica gel with a nonpolar liquid phase, such as a hydrocarbon or silicone. Bonded liquid phases, such as the ones described for HPLC [OP-38], can also be used.

To prepare a column using coated silica gel, the liquid mobile and stationary phases are shaken together in a separatory funnel to saturate each phase with the other and the layers are separated. The stationary phase is then stirred with the silica gel and the coated support is made into a slurry with the saturated mobile phase. The column is slurry packed, usually with stirring to make it more uniform and to remove air bubbles. A separation is carried out by eluting with the saturated mobile phase, essentially as described previously for normal-phase column chromatography.

Thin-Layer Chromatography

Principles and Applications

Like column chromatography, thin-layer chromatography (TLC) utilizes a solid adsorbent as the stationary phase and a liquid solvent as the mobile phase, but the mobile phase creeps *up* the adsorbent layer by capillary action rather than filtering down through it by gravity. A *TLC plate* consists of a thin layer of the adsorbent on an appropriate *backing* (solid support) made of plastic, aluminum, or glass. The sample, or several samples, are dissolved in a suitable solvent and applied near the lower edge of the TLC plate as small spots. The plate may also be spotted with a selection of standard solutions for comparison. The TLC plate is *developed* by immersing its lower edge in a suitable mobile phase, the *developing solvent*. As this solvent moves up the adsorbent layer, it carries with it the components of each spot, which are separated by the mechanism described in OP-21a for column chromatography.

Although TLC is not useful for separating large quantities of material, it is much faster than column chromatography and it can be carried out with very small sample volumes, so that little is wasted. TLC provides better separation than the related technique of paper chromatography [OP-23], and it can be applied to a wider range of organic compounds. Although TLC is categorized under separation operations in this book, it is also used for qualitative and quantitative analysis of organic compounds.
Applications of TLC include the following:

- To identify unknown substances and unknown components of mixtures.
- To monitor the course of a reaction and assess the purity of its product by comparing the relative amounts of product, reactants, and by-products on successive chromatograms.
- To determine the best solvent for a column chromatography separation (see *Choosing a Developing Solvent*).
- To determine the composition of each fraction from a column chromatography separation, so that fractions containing the same component can be detected and combined.
- To determine whether a substance purified by recrystallization or another method still contains appreciable amounts of impurities.

Experimental Considerations

Adsorbents. The most commonly used adsorbents for TLC are silica gel, alumina, and cellulose. The adsorbent is more finely divided than that used in column chromatography, and it is provided with a *binder*, such as polyacrylic acid, to make it stick to the backing. It may also contain a *fluorescent indicator*, which makes most spots visible under ultraviolet light.

TLC Plates. Small "do-it-yourself" TLC plates can be prepared by dipping glass microscope slides into a slurry of the adsorbent and binder in a suitable solvent and allowing the solvent to evaporate. Such plates may give inconsistent results because of variations in the thickness of the adsorbent layer. More uniform TLC plates measuring 20 × 20 cm or larger are commercially available with a wide variety of adsorbents, backings, and layer thicknesses. For example, a typical TLC plate suitable for use in undergraduate laboratories has a flexible plastic backing coated with a 200-μm-thick layer of silica gel mixed with a binder, and possibly a fluorescent indicator as well. The adsorbent layer of a TLC plate is easily damaged, so it is important to avoid unnecessary contact with its coated surface and to protect the plate from foreign materials. A TLC adsorbent will pick up moisture when exposed to the atmosphere, making it less active; TLC plates can be *activated* by heating them in a 110°C oven for an hour or so.

Spotting. A TLC plate is prepared for development by applying solutions of the sample(s) to be analyzed and any reference standards as small spots; this process is known as *spotting*. The sample is dissolved in a suitable solvent to make an approximately 1% solution. If possible, the solvent should be quite nonpolar and have a boiling point in the 50–100°C range. Column chromatography fractions and other solutions can often be used as is, if the solute is present at a concentration in the 0.2–2.0% range.

It is best to wear thin disposable gloves while spotting a TLC plate, because if you touch the surface of the adsorbent with your bare fingers, your fingerprints may hinder development or obscure developed spots. Position the spots accurately, because incorrectly placed spots may run into one another or onto the edge of the adsorbent layer. This can be done with a transparent plastic ruler supported just above the surface of the plate so that it doesn't touch the adsorbent. Mark the starting line with a pencil on both edges, about 1.5 cm from the bottom of the plate (or 1.0 cm for a microscope-slide plate), and position the spots along the starting line at least 1.5 cm from each edge of the plate and 1.0 cm from each other. Thus a 10 × 10-cm TLC plate can accommodate up to eight spots, as shown in Figure C20.

Large, diffuse spots spread out too much for accurate results, so each spot should be as small and concentrated as possible. The spots are best applied with a microliter syringe or a capillary micropipet. Capillary micropipets such as Drummond Microcaps are commercially available, but suitable micropipets can also be prepared by heating an open-ended melting-point capillary in the middle over a small flame, drawing it out to form

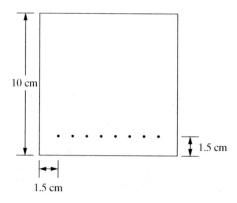

Figure C20 Spotted 10 × 10-cm TLC plate

a fine capillary about 4–5 cm long, allowing the tube to cool, and then scoring and snapping it apart in the middle to form two micropipets (see Figure C21).

To spot a TLC plate with a micropipet, dip the narrow tip into the solution to draw in a small amount of liquid, and then gently touch the tip to the surface of the TLC plate, at the proper location, for only an instant. Be careful not to dig a hole in the adsorbent surface, because this will obstruct solvent flow and distort the chromatogram. Make several successive applications at each location, letting the solvent dry each time, to form a spot 1–3 mm in diameter. It may be worthwhile to try one, two, and three applications of a sample at three separate locations on the TLC plate to determine which quantity gives the best results. Too much solution can result in "tailing" (a zone of diffuse solute following the spot), "bearding" (a zone of diffuse solute preceding the spot), and overlapping of components. Too little solution makes it difficult to detect some of the components. Capillary micropipets can be reused a few times, but a different micropipet should be used for each different solution to avoid cross-contamination.

To spot a TLC plate with a microliter syringe, deliver about 1 µL of solution with each application and make two to three applications for each spot, letting the solvent dry between applications. Take care not to touch the adsorbent surface with the syringe needle. Before you fill the syringe with a different solution, rinse it with a suitable solvent and remove excess solvent by pumping the plunger gently a few times. After the last application, clean the syringe with more solvent, remove the plunger, and let it dry.

Choosing a Developing Solvent. Solvents that are suitable eluants for column chromatography are equally suitable as TLC developing solvents; the eluotropic series in Table C4 (OP-21) may help you choose a solvent for a particular application. A quick way to find a suitable solvent is to spot a TLC plate with the sample, applying as many spots as you have solvents to test (you can use a grid pattern with the spots 1.5–2.0 cm apart), and then apply enough solvent directly to each spot to form a circle of solvent 1–2 cm in diameter. Mark the circumference of each circle before the solvent dries. A solvent whose chromatogram (after visualization) shows well-separated rings, with the outermost ring about 50–75% of the distance from the center to the solvent front, should be satisfactory. It is preferable to use the least polar solvent that gives good separation. Hexane, toluene, dichloromethane, and methanol or ethanol (alone or in binary combinations) are suitable for most separations. If no single solvent is suitable, choose two mutually miscible solvents whose outermost rings bracket the 50–75% range (one with too little solute migration, the other with too much), and test them in varying proportions.

Development. TLC plates are developed by placing them in a *developing chamber* containing the developing solvent. A paper wick can be used to help saturate the air in the developing chamber with solvent vapors, which improves reproducibility and increases the rate of development. The developing chamber can be a jar with a screw-cap lid, a beaker covered with plastic film or aluminum foil, or a commercial developing tank with a lid. You should use the smallest available container that will accommodate the TLC plate, since a larger container takes longer to fill with solvent vapors. Development should be carried out in a place away from direct sunlight or

You can practice your spotting technique on a used or damaged TLC plate.

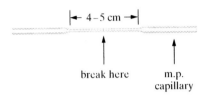

Figure C21 Drawing out a capillary micropipet (not to scale)

drafts to prevent temperature gradients. It may take 20 minutes or more to develop a 10 × 10-cm TLC plate; a microscope-slide plate can often be developed in 5–10 minutes. The solvent should not be allowed to reach the top edge of the plate, since the spots will spread by diffusion once the solvent has stopped advancing. When the *solvent front* — the boundary between the wet and dry parts of the adsorbent — is within 5 mm or so of the top of the plate, remove the TLC plate from the developing chamber and mark the solvent front with a pencil before the plate has had time to dry. Outline any colored spots with a pencil and let the plate dry, preferably under a hood.

Visualization.   If the spots are colored, they can be observed immediately; otherwise, they must be *visualized* (made visible) by some method. The simplest way to visualize many spots is to observe the TLC plate under ultraviolet light. Hand-held ultraviolet lamps used for this purpose may have a switch to select either short-wave (254 nm) or long-wave (365 nm) radiation. If the TLC adsorbent contains a fluorescent indicator, compounds that quench fluorescence will show up as dark spots on a light background when the plate is irradiated with 254-nm UV light. Fluorescent compounds will produce bright spots when irradiated with UV light of an appropriate wavelength. Mark the center of each spot immediately with a pencil, because the spots will disappear when the ultraviolet light is removed. If a spot is irregular, mark its center of concentration — the midpoint of its most densely shaded region — instead. It is also a good idea to outline each spot with your pencil.

See J. Chem. Educ. **1985**, *62, 156 for an alternative method of visualizing TLC plates with iodine.*

Another general visualization procedure is to place the *dry* plate in a closed chamber, such as a wide-mouthed jar with a screw-cap lid, add a few crystals of iodine, and heat the chamber gently (preferably on a steam bath) so that the iodine vapors sublime onto the adsorbent. Most organic compounds, except saturated hydrocarbons and halides, form brown spots with iodine vapor. Unsaturated compounds may show up as light spots against the dark background. The iodine color fades in time, so mark the spots shortly after visualization. It is important that the TLC plate be completely dry; otherwise the residual solvent will pick up the iodine color, resulting in a dark background that may obscure the spots.

See J. Chem. Educ. **1996**, *73, 358 for a description of the cotton-ball procedure.*

Spots can also be made visible by applying a visualizing reagent to the TLC plate. The visualizing reagent can be applied by spraying it onto the *dry* plate, dipping the plate into the reagent, or wiping the plate with a cotton ball that has been saturated with a noncorrosive reagent. A 20% (mass/volume) solution of phosphomolybdic acid in ethanol can be used to visualize most organic compounds; the spots appear after the plate is heated with a heat gun or in an oven. Other visualizing reagents are used for specific classes of compounds, such as ninhydrin reagent for amino acids and 2,4-dinitrophenylhydrazine reagent for aldehydes and ketones. All spraying should be done under a hood in a "spray box," which can be made from a large cardboard box with the top and one side removed. A thin spray is applied to the TLC plate from about two feet away, using an aerosol can or spraying bottle containing the visualizing reagent. Large plates are sprayed by crisscrossing them with horizontal and vertical passes. As for the previous method, the TLC plate must be completely dry before the visualizing reagent is applied, since residual solvent may otherwise inhibit the visualization reaction.

Analysis. The ratio of the distance a component travels up a TLC plate to the distance the solvent travels is called its R_f *value.*

$$R_f = \frac{\text{distance traveled by spot}}{\text{distance traveled by solvent}}$$

The R_f (ratio to front) value of a spot is determined by measuring the distance from the starting line to the center of the spot and dividing that value by the distance from the starting line to the solvent front, both distances being measured along a line extending from the starting point of the spot to its final location. If the spot is irregular, its center of concentration (the midpoint of its most densely shaded region) is used instead. The R_f value of a substance with a given mobile and stationary phase depends on the polarity of its functional groups and other structural features, so it is a physical property of the substance that can be used in its identification. R_f values for some compounds have been reported in the literature, using specified mobile and stationary phases. Reported R_f values alone can seldom be used to establish the identity of a substance, however, since they depend on a number of factors that are difficult to standardize, such as the sample size, the thickness and activity of the adsorbent, the purity of the solvent, and the temperature of the developing chamber. The only way to be reasonably sure that a TLC unknown is identical to a known compound is to spot a solution of the known compound on the same TLC plate as the unknown. Even then, the identity of the unknown may have to be confirmed by an independent method. An unknown substance or mixture is often analyzed on the same plate with a series of standard solutions, each containing a substance that may be identical to the unknown, or to one of its components if it is a mixture.

Directions for Preparing Microscope-Slide TLC Plates

Several students should work together so that they can use the same slurry. For each TLC plate, clean a microscope slide with detergent and water, and then rinse it with distilled water and 50% aqueous methanol. After a slide has been cleaned, do not touch the surface that will be coated; hold the slide by its edges or at the top. *Under the hood,* measure 100 mL of dichloromethane into a 4-oz (125-mL) screw-cap jar, add 35 g of Silica Gel G with vigorous stirring or swirling, and shake the capped jar vigorously for about a minute to form a smooth slurry. Stack two clean microscope slides back to back, holding them together at the top. Without allowing the slurry to settle (shake it again, if necessary), dip the stacked slides into the slurry for about 2 seconds, using a smooth, unhurried, paddle-like motion (see Figure C22) to coat them uniformly with the adsorbent. Immerse the slides deeply enough so that only the top 1 cm or so remains uncoated.

    Touch the bottom of the stacked slides to the jar to drain off the excess slurry and let them air-dry for a minute or so to evaporate the solvent. Then separate them and wipe off the excess adsorbent from the edges with a tissue paper. Repeat with more slides, as needed; if the slurry becomes too thick, dilute it with dichloromethane. Activate the coated slides by heating them in a 110°C oven for 15 minutes. Slides having streaks, lumps, or thin spots in the

Take Care! Avoid contact with dichloromethane and do not breathe its vapors.

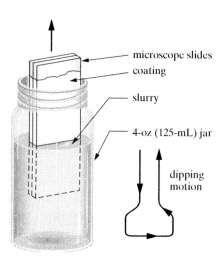

Figure C22 Dipping a pair of stacked microscope slides

coating should be wiped clean and redipped. If the coating is too fragile, try adding some methanol (up to one-third by volume) to the slurry. Coated slides can be stored in a microscope-slide box inside a dessicator.

General Directions for Thin-Layer Chromatography

 Standard Scale and Microscale

Equipment and Supplies

TLC plate(s)
developing chamber
paper wick
developing solvent
pencil and ruler
capillary micropipets or microliter syringe
solutions to be spotted
visualizer (sprayer, iodine chamber, UV lamp, etc.)

Obtain a beaker, screw-cap jar, or another container large enough to hold the TLC plate. A 4-oz screw-cap jar is suitable for a microscope-slide TLC plate, a 400-mL beaker will hold a 6.7×10-cm plate, and a 1-L beaker will hold a 10×10-cm plate. If necessary, prepare and insert a paper wick made from filter paper or chromatography paper. The wick can be prepared by cutting a rectangular strip of paper 3–5 cm wide and long enough to extend in a "U" down one side of the developing chamber, along the bottom, and up the opposite side (see Figure C23). Pour in enough developing solvent to form a liquid layer about 5 mm deep on the bottom of the developing chamber. Cover the chamber with a screw cap, a square of plastic food wrap, or another suitable closure. Tip the chamber and slosh the solvent around to soak the wick with developing solvent; then put it in a place

The wick may not be necessary if the developing chamber is allowed to stand for an hour or more after the developing solvent has been added.

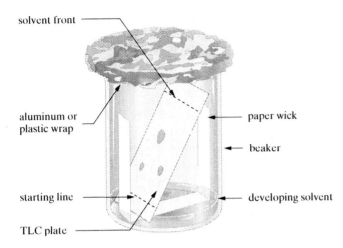

solvent front

aluminum or plastic wrap

paper wick

beaker

starting line

developing solvent

TLC plate

Figure C23 Development of a TLC plate

where it can sit undisturbed away from drafts and direct sunlight. Let the developing chamber stand for 30 minutes or more to saturate the atmosphere inside the chamber with solvent vapors.

Mark the starting line on a TLC plate and spot it with (1) the solution(s) to be analyzed and (2) any standard solutions required, making spots 1–3 mm in diameter. Except on a microscope-slide TLC plate, which can accommodate up to three evenly spaced spots, the spots should be at least 1.0 cm apart and the outermost spots should be 1.5 cm from the edges of the plate. When the spots are dry, place the TLC plate in the developing chamber, with its spotted end down, so that it leans *across* the wick (perpendicular to it) with its top against the glass wall of the chamber. No part of a plastic- or aluminum-backed TLC plate should touch the exposed part of the wick because solvent can diffuse onto the adsorbent at that point. Cover the developing chamber without delay and do not move it during development.

Observe the development frequently, and when the solvent front is within about 5 mm of the top of the plate, remove the plate from the developing chamber. Before the solvent evaporates, use a pencil to trace a line along the solvent front and mark the centers (or centers of concentration) of any visible spots. Let the plate dry thoroughly and visualize the spots (if necessary) by one of the methods described previously, marking the center (or center of concentration) of each spot with a pencil. It is advisable to outline each spot with your pencil as well. Keeping your ruler parallel to the edges of the plate along the line of development, measure the distance in millimeters from the starting line to the solvent front and from the starting line to the center of each spot. Calculate the R_f value for each spot. If requested, make a permanent record of the chromatogram by photocopying or photographing it.

The distance from the starting line to the solvent front may vary, so measure that distance for each spot along an imaginary vertical line that goes through the spot.

Summary

1. Put wick and developing solvent in developing chamber, cover, and let stand.
2. Obtain or prepare TLC plate.
3. Spot TLC plate with solutions of the unknown(s) and any standards.
4. Place TLC plate in developing chamber, cover, and observe.
5. Remove TLC plate when solvent front nears top of plate.
6. Mark solvent front and visible spots, let plate dry.

7. Visualize and mark spots, as needed.
8. Measure R_f values.
9. Clean up, dispose of solvent as directed.

When Things Go Wrong

If the spots on your developed TLC plate are so large or irregular that you can't measure their R_f values accurately, you probably used too much sample to spot the plate. Spot another TLC plate, applying smaller spots or making fewer successive applications, or both. (See *Spotting.*) You may also have let the plate develop too long so that the solvent front reached the top end of the plate. Spot and develop a new plate, taking it out of the developing solvent when the solvent front is 0.5–1 cm from the top of the plate. (See *Development.*) If you used a "do-it-yourself" plate it may have an irregular surface; if so prepare another plate, more carefully this time. (See "Directions for Preparing Microscope-Slide TLC Plates.")

If the spots for some components are too close together to identify from their R_f values, try another developing solvent or solvent mixture. Be sure that the solvents are not contaminated by water or other impurities. (See *Choosing a Developing Solvent.*)

It's also possible that the plate has an organic binder, in which case a different type of TLC plate or a different visualization method should be used.

If you used iodine vapor for visualization and the entire TLC plate has become quite dark, you probably didn't let it dry long enough. If you can't locate all of the expected spots, you will have to prepare, develop, thoroughly dry, and visualize another plate. (See *Visualization.*)

If you don't see the expected spots on your developed TLC plate, think about what you might have done wrong. Were your spots so small or diffuse that they could no longer be seen after development? (See *Spotting.*) Did you use too much developing solvent or draw your starting line too low on the TLC plate, so that the solvent covered the starting line and washed out the components of the spots? (See *Development.*) If you used a UV lamp to visualize the spots, did you irradiate the wrong side of the TLC plate or use the wrong UV wavelength? (See *Visualization.*) If you put your TLC plate in an iodine-vapor chamber, did you set the plate aside so that the iodine spots faded before you examined it? (See *Visualization.*) If you used a visualizing reagent such as phosphomolybdic acid, did you fail to dry the plate completely before applying the reagent, or didn't you heat the plate long enough for the spots to appear? (See *Visualization.*) If you think you may have done (or not done) any of these things, read the indicated section carefully, then prepare, develop, and visualize another plate, taking steps to correct any possible mistakes.

OPERATION 23   Paper Chromatography

Principles and Applications

Paper chromatography is similar to thin-layer chromatography [OP-22] in practice, but quite different in principle. As for TLC, the paper is spotted with solutions of the sample and standards, and the chromatogram is developed with a suitable mobile phase. Although paper consists mainly of cellulose, the stationary phase is not cellulose itself but the water that is adsorbed by it. Chromatography paper can adsorb up to 22% water, and the developing solvents usually contain enough water to keep it saturated. During development, a comparatively nonpolar mobile phase seeps up through the cellulose fibers, partitioning the solutes between the bound water and the mobile phase. Paper chromatography thus operates by a liquid–liquid partitioning process rather than by adsorption on the surface of a solid.

Since only polar compounds are appreciably soluble in water, paper chromatography is most frequently used to separate polar substances such as amino acids and carbohydrates. Manufactured chromatography paper is quite uniform, and the activity of cellulose doesn't vary as much as the activity of most TLC adsorbents, so R_f values obtained by paper chromatography may be more reproducible than those from thin-layer chromatography. However, the resolution of spots is often poorer and the development times are usually much longer. Nevertheless, paper chromatography is a useful analytical technique that can accomplish a variety of separations.

Experimental Considerations

Since many of the experimental aspects of paper chromatography are similar or identical to those for thin-layer chromatography, you should read the appropriate parts of OP-22 for additional information about experimental techniques.

Paper. Various grades of chromatography paper, such as Whatman #1 Chr, are manufactured in rectangular sheets, strips, and other convenient shapes. For good results, the chromatography paper must be kept clean. The bench top or other surface on which the paper is handled should be covered with a sheet of freezer paper (coated side up) or some other liner. The chromatography paper should be held only by the edges or along the top; alternatively, an extra strip of paper can be left on one or both ends of the chromatography paper, used for handling it, and then cut off just before development. The paper should be cut so that its grain is parallel to the direction of development. For most purposes, the paper should be 10–15 cm high (in the direction of development) and wide enough to accommodate the desired number of spots.

Spotting. As for thin-layer chromatography, the substance(s) to be analyzed should be dissolved in a suitable solvent, usually at a concentration of about 1% (mass/volume). The starting line should be marked with a pencil about 2 cm above the bottom edge; the positions of the spots can also be marked lightly with a pencil. The spots tend to spread out more with paper chromatography than with TLC, so they should be applied farther apart—about 1.5–2.0 cm from each other and 2.0 cm from the edges of the paper, as shown in Figure C24. Spotting can be performed by the same techniques as for TLC using a capillary micropipet or syringe (see OP-22), but a round wooden toothpick is adequate for most purposes. Each spot should be 2–5 mm in diameter. It should be made using several consecutive applications, allowing the solvent to dry after each application. Use no more than 10 μL of solution, in all, per spot; too little is better than too much. You

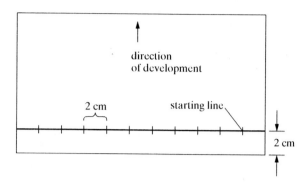

Figure C24 11 × 22-cm chromatography paper marked for spotting

can practice your spotting technique on a piece of filter paper before attempting to spot the chromatogram.

Developing Solvents.   As a rule, paper chromatography developing solvents contain water, which is needed to maintain the composition of the aqueous stationary phase on the cellulose. They also contain an organic solvent and (if necessary) one or more additional components to increase the solubility of the water in the organic solvent or provide an acidic or basic medium. A developing solvent for paper chromatography may be prepared by saturating the organic solvent(s) with water in a separatory funnel, separating the two phases, and using the organic phase for development. Alternatively, a monophase (single-phase) mixture with essentially the same composition as the organic phase of the saturated mixture can be used. Some typical monophase solvent mixtures are listed in Table C5. Most solvent mixtures should be made up fresh each time they are used and not kept for more than a day or two.

Development.   Narrow paper strips accommodating two or more spots can be developed in test tubes, bottles, cylinders, or Erlenmeyer flasks. A wider sheet can be rolled into a cylinder and developed in a beaker. A typical paper chromatogram is developed in much the same way as for TLC, but more time is required to saturate the developing chamber with solvent vapors.

Visualization.   Visualization of spots by ultraviolet light is quite useful in paper chromatography because paper fluoresces dimly in a dark room and many organic compounds will quench its fluorescence, yielding dark spots on a light background. Paper chromatograms can also be visualized by

Table C5 Some monophase solvent mixtures for paper chromatography

Solvents	Composition
2-propanol/ammonia/water	9 : 1 : 2
1-butanol/acetic acid/water	12 : 3 : 5
phenol/water	500 g phenol : 125 mL water
ethyl acetate/1-propanol/water	14 : 2 : 4

Note: Ratios are by volume unless indicated otherwise.

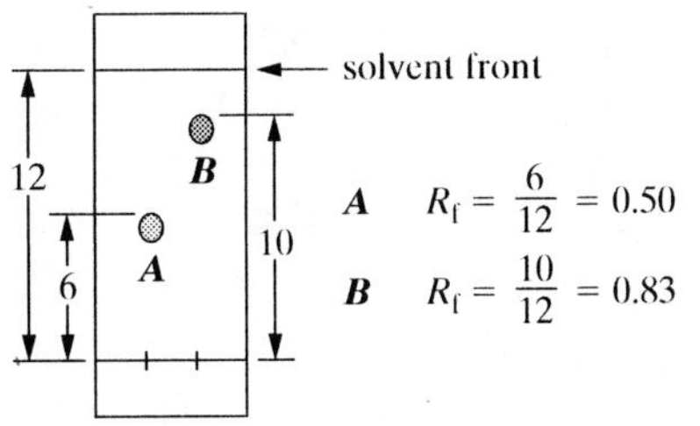

$$R_f = \frac{\text{distance to leading edge of spot}}{\text{distance to solvent front}}$$

Figure C25 Measuring R_f values on a paper chromatogram

spraying them or by dipping them into a solution of a suitable visualizing reagent.

Analysis.   A substance responsible for a spot on a developed chromatogram is characterized by its R_f value, which is determined as illustrated in Figure C25. Spot migration distances are customarily measured from the starting line to the *front* of each spot, rather than to its center as for TLC. It is usually necessary to run one or more standards along with an unknown to identify the unknown. Whenever possible, the solvent and concentration should be the same for the standard as for the unknown.

General Directions for Paper Chromatography

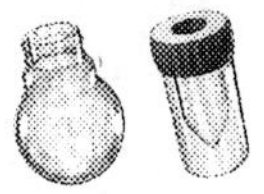

Standard Scale and Microscale

Equipment and Supplies

> chromatography paper
> developing chamber
> developing solvent
> pencil and ruler
> capillary micropipets (or toothpicks, etc.)
> solutions to be spotted
> visualizer (sprayer, UV lamp, etc.)

The following procedure can be used when up to 13 spots are to be applied. For just a few spots, a strip of chromatography paper can be spotted, folded in the middle (or hung from a wire embedded in a cork), and developed in a test tube, a jar, or another appropriate container.

Add enough developing solvent to a 600-mL beaker (or another suitable developing chamber) to provide a liquid layer about 1 cm deep (see Figure C26). Cover the chamber tightly with plastic food wrap (or cap it), slosh the solvent around in it for about 30 seconds, and then put it in a place where it can sit undisturbed away from drafts and direct sunlight. Allow sufficient time (usually an hour or more) for the solvent to saturate the developing chamber. While the developing chamber is equilibrating, obtain a sheet of chromatography paper and cut it to form an 11 × 22-cm rectangle (or another appropriate

(cover omitted for clarity)

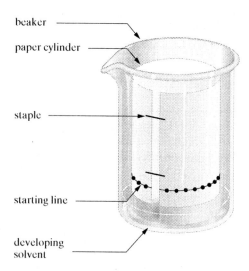

beaker

paper cylinder

staple

starting line

developing
solvent

Figure C26 Developing a paper chromatogram

size). Without touching the surface of the paper with your fingers, use a pencil to draw a starting line 2 cm from one long edge (the bottom edge) and lightly mark the positions for spots with a pencil, spacing them about 2 cm from each side and 1.5–2.0 cm apart. Spot the paper with all of the solutions to be chromatographed; then roll it into a cylinder and staple the ends together, leaving a small gap between them. Uncover the developing chamber, place the paper cylinder inside (spotted end down), and cover it without delay. The paper must not touch the sides of the chamber and the chamber should not be moved during development. Development may take an hour or more, depending on the developing solvent and the distance traveled.

When the solvent front is a centimeter or less from the top of the paper, remove the cylinder and separate its edges; then accurately draw a line along the entire solvent front with a pencil. If any spots are visible at this time, outline them with a pencil (carefully, to avoid tearing the wet paper), as they may fade in time. Again bend the paper into a cylinder and stand it on edge to air dry, preferably under a hood. When it is completely dry, visualize the spots by an appropriate method, if necessary, and outline them with a pencil. Keeping your ruler parallel to the edges of the paper along the line of development, measure the distance in millimeters from the starting line to the solvent front and from the starting line to the leading edge of each spot. Calculate the R_f value of each spot

The distance from the starting line to the solvent front may vary, so measure that distance for each spot along an imaginary vertical line that goes through the spot.

Summary

1. Add developing solvent to developing chamber, cover, and let stand.
2. Obtain chromatography paper and cut it to size.
3. Spot chromatography paper with solutions, roll into cylinder.
4. Place paper in developing chamber, cover, and observe.
5. Remove paper chromatogram when solvent front nears top of paper.

6. Mark solvent front and any visible spots, let chromatogram dry.
7. Visualize and mark spots, as needed.
8. Measure R_f values.
9. Clean up, dispose of solvent as directed.

When Things Go Wrong

Most of the things that go wrong during TLC analysis can also go wrong during a paper chromatography analysis (with minor modifications), so you can refer to *When Things Go Wrong* in OP-22 for help.

D. Washing and Drying Operations

During various procedures it may be necessary to wash or otherwise clean a substance or solution to remove impurities, or to dry it by removing traces of water or another liquid. The method used depends on the physical state of whatever is being washed or dried. Liquids can be washed [OP-24] by shaking them with an appropriate wash liquid, and dried [OP-25] by mixing them with a desiccant (drying agent). Solids can be washed [OP-26a] with an appropriate solvent, which is removed by filtration, and dried [OP-26b] by leaving them in an oven or a dry atmosphere. Air and other gases can be cleaned [OP-27a] by passing them through a chemically reactive wash liquid or a filtering agent such as cotton, and dried [OP-27b] by contact with a desiccant.

OPERATION 24

Washing Liquids

In practice, the process of washing liquids is identical to liquid–liquid extraction [OP-18a], but its purpose is different. *Extraction* separates a desired substance (such as the product of a reaction) from an impure mixture; *washing* removes impurities from a desired substance. The liquid being washed may be a neat liquid (one without solvent) or a solvent containing the desired substance. In either case, the substance must not dissolve appreciably in the wash liquid or it will be extracted along with the impurities.

The *wash liquid* is usually water or an aqueous solution, although organic solvents such as diethyl ether can be used to remove low-polarity impurities from aqueous solutions containing polar solutes. Both water and saturated aqueous sodium chloride remove water-soluble impurities, such as salts and polar organic compounds, from organic liquids. Saturated sodium chloride is frequently used for the last washing before a liquid is dried [OP-25] because it removes excess water from the organic liquid by the salting-out effect described in OP-18b. It is preferred to water in some other cases, because it helps prevent the formation of emulsions at the interface between the liquids.

Some aqueous wash liquids contain chemically reactive solutes that convert water-insoluble impurities to water-soluble salts, which then dissolve in the wash solvent. Aqueous solutions of bases remove acidic impurities, as illustrated for sodium bicarbonate:

$$NaHCO_3 + HA \text{ (acidic impurity)} \rightarrow Na^+ A^- \text{ (soluble salt)} + H_2O + CO_2$$

Aqueous solutions of acids remove alkaline impurities, as illustrated for hydrochloric acid:

$$HCl + B \text{ (basic impurity)} \rightarrow BH^+Cl^- \text{ (soluble salt)}$$

When a chemically reactive wash liquid is used, it is usually advisable to perform a preliminary washing with water or aqueous NaCl to remove most of the water-soluble impurities. This may prevent a potentially violent reaction between the reactive wash liquid and the impurities. Sometimes a follow-up washing with water or aqueous NaCl is used to remove traces of the chemically reactive solute from the product.

The effectiveness of washing with a given total volume of wash liquid increases if it is carried out in several steps. Unless otherwise indicated in the experimental procedure, a liquid should be washed in two or three stages using equal volumes of wash liquid for each stage, and the total volume of wash liquid should be roughly equal to the volume of the liquid being washed. For example, if you are washing 6 mL of a liquid, you can use two 3-mL or three 2-mL portions of the wash liquid. When several different wash liquids are used in succession, one or two washings with each wash liquid may be sufficient, but the total volume of all of the wash liquids should usually equal or exceed the volume of the liquid being washed.

The procedure for washing liquids is essentially the same as that for liquid–liquid extraction [OP-18a], except that the wash liquid is discarded and the liquid being washed is saved. As for extraction, it is important to save *both* layers until you are absolutely certain you are working with the right layer. Refer to the general directions in OP-18a for illustrations and experimental details.

Rule of Thumb: *Total volume of wash liquid ≈ volume of liquid being washed.*

General Directions for Washing Liquids

Standard Scale

Combine the wash liquid and the liquid being washed in a separatory funnel. If the wash liquid contains sodium carbonate, sodium bicarbonate, or another reactive solute that generates a gas, use a glass stirring rod to stir it vigorously with the liquid being washed until gas evolution subsides. Otherwise, a pressure buildup might cause the stopper to pop out and your product to spray all over the lab. Then stopper the separatory funnel and shake it, very gently at first, with frequent venting. When you no longer hear a "whoosh" of escaping vapors upon venting, shake the separatory funnel more vigorously for 1–2 minutes, then set it on a support until the layers separate sharply.

If the liquid being washed is *less* dense than the wash liquid (for example, if it is a diethyl ether solution), drain the lower layer (the wash liquid) into a flask or beaker after each washing. Then add the next portion of wash liquid to the liquid being washed, which remains in the separatory funnel, and repeat the washing as needed. Set the wash liquid aside for later disposal.

If the liquid being washed is *more* dense than the wash liquid (for example, if it is a dichloromethane solution), drain the lower layer into a flask and pour the wash liquid out of the top of the separatory funnel into another container after each washing. Return the lower layer to the separatory funnel, add the next portion of wash liquid, and repeat the washing as needed. Set the wash liquid aside for later disposal.

Microscale

Use a calibrated Pasteur pipet to transfer the desired volume of wash liquid to a conical vial or conical centrifuge tube containing the liquid being washed. If the wash liquid contains sodium carbonate, sodium bicarbonate, or another reactive solute that generates a gas, use a thin glass stirring rod to stir it vigorously with the liquid being washed until gas evolution subsides. Then cap the washing container securely, shake it gently 5 to 10 times, and unscrew the cap slightly to allow gases to escape. Shake it more vigorously about 100 times, venting occasionally, then loosen the cap and let it stand until the layers separate sharply.

If the liquid being washed is *less* dense than the wash liquid (for example, if it is a diethyl ether solution), use a Pasteur pipet to transfer the lower layer (the wash liquid) to another container after each washing. Then add the next portion of wash liquid to the liquid being washed, which remains in the washing container, and repeat the washing as needed. Set the wash liquid aside for later disposal. After the last washing, try to remove all of the wash solvent from the liquid being washed.

If the liquid being washed is *more* dense than the wash liquid (for example, if it is a dichloromethane solution), use a Pasteur pipet to transfer the lower layer to an appropriate container (**A**) and transfer the wash liquid to a different container (**B**) after each washing. Return the lower layer from **A** to the washing container, add the next portion of wash liquid, and repeat the washing as needed. Set the wash liquid aside for later disposal. After the last washing, try not to include any of the wash solvent when you remove the lower layer.

When Things Go Wrong

Most of the things that go wrong during a liquid–liquid extraction can also go wrong when you are washing a liquid, so you can refer to *When Things Go Wrong* in OP-18 for help.

OPERATION 25

Drying Liquids

When an organic substance (or a solution containing it) is extracted from an aqueous reaction mixture, washed with an aqueous wash liquid, steam distilled from a natural product, or comes into contact with water in some other way, the substance or its solution will retain traces of water that must be removed before such operations as evaporation and distillation are carried out. Organic liquids and solutions can be dried in bulk by allowing them to stand in contact with a *drying agent*, which is then removed by decanting or filtration. They can also be dried by passing them through a *drying column*, such as a Pasteur pipet packed with a suitable drying agent.

Drying Agents

Most drying agents are anhydrous (water-free) inorganic salts that form hydrates by combining chemically with water. For example, a mole of anhydrous magnesium sulfate can combine with up to seven moles of water to form hydrates of varying composition.

$$MgSO_4 + nH_2O \rightleftharpoons MgSO_4 \cdot nH_2O \quad (n = 1-7)$$

The effectiveness and general applicability of a drying agent depend on the following characteristics:

- Speed—how fast drying takes place
- Capacity—the amount of water absorbed per unit of mass
- Intensity—the degree of dryness attained
- Chemical inertness—unreactivity with substances being dried
- Ease of removal

Ideally, a drying agent should be very fast and have both a high capacity and a high intensity. It should not react with (or dissolve in) a substance being dried, and it should be easy to remove when drying is complete. Table D1 summarizes the properties of some common drying agents. As you can see in the table, there is no ideal drying agent, but anhydrous magnesium sulfate is perhaps the best all-around drying agent.

The choice of a drying agent depends, in part, on the properties of the liquid being dried and the degree of drying required. Solvents such as diethyl ether and ethyl acetate retain appreciable quantities of water, so their solutions are often washed [OP-24] with saturated aqueous sodium chloride to salt out [OP-18b] some of the water before further drying. They should then be dried by a drying agent with a relatively high capacity, such as magnesium sulfate or sodium sulfate. Powdered anhydrous magnesium sulfate tends to adsorb organic solutes onto its surface, so it should always be washed with a suitable pure solvent to recover adsorbed material. Granular anhydrous sodium sulfate doesn't adsorb much of the substance being dried and is easy to remove, so it is often used when thorough drying isn't necessary.

Relatively nonpolar solvents, such as petroleum ether and dichloro-methane, retain little water and can be dried with lower capacity drying agents such as calcium sulfate (Drierite) or calcium chloride. Calcium chloride reacts with many organic compounds that contain oxygen or nitrogen, including alcohols, aldehydes, ketones, carboxylic acids, phenols, amines, amides, and some

Table D1 Properties of commonly used drying agents

Drying agent	Speed	Capacity	Intensity	Comments
magnesium sulfate	fast	medium	medium	good general drying agent, suitable for nearly all organic liquids
calcium sulfate (Drierite)	very fast	low	high	fast and efficient, but low capacity; may contain a blue indicator that turns pink when hydrated
sodium sulfate	slow	high	low	good for predrying, easy to remove; loses water above 32.4°C
calcium chloride	slow to fast	low to medium	medium	removes traces of water quickly, larger amounts slowly; reacts with many organic compounds
silica gel	medium	medium	high	good general drying agent, more expensive than most
potassium carbonate	fast	low	medium	cannot be used to dry acidic compounds
potassium hydroxide	fast	very high	high	used to dry amines, reacts with many other compounds; caustic

esters, so it should not be used to dry such compounds or their solutions. To dry a dichloromethane solution that will later be evaporated, it may be sufficient to filter the solution through a cotton plug or a layer of sodium sulfate to remove water droplets (if there are any). The remaining water forms an azeotrope (see OP-32) with dichloromethane and evaporates along with it.

Drying agents come in different particle sizes, as indicated by their mesh numbers. Finer particles dry liquids more effectively, but coarser particles make it easier to separate the dried liquid. Anhydrous magnesium sulfate is usually finely powdered, so it should be removed by gravity filtration [OP-15]. Anhydrous sodium sulfate can be obtained in coarse particles like those of granulated sugar, and calcium chloride in coarse granules with mesh numbers of 20 or lower, so they are often removed by decanting (see OP-15).

The mesh number of a granular substance is the number of openings per linear inch in the finest screen that will allow the substance's particles to pass through. Thus the finer the particles, the higher the mesh number.

Small quantities of neat liquids should be dissolved in a *carrier solvent*, such as diethyl ether or dichloromethane, before drying. Otherwise, you may lose much of your product by adsorption on the drying agent. After drying, the solvent is removed by evaporation [OP-19].

Predrying

Saturated aqueous sodium chloride is often used to predry a wet solution obtained by extraction, especially when the extraction solvent is diethyl ether. Most other common extraction solvents, including dichloromethane and all hydrocarbons, dissolve little water and do not require predrying. For example, when diethyl ether is used to extract an aqueous solution, the resulting ether layer contains about 1.5% by mass of dissolved water. Washing it with saturated aqueous sodium chloride draws out most of the dissolved water from the ether solution, in part because dilution of a concentrated salt solution is energetically favorable. This reduces the amount of drying agent needed to dry it efficiently. The washing is carried out as described in OP-24, using a volume of saturated sodium chloride that is 50–100% that of the ether solution being washed.

Because of its high capacity, anhydrous sodium sulfate is sometimes used to predry very wet solutions that are later dried with a high-intensity drying agent such as calcium sulfate.

 Drying Liquids in Bulk

Amount of Drying Agent.   The amount of drying agent needed for bulk drying depends on the capacity and particle size of the drying agent and on the amount of water present. Usually about 1 g of drying agent should be used per 25 mL of liquid, or 40 mg per milliliter. More may be needed if the drying agent has a low capacity or large particle size or if the liquid has a high water content. It is best to start with a small amount of drying agent and then add more if necessary, since using too much results in excessive losses by adsorption of liquid on the drying agent. For microscale work, you can usually start with the amount of drying agent that will fit in the V-shaped groove of a Hayman-style microspatula, and then add more if most of the drying agent clumps up or shows other evidence of hydration, as described next. The appearance of the drying agent when the drying time is up often suggests whether more drying agent is needed. As they become hydrated,

Rule of Thumb: Use about 1 g of drying agent per 25 mL of liquid (40 mg/mL).

sodium sulfate and magnesium sulfate particles clump together, calcium chloride displays a glassy surface appearance, and blue indicating Drierite changes color to pink. If, after the initial drying period, most of a drying agent has changed as described, more drying agent should be added or the spent drying agent should be removed and replaced with fresh drying agent.

Spent drying agent contains hydrates that reduce drying efficiency.

Drying Time. The time required for bulk drying depends on the speed of the drying agent and the amount of water present. Most drying agents attain at least 80% of their ultimate drying capacity within 15 minutes, so longer drying times are seldom necessary—5 minutes is usually sufficient for magnesium sulfate or Drierite, while 15 minutes is recommended for calcium chloride and sodium sulfate. When more complete drying is required, it is better to replace the spent drying agent or use a more efficient drying agent than to extend the drying time.

Removal of the Drying Agent. A drying agent should be removed as completely as possible when the drying period is over. For standard scale work, most drying agents are removed by gravity filtration through a coarse fluted filter paper. A coarse-grained drying agent, such as granular calcium chloride, sodium sulfate, or Drierite, can sometimes be removed by carefully decanting the liquid, but granular calcium chloride often contains a fine powder that requires filtration. For microscale work, the liquid being dried can be separated from the drying agent with a filter-tip pipet [OP-6] or by passing the liquid through a glass-wool plug in a filtering pipet [OP-15].

When a solution (an ether extract, for example) is being dried, the spent drying agent should be washed with a small amount of the pure solvent to recover adsorbed solute that would otherwise be discarded with the drying agent. When a neat liquid is being dried, the drying agent can be washed with dichloromethane or diethyl ether and the solvent evaporated to recover the adsorbed liquid.

General Directions for Bulk Drying

 ## Standard Scale and Microscale

Equipment and Supplies

> Erlenmeyer flask, conical vial, or other drying container
> drying agent
> funnel and fluted filter paper (standard scale)
> filter-tip pipet or filtering pipet (microscale)

If you are drying a small amount of a neat liquid, it is advisable to dissolve the liquid in a carrier solvent, such as diethyl ether or dichloromethane, and evaporate the solvent after drying. If the liquid to be dried contains water droplets or a separate aqueous layer, remove the water using a Pasteur pipet or a separatory funnel. For microscale work you can pass the wet liquid through a filtering pipet [OP-15] containing a cotton plug, using a rubber bulb to force the last droplets through.

Select an Erlenmeyer flask, a conical vial, or another suitable container that will hold the liquid with plenty of room to spare, and add the liquid.

Measure out the estimated quantity of a suitable drying agent, being sure to cap its original container tightly. Protect it from atmospheric moisture until you are ready to use it. Add the drying agent to the liquid, stopper or cap the container, and swirl or shake the container for a few seconds. Let it dry for 5–15 minutes, using the longer drying time for calcium chloride, sodium sulfate, and other slow drying agents. Swirl, stir, or shake the mixture occasionally during the drying period. If a second (aqueous) phase forms during drying, remove it with a Pasteur pipet and add more drying agent.

When the drying period is over, examine the drying agent carefully. If most of it is spent, add more drying agent (or remove it and replace it with fresh drying agent), and continue drying for 5 minutes or more. When drying appears to be complete, separate the dry liquid from the drying agent by one of the following methods. For standard scale work, filter it by gravity [OP-15] through fluted filter paper. (With a coarse-grained drying agent, you may be able to decant the liquid into another container, leaving the solid behind.) For microscale work, withdraw the liquid with a dry filter-tip pipet and transfer it to another container, or filter it by gravity through a filtering pipet containing glass wool.

If you are drying a solution, rinse the drying agent with a small volume of the pure solvent and combine the rinse liquid with the dried liquid. If you are drying a neat liquid, you can rinse the drying agent with a suitable solvent, combine it with the dried liquid, and evaporate the solvent.

Summary

1. Select drying agent and measure estimated quantity needed.
2. Add drying agent to liquid in drying container.
3. Stopper container and mix contents, set aside.
4. Shake, stir, or swirl occasionally until drying time is up.
   IF drying agent is spent or aqueous phase separates, GO TO 5.
   IF not, GO TO 6.
5. Remove aqueous phase or spent drying agent if necessary; add fresh drying agent. GO TO 3.
6. Separate liquid from drying agent.
   IF liquid being dried is a neat liquid, GO TO 7.
   IF liquid being dried is a solution, GO TO 8.
7. (Optional) Wash drying agent with appropriate solvent, combine solvent with dried liquid, evaporate solvent. GO TO 9.
8. Wash drying agent with fresh solvent, combine with dried solution.
9. Clean up, dispose of spent drying agent as directed.

When Things Go Wrong

If most or all of the drying agent you added has clumped together or (in the case of calcium chloride) has a glassy appearance, try adding roughly half as much fresh drying agent as you used initially (this assumes that you followed the rule of thumb given previously). If, after swirling and standing, the added drying agent has about the same appearance as the drying agent already there, decant (see OP-15) or withdraw (with a Pasteur pipet) the liquid from the spent drying agent and transfer it to another container; it doesn't matter if a little drying agent goes with the liquid. Then add fresh drying agent. After swirling and standing, most of the fresh drying agent should have the same granular or powdery appearance as it has in the dry form; if not, repeat this process until it does. If you have to add a large amount of drying agent to get to this point, you may want to add more solvent (if you're drying a solution) to minimize product losses. If you're drying a neat (undiluted) liquid, you can add a low-boiling solvent such as dichloromethane and then evaporate it after the drying agent has been removed.

   If after you add the drying agent you observe a bubble or layer of water (usually cloudy), you may not have separated the organic layer cleanly from the aqueous layer during a previous extraction or washing process, or you may not have dried your glassware carefully. Remove the water with a Pasteur pipet or a filter-tip pipet [OP-6] (try not to remove any of the organic layer), and then add more drying agent.

   If you record an infrared spectrum of your dried product and it shows a broad absorption band centered near 3500 cm^{-1}, it is probably not dry enough, because that is the usual frequency of the O—H stretching vibration of water. Similarly, a ^{1}H NMR singlet at 1.55 ppm in $CDCl_3$ may indicate the presence of dissolved water. Unless there is another good explanation for such results, you should redry your product with fresh drying agent for at least 10 minutes and then repeat the analysis.

The location of the water band can vary with the sampling method used. In a CCl_4 solution it occurs around 3700 cm^{-1}.

 ## Drying Columns

Drying columns are seldom practical for standard scale drying in undergraduate laboratories because of the relatively large quantity of drying agent required. A microscale drying column can be constructed by supporting the drying agent on a cotton or glass-wool plug inside a Pasteur pipet (Figure D1). When thorough drying isn't necessary, granular anhydrous sodium sulfate is the preferred drying agent. This form of sodium sulfate comes in relatively large grains, allowing liquids to flow through easily, and a sodium sulfate column can absorb relatively large amounts of water without plugging up. The powdered form of anhydrous magnesium sulfate isn't suitable for a drying column because liquids won't flow through it, but a granular form of the drying agent may be used for this purpose. Although magnesium sulfate is a more efficient drying agent than sodium sulfate, a magnesium sulfate column is more likely to plug up if the liquid being dried is quite wet. So it is important to make sure that the liquid doesn't contain any visible water droplets, such as those resulting from imperfect separation during an extraction.

General Directions for Using a Drying Column

 ## Microscale

Equipment and Supplies

Two $5\frac{3}{4}$-inch Pasteur pipets
latex rubber bulb
cotton
glass wool (optional)
applicator stick or stirring rod

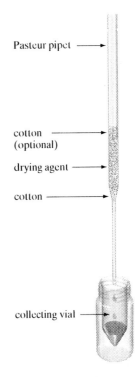

Figure D1 A drying column

drying agent

collecting container

If you are drying a neat liquid, dissolve it in approximately 2 mL of a carrier solvent, such as diethyl ether or dichloromethane. (You will evaporate the solvent after drying.) To prepare a drying column, obtain a $5\frac{3}{4}$-inch Pasteur pipet and use an applicator stick or a thin glass stirring rod to gently tamp a small ball of glass wool into the column until it lodges where the pipet narrows. Add enough anhydrous sodium sulfate or *granular* anhydrous magnesium sulfate (usually about 0.7–1 g) through the top of the Pasteur pipet to provide a 2–3-cm layer of the drying agent. If the liquid to be dried contains visible water droplets or a separate aqueous layer, remove most of the water with a Pasteur pipet, and then tamp a loose plug of cotton into the drying column so that it rests on top of the layer of drying agent. Clamp the drying column vertically to a ring stand, place a suitable collecting container under the outlet, and transfer the liquid being dried to the column with a Pasteur pipet (preferably filter-tipped). If the liquid flow stops or becomes very slow, attach a latex rubber bulb to the top of the column and squeeze it just enough to force the liquid *slowly* through the column. If that doesn't work, obtain a pipet pump, extend its plunger, attach it to the top of the column, and rotate the thumb wheel enough to force the liquid *slowly* through the column. When all of the liquid has passed through the drying agent, add about 0.5 mL of fresh solvent (the solvent present in the liquid being dried) to the top of the drying column and let it drain into the container holding the dried liquid. Use a latex rubber bulb or a pipet pump to gently expel any liquid remaining on the column into the container.

When it is dry, remove the drying agent by inverting the column in a beaker and tapping it gently, and remove the plug by snagging it with a length of copper wire bent to form a "J" at one end. If that doesn't work, attach a rubber hose from an air line to the column at its narrow end (be careful you don't break the tip), hold the open end over a beaker, and *slowly* open the air valve until the air blows out the drying agent.

Summary

1. Select drying agent and measure quantity needed.
2. Prepare drying column containing drying agent supported on glass-wool plug.
3. Clamp drying column to ring stand over collecting container.
4. Transfer liquid to column with Pasteur pipet, drain into collecting container.
5. Rinse drying agent with fresh solvent, combine with dried liquid.
6. Clean up, dispose of spent drying agent as directed.

OPERATION 26

Washing and Drying Solids

Solids that have been collected by vacuum filtration [OP-16] tend to retain traces of impurities on the surfaces of their crystals, which can be washed with an appropriate solvent. Solids obtained by other operations such as evaporation of a solvent may also benefit from careful washing.

Solids that have been separated from a reaction mixture or isolated from other sources usually retain traces of water or other solvents used in the separation. The solvent can be removed by a number of drying methods, depending on the nature of the solvent, the amount of material to be dried, and the melting point and thermal stability of the solid compound.

a. Washing Solids

Solid products obtained from a reaction or a recrystallization [OP-28] operation are ordinarily collected by vacuum filtration [OP-16] and washed directly on the filter. The wash solvent should be chosen carefully, because one in which the solid is appreciably soluble will reduce the amount of product you recover. To minimize losses due to solubility, the wash solvent should ordinarily be chilled in ice water before use.

The solvent from which a solid was originally filtered is usually a suitable wash solvent, since the solid shouldn't be very soluble in it—otherwise there wouldn't be much solid to filter. For example, a solid that was recrystallized [OP-28] from 95% ethanol can be washed with cold 95% ethanol. Since most organic compounds are less soluble in water than in ethanol, using a wash solvent that contains a higher percentage of water, such as 50% ethanol, should reduce its solubility even further (but make sure that the solid isn't one of the few organic compounds that is more soluble in water than in ethanol). If you filter a solid from a mixture of two solvents, as in a mixed-solvent recrystallization [OP-28b], wash it with the solvent in which it is *least* soluble or with an appropriate mixture of the two solvents. For example, aspirin can be recrystallized from a mixture of ethanol and water that contains about twice as much water as ethanol, so it can be washed with an ethanol–water mixture that contains at least that much water, or with water alone.

Solids that are filtered and washed are ordinarily dried afterward, so if a solid was originally filtered from a high-boiling solvent, washing it with a lower boiling solvent having similar properties will help it dry faster. For example, a solid that was filtered from a toluene (bp = 111°C) solution can be washed with low-boiling petroleum ether (bp = ~35–60°C), since the components of petroleum ether, like toluene, are hydrocarbons. Of course the wash solvent must be miscible with the original solvent, so never use petroleum ether, for example, to wash a solid that was filtered from an aqueous solution.

When a solid contains appreciable amounts of one or more impurities, it may be advantageous to wash it in a separate container before filtration. For example, the alcohol triphenylmethanol is contaminated by the hydrocarbon biphenyl when it is prepared as described in Experiment 30. To remove the impurity, a hydrocarbon solvent such as high-boiling petroleum ether can be added to a flask containing the impure triphenylmethanol, which is then rubbed and ground against the side of the flask with the tip of a glass stirring rod to remove the biphenyl. This rubbing and grinding process is called *trituration*.

To wash a solid that has been collected in a Hirsch or Buchner funnel, first cool the solvent in ice water for 10 minutes or more. With the vacuum turned off, add enough of the solvent to completely cover the solid. Stir the mixture *gently* with a spatula or stirring rod to suspend the solid in the liquid, being careful not to disturb the filter paper. If you have only a few tenths of a gram of product, it may be advisable to omit the stirring because of the likelihood of product loss. Without delay, turn on the vacuum to drain the wash liquid. The washing may be repeated with fresh chilled solvent, if desired. After the last washing, leave the vacuum turned on for several minutes to partially dry the solid. If you experience difficulties while washing a solid by vacuum filtration, refer to *When Things Go Wrong* in OP-16.

To wash a solid by trituration, place it in a glass container such as a test tube, beaker, or Erlenmeyer flask and add enough of the wash solvent (preferably chilled) to completely cover the solid, or the amount of solvent specified by the procedure you are following. Using the tip of a glass stirring rod, rub and grind the solid against the sides of the container for several minutes. It is important to grind the solid finely to increase the amount of its surface area exposed to the solvent. Then remove the solvent by vacuum filtration [OP-16].

b. Drying Solids

Solids that have been collected by vacuum filtration [OP-16] are usually air dried on the filter by leaving the vacuum on for a few minutes after filtration is complete. Unless the solvent is very volatile, further drying is required. Comparatively volatile solvents can be removed by simply spreading the solid on a watch glass or evaporating dish (covered to keep out airborne particles) and placing the container in a location with good air circulation, such as a hood, for a sufficient period of time (Figure D2). Clamping an inverted funnel above the watch glass or evaporating dish and passing a gentle stream of dry air or nitrogen over it will accelerate the drying rate.

A very wet solid can be partially dried by transferring it to a filter paper on a clean surface and blotting it with another filter paper to remove excess solvent. The solid is then rubbed against the filter paper with the blade of a flat-bladed spatula until it is finely divided and friable, using fresh filter paper if necessary. It should then be dried completely by one of the methods described next. Some of the solid will adhere to the filter paper, so this method is not recommended for microscale work.

Many wet solids can be spread out in a shallow ovenproof container and dried in a laboratory oven set at 110°C or another suitable temperature. The expected melting point of the solid should be at least 20°C above the oven temperature and the solid should not be heat sensitive or sublime readily at the oven temperature. It isn't unusual for a student to open an oven door and discover that the product he or she worked many hours to prepare has just turned into a charred or molten mass, or disappeared entirely. Aluminum weighing dishes and other commercially available containers made of heavy aluminum foil are usually suitable for oven drying because the aluminum conducts heat well, cools quickly, and isn't likely to burn your fingers. Aluminum reacts with acids and bases, so acidic and basic solids (or solids that may be wet with acidic or basic liquids) should be oven-dried in Pyrex or porcelain containers, such as watch glasses or evaporating dishes.

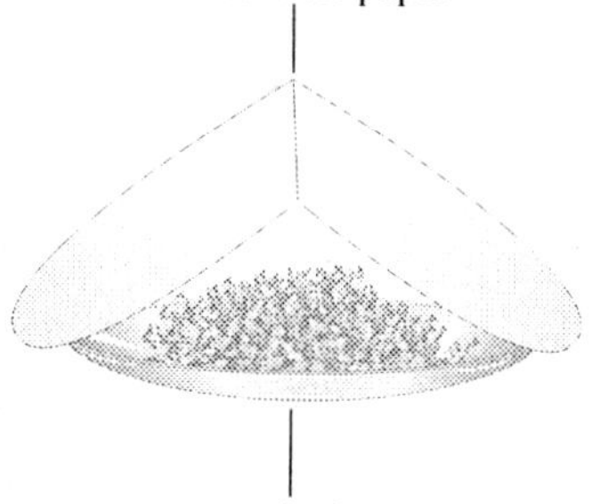

Figure D2 Covered watch glass for drying solids

Take Care! Don't blow the crystals away.

If you are not sure whether your product can be oven-dried safely, consult your instructor.

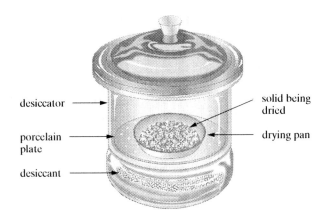

Figure D3 Commercial desiccator

A *vacuum oven* combines the use of heat with low pressure for very fast, efficient drying. A simple vacuum oven can be constructed by clamping a sidearm test tube horizontally and connecting it to a vacuum source (see *J. Chem. Educ.* **1988**, *65,* 460). The sample, in an open vial, is inserted in the sidearm test tube, which is stoppered and heated with a heat lamp while the vacuum is turned on.

When time permits, the safest way to dry a solid is to leave it in a *desiccator* overnight or longer. A desiccator consists of a tightly sealed container partly filled with a *desiccant* (drying agent) that absorbs water vapor, creating a moisture-free environment in which the solid should dry thoroughly. Desiccators such as the one shown in Figure D3 are available commercially. A simple "homemade" desiccator for drying small amounts of solid can be constructed using an 8-oz (~250-mL) wide-mouth jar with a screw cap (see Figure D4). Its size is well suited to microscale work, since several samples in small vials can be dried at the same time, but it is also useful for standard scale work. Enough of a solid desiccant (about 50 mL, measured in a graduated beaker) is added to form a 1-cm layer of desiccant on the bottom. The wet solid, in an appropriate container, is set inside the desiccator, which is capped and allowed to stand undisturbed until drying is complete. If there is any danger of the container tipping over, a wire screen can be cut to fit on top of the desiccant layer and provide a more stable surface. Except when a product is being added or removed, the desiccator must be kept tightly closed at all times to keep the dessicant active.

A similar desiccator with a polyethylene storage rack is available commercially.

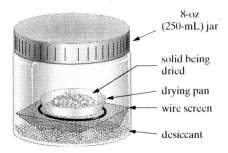

Figure D4 "Homemade" desiccator

Anhydrous calcium chloride is a good (if rather slow-working) desiccant because it is inexpensive and has a high water capacity. Drierite (anhydrous calcium sulfate) is faster and more efficient than calcium chloride but has a much lower capacity. A combination of calcium chloride with a small amount of indicating (blue) Drierite works better than either desiccant separately. The blue Drierite removes traces of moisture that calcium chloride cannot, and it turns pink when the desiccant is spent and needs to be replaced. Unless your instructor indicates otherwise, spent desiccant should be placed in a designated container for reactivation. Drierite can be reactivated by heating it in a 225°C oven overnight; calcium chloride should be heated overnight at 250–350°C.

General Directions for Drying Solids in an Oven

 Standard Scale and Microscale

Obtain a wide, shallow, ovenproof container, such as a small evaporating dish or an aluminum weighing dish. Weigh it and label it to prevent mix-ups. Spread the solid on the bottom of the container in a thin, uniform layer and place it in the oven, preferably where it is well separated from other containers. After 30 minutes or so, remove the container (wear gloves when handling a glass container), let it cool to room temperature, and weigh it. Then put it back in the oven for 5–10 minutes, let it cool, and weigh it again. If the mass has decreased by 1% or more, repeat this process until the mass does not change significantly between weighings.

General Directions for Drying Solids in a Desiccator

 Standard Scale and Microscale

If the sample will remain in the desiccator until the next lab period, you can dry it in a tared, labeled storage vial with the cap removed. (At the beginning of the next lab period, you should cap the vial and weigh it.) If the sample will only be in the desiccator overnight or for a day or two, spread it out in a shallow tared container, such as a plastic or aluminum weighing dish or a square drying tray made by folding a square of heavy aluminum foil. Label the container with the name of the compound, your name or initials, and the tare mass. Set the container inside a desiccator provided with fresh desiccant, taking care to place it securely so that it won't tip over. With a homemade desiccator, vials can be pushed down into the desiccant layer and low containers can be set on a rack made of wire screen or another material (see Figure D4). A commercial desiccator has a porcelain plate or another kind of rack to hold the samples. Cap or cover the desiccator securely and let it stand overnight or longer (preferably longer). Remove and weigh the container with the sample in it, then return it to the desiccator and reweigh it after an hour or so. If the mass has decreased by 1% or more, return the container to the desiccator and leave it there until the mass does not change significantly between weighings.

When Things Go Wrong

If a solid being dried in an oven begins to discolor, it is probably beginning to decompose. Dry it at a lower temperature or by a different method.

If a solid being dried in an oven is decreasing in volume, it is probably undergoing sublimation (conversion from a solid to a vapor). Dry it by a different method.

If a solid being dried in a desiccator doesn't dry overnight or after several days, check the appearance of the desiccant. Calcium chloride granules and some other desiccants clump together when exhausted, and indicating Drierite turns pink. If the desiccant appears to be exhausted, remove the old desiccant, clean and dry the inside of the desiccator, and add fresh desiccant. Then resume drying.

Cleaning and Drying Gases

OPERATION **27**

Air and nitrogen are often used to evaporate [OP-19] a solvent from a desired product, nitrogen and other inert gases are used to provide an inert atmosphere [OP-13] for reactions of oxygen-sensitive compounds, and reactive gases such as carbon dioxide are used in certain chemical syntheses. Gases must ordinarily be clean and dry for such applications. Gases can be dried using desiccants like the ones described in OP-26 for drying solids.

Many gases, such as nitrogen, are available in cylinders and can be purchased in a form that is pure enough for most applications. Other gases, especially compressed air obtained from a laboratory air line, may have to be cleaned and dried before use. For most purposes, air from an air line and other impure gases can be cleaned and dried by passing the gas slowly through a standard scale drying tube or a U-tube filled with a suitable desiccant (drying agent) and plugged with a layer of cotton at both ends (read about the use of drying tubes in OP-12). Indicating silica gel and granular alumina are very efficient desiccants; indicating Drierite and calcium chloride are satisfactory for many purposes. Silica gel has the advantage of being chemically inert but calcium chloride is cheaper and easier to use. The cotton plugs remove most particles, grease, and other impurities from the air line. One end of the drying tube should have a connector that is inserted in a rubber or plastic tube going to the air line; the other end should have a connecter that is inserted in a similar flexible tube going to a Pasteur pipet or similar gas delivery tube. If necessary, you can make a connector by inserting a short length of fire-polished glass tubing (see OP-3) into a one-hole rubber stopper of a size that will fit in the drying tube.

When it is important that a gas be very clean and dry, it can be bubbled through a gas-washing bottle such as the one in Figure D5. The bottle is partly filled with concentrated sulfuric acid or another suitable liquid. The gas is bubbled into the acid through the long glass tube and exits through

Take Care! Avoid contact with sulfuric acid and use it under a fume hood

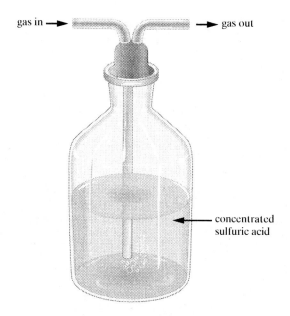

gas in →   → gas out

concentrated
sulfuric acid

Figure D5 Gas-washing bottle

the short tube, from which it is conveyed through a flexible tube to wherever it is needed. To make sure that the acid isn't carried with the air stream into a reaction flask or another vessel, it's a good idea to attach an empty trap (see Figure C4, OP-16) at the outlet of the gas washing bottle. The acid should be replaced after extended use. Note that many reactive gases, such as ammonia, can't be dried in sulfuric acid.

E. Purification Operations

Once a product has been separated from a reaction mixture it must ordinarily be *purified* to remove any residual impurities. Solids are usually purified by recrystallization [OP-28], but solids with high vapor pressures can be purified by sublimation [OP-29] and a few low-melting solids can be purified by distillation [OP-30b]. Most liquid products of reactions are purified by simple distillation [OP-30a]. High-boiling liquids that may decompose during an ordinary distillation can be purified by vacuum distillation [OP-31]. When a liquid product contains a substantial amount of liquid impurity having a boiling point close to that of the product itself, it should be purified by fractional distillation [OP-32]. Both liquid and solid products can sometimes be purified by column chromatography [OP-21], which was described in section C.

Recrystallization

OPERATION 28

The simplest and most widely used operation for purifying organic solids is *recrystallization*. Recrystallization is so named because it involves dissolving a solid that (in most cases) had originally crystallized from a reaction mixture or another solution, and then causing it to *again* crystallize from solution. In a typical recrystallization procedure, the crude solid is dissolved by heating it in a suitable *recrystallization solvent*. The hot solution is then filtered by gravity and the filtrate is allowed to cool to room temperature or below, whereupon crystals appear in the saturated solution and are collected by vacuum filtration. The crystals are ordinarily much purer than the crude solid because most of the impurities either fail to dissolve in the hot solution, from which they are separated by gravity filtration or transfer, or remain dissolved in the cold solution, from which they are separated by vacuum filtration or centrifugation.

A table in the front endpapers gives the properties of common solvents, most of which can be used for recrystallization. See also Table E1 in OP-28c.

Recrystallization is based on the fact that the solubility of a solid in a given solvent increases with the temperature of the solvent. Consider the recrystallization from boiling water of a 5.00-g sample of salicylic acid contaminated by 0.25 g of acetanilide. The solubility of salicylic acid in water at 100°C is 7.5 g per 100 mL, so the amount of water required to just dissolve 5.00 g of salicylic acid at the boiling point of water is 67 mL.

$$5.00 \text{ g} \times \frac{100 \text{ mL}}{7.5 \text{ g}} = 67 \text{ mL of water}$$

All of the acetanilide impurity will also dissolve in the boiling water. If the solution is cooled to 20 °C, at which temperature the solubility of salicylic acid is only 0.20 g per 100 mL, about 0.13 g of salicylic acid will remain dissolved.

$$67 \text{ mL} \times \frac{0.20 \text{ g}}{100 \text{ mL}} = 0.13 \text{ g of salicylic acid}$$

The dissolved salicylic acid will end up in the filtrate during the vacuum filtration; the remaining 4.87 g will crystallize from solution (if sufficient time is allowed) and will be collected on the filter. The solubility of acetanilide in water is 0.50 g per 100 mL at 20 °C, so up to 0.35 g of acetanilide can dissolve in 67 mL of water at 20 °C.

$$67 \text{ mL} \times \frac{0.50 \text{ g}}{100 \text{ mL}} = 0.35 \text{ g of acetanilide}$$

This means that all 0.25 g of acetanilide in the crude product should remain in solution and end up in the filtrate. Therefore, under ideal conditions, the recrystallization should yield a 97% recovery of salicylic acid uncontaminated by acetanilide.

The recovery could be increased by cooling the mixture in an ice/water bath to further lower the solubility of salicylic acid, but even then, some of the salicylic acid would remain in solution. You can never recover all of your product after a recrystallization, but by allowing plenty of time for the product to crystallize and making careful transfers, you should be able to minimize losses.

This is a simplified description of a rather complex process; a number of factors may bring about results different from those calculated:

- Crystals of the desired solid may adsorb impurities on their surfaces or trap them within their crystal lattice.
- The solubility of a solute in a saturated solution of a different solute may not be the same as its solubility in the pure solvent.
- Using only enough recrystallization solvent to dissolve a solid can result in premature crystallization, so additional solvent may be added to prevent this.

a. Recrystallization from a Single Solvent

Experimental Considerations

Most experimental procedures that involve recrystallization specify a suitable recrystallization solvent in the directions. If the solvent isn't specified, see Section c, "Choosing a Recrystallization Solvent."

In its simplest form, the recrystallization of a solid is carried out by dissolving the impure solid in the hot (usually boiling) recrystallization solvent and letting the resulting solution cool to room temperature or below to allow crystallization to occur. Additional steps, such as filtering or decolorizing the hot solution, may also be necessary. Sometimes it's desirable to collect a second or third crop of crystals by concentrating (see OP-19) the *mother liquor*—the liquid from which the crystals are filtered—from the previous crop. These crystals will contain more impurities than the first crop and may require recrystallization from fresh solvent. A melting-point determination [OP-33] or TLC analysis [OP-22] can be used to assess the purity of a recrystallized solid.

Recrystallization of a gram or more of solid is usually carried out in an Erlenmeyer flask, which—depending on the boiling point of the solvent—can be heated using a steam bath or hot plate. If necessary, the hot solution is filtered by gravity through fluted filter paper to remove insoluble impurities. After crystallization is complete, the product is collected by vacuum filtration with a Buchner funnel.

Recrystallization of about 0.1 g to 1 g of solid can be carried out using test tubes or small (10–25-mL) Erlenmeyer flasks. A sand bath or an aluminum block [OP-7a] is generally used for heating. If necessary, the hot solution is filtered using a preheated filter-tip pipet or filtering pipet, and the crystals are collected by vacuum filtration on a Hirsch funnel.

For quantities smaller than about 0.1 g, a specialized crystallization tube called a Craig tube can be used. A Craig tube looks much like a small test tube except that it widens at the top, where it has a ground-glass inner surface that accommodates a glass or Teflon plug (see Figure E3). The impure solid is dissolved in the recrystallization solvent and allowed to crystallize in the Craig tube. Then the Craig tube and its plug are inverted in a centrifuge tube and spun in a centrifuge [OP-17], forcing liquid past the plug into the centrifuge tube and leaving the solid behind on the plug. This procedure reduces the number of transfers needed for recrystallization and therefore reduces product losses. But Craig tubes are fragile and expensive, and they can easily roll off a bench top and break, so they must be handled with care.

*An alternative method for purifying 10–100 mg of solid, which uses a Pasteur pipet as the recrystallization vessel, is described in J. Chem. Educ. **1988**, 65, 460.*

Dissolving the Impure Solid.

The impure solid is usually dissolved by heating it with a sufficient amount of recrystallization solvent at the boiling point of the solvent. Except with very small amounts of solvent, the recrystallization solvent should be kept at or near the boiling point before it is added to the solid. If you know the solubility of the solid substance at the boiling point of the solvent you are using, you can calculate the approximate volume of boiling solvent that will be needed to dissolve it. Otherwise, you will have to determine the necessary amount of recrystallization solvent by trial and error; adding a measured amount of the hot solvent, boiling the mixture to see if it dissolves, adding more solvent if it doesn't, and continuing to add and boil fresh portions of hot solvent until it eventually goes into solution. Since the solid is most soluble at the boiling point of the solvent, it is important to bring the mixture back to the boiling point after each solvent addition and to boil it for a minute or more before making the next addition. Most solids dissolve fairly rapidly in a boiling solvent, but some do not. Slow-dissolving solids will usually dissolve when heated under reflux [OP-7c] (to prevent solvent loss) for 5–10 minutes after each addition of fresh solvent.

Filtering the Hot Solution.

Some impurities in a substance being crystallized may be insoluble in the boiling solvent and should be removed after the desired substance has dissolved. Don't mistake such impurities for the substance being purified and add too much solvent in an attempt to dissolve them, because excess solvent will reduce the yield of crystals and may even prevent the substance from crystallizing at all. If, after most of the solid has dissolved, addition of another portion of hot solvent doesn't appreciably reduce the amount of solid in the flask, that solid is probably an impurity—particularly if it is different in appearance from the solid that dissolved.

Excess solvent can be removed by evaporation [OP-19] if necessary.

🌑 For most standard scale recrystallizations, you can remove undissolved impurities by filtering the hot solution through coarse fluted filter paper, using the procedure for gravity filtration [OP-15]. To help prevent premature crystallization, you should use at least 10% more recrystallization solvent than the minimum amount needed to dissolve the crude solid, and carry out the filtration as rapidly as possible. You should also preheat the filtration apparatus by, for example, setting the funnel on an Erlenmeyer flask containing the boiling recrystallization solvent so that it is heated by hot solvent vapors; this flask is used as the collecting flask after unused solvent has been removed. (Alternatively, you can preheat the funnel in an oven or invert it inside a large beaker set on a steam bath.) If a few crystals form on the filter paper or in the funnel stem during filtration, dissolve them by pouring a small amount of hot recrystallization solvent over them. If a relatively large quantity of solid crystallizes in the filter or funnel stem, scrape it into the filtrate and redissolve it by adding about 10% more recrystallization solvent and heating the mixture to boiling. Then refilter the hot solution and dissolve any precipitate or cloudiness that forms in the collecting flask by heating it before you set it aside to cool.

If the particles of the solid impurity are relatively large, you may be able to remove them by letting them settle to the bottom of the recrystallization container and then decanting (pouring) the liquid into another container without disturbing the solid.

🧂 In microscale work, filtering the hot solution is complicated by the fact that small quantities of liquid and small-scale filters cool more rapidly than larger ones, making premature crystallization much more likely than in a standard scale recrystallization. To prevent it, you should dilute the hot solution with about 30–50% more recrystallization solvent than was needed to dissolve the solid, and preheat everything that will contact the solution when it is being filtered. For relatively small amounts of liquid, you can filter the diluted solution with a shortened filter-tip pipet—one having all but ~5 mm of the capillary tip cut off (see OP-6). Preheat the filter-tip pipet by drawing in and expelling several portions of the boiling recrystallization solvent (*not* the solution you are filtering). Without delay, use it to transfer the hot solution, while it is just below its boiling point, to another container such as a test tube or Craig tube. To remove the excess solvent, continuously twirl the rounded end of a microspatula in the tube as you boil the resulting solution (see Figure C13b, OP-19) in a heating block or sand bath. Stop boiling when solid begins to form on the spatula just above the liquid level, indicating that the solution is near the saturation point. Then set the solution aside to cool and crystallize as described later.

Filtration of the hot solution can also be accomplished with a similarly shortened filtering pipet (see OP-15). Clamp the preheated filtering pipet over the crystallization container (test tube, Craig tube, etc.). Add 30–50% more recrystallization solvent than was needed to dissolve the solid and heat the solution to boiling, and then use a preheated Pasteur pipet to transfer the recrystallization solution to the filtering pipet. Work fast, as the pipets and their contents will cool down rapidly. If necessary, you can increase the filtration rate through the filtering pipet by applying gentle pressure with a rubber bulb or pipet pump. Boil off the excess solvent, as described previously, before cooling.

If premature crystallization occurs in either kind of filtration apparatus, you may be able to redissolve the crystals in the solvent by blowing hot air from a heat gun over the pipet while its outlet is over the crystallization container, adding a little more solvent if necessary. Otherwise, return the contents of the pipet to the hot recrystallization mixture, add more recrystallization solvent, and try again. If necessary, use a freshly prepared and preheated filtering device to complete the filtration.

Removing Colored Impurities.

If a crude sample of a compound known to be white or colorless yields a recrystallization solution with a pronounced color, activated carbon (Norit) can often be used to remove the colored impurity. Pelletized Norit, which consists of small cylindrical pieces of activated carbon, is usually preferable to finely powdered Norit. Powdered Norit, although it is somewhat more efficient than the pelletized form, obscures the color and is difficult to filter out completely.

To use pelletized Norit, let the boiling recrystallization solution cool down for a minute or so, then stir in a *small* amount of the Norit. Unless otherwise directed, start with about 20 mg of Norit for microscale work and 0.1 g or so for standard scale work. Stir or swirl the mixture for a few minutes, keeping it hot enough to prevent your product from crystallizing but not boiling it. Then let it settle and observe the color of the solution. If much color remains, you can add more pelletized Norit and repeat the process. Avoid adding too much, because any excess Norit can adsorb your product as well as the impurities, and some color may remain no matter how much you use. Then heat the solution just to boiling and separate it from the Norit by one of the methods described in the section "Filtering the Hot Solution." Note that even pelletized Norit contains some fine particles that can plug up a filter-tip pipet, so for microscale work a filtering pipet may be preferable.

Powdered Norit can be used in much the same way as pelletized Norit, except that less is needed and you will not be able to see whether the decolorization was successful until the solution is filtered. If Norit particles pass through the filter during gravity filtration of the hot solution, they can be removed (with some product loss) by vacuum filtration through a bed of a filtering aid such as Celite. To prepare such a bed, mix the filtering aid with enough low-boiling solvent (such as diethyl ether or dichloromethane) to form a thin slurry, then pour it onto the filter paper in a Buchner or Hirsch funnel, with the vacuum turned on, until it forms a layer about 3 mm thick. When the bed is dry, remove the solvent from the filter flask. Quickly filter the hot solution under vacuum and wash the filtering aid with a small amount of hot recrystallization solvent. Turn off the vacuum without delay to prevent evaporation of the filtrate. Redissolve any solid that forms in the filtrate by heating it and, as necessary, adding more hot solvent. Then let it cool and crystallize as described later.

Inducing Crystallization.

If no crystals form after a hot recrystallization solution is cooled to room temperature, the solution may be supersaturated. If so, crystallization can often be induced by one or more of the following methods:

- Dip the end of a glass stirring rod into the liquid; then remove it and let the solvent evaporate to leave a thin coating of the solid. Reinsert the glass rod into the liquid and stir gently.

Take Care! Never add Norit to a solution at or near the boiling point—it may boil up violently.

A glass rod that has not been fire polished works best, but it will also scratch the glass, so don't use one without your instructor's permission.

- Rub the tip of a glass stirring rod against the inside of the recrystallization container, just above the liquid surface, for a minute to two. Use an up and down motion with the rod tip just touching the liquid on the downstroke. If you are using a Craig tube, rub gently so as not to scratch or break the tube.
- Cool the solution in an ice/water bath, then continue rubbing as described previously for several minutes.
- If any *seed crystals* (crystals of the pure compound) are available, drop a few into the solution with cooling and stirring.

If crystals still do not form, you may have used too much recrystallization solvent. In that case, try one or both of the following procedures, in order:

- Concentrate the solution by evaporation [OP-19] until it becomes cloudy or crystals appear, heat it until the cloudiness or crystals disappear (add a little more recrystallization solvent if necessary), and then let it cool. If necessary, use one or more of the previous methods to induce crystallization.
- Heat the solution back to boiling and add another solvent that is miscible with the first, and in which the compound should be less soluble (for example, try adding water to an ethanol solution). Add just enough of the second solvent to induce cloudiness or crystal formation at the boiling point; then add enough of the original solvent to cause the cloudiness or crystals to disappear from the boiling solvent, and let it cool.

As a last resort, remove all of the solvent by evaporation and try a different recrystallization solvent—but see your instructor for advice first.

Cooling the Hot Solution. The size and purity of the crystals formed depends on the rate of cooling; rapid cooling yields small crystals and slow cooling yields large ones. The medium-sized crystals obtained from moderately slow cooling are usually the best, because larger crystals tend to *occlude* (trap) impurities, while smaller ones adsorb more impurities on their surfaces and take longer to filter and dry. You can reduce the rate of cooling by setting the recrystallization container on a surface that is a poor heat conductor and inserting it inside or covering it with another container. For standard scale work, the flask can be set on the bench top and its mouth covered with a watch glass. For slower cooling, a large beaker can be inverted over the flask and some paper towels or other insulating material placed under it. For microscale work, a test tube can be supported in a small Erlenmeyer flask and a beaker inverted over the apparatus. A Craig tube, with its plug inserted, can be supported in a 25-mL Erlenmeyer flask, which is set inside a 100-mL beaker; if desired, a larger beaker can be inverted over this assembly.

You can increase the yield of crystals somewhat by cooling the mixture in an ice/water bath once a good crop of crystals is present, but their purity may decrease slightly as a result. Cooling the mixture before well-formed crystals are present may result in small, impure crystals that take longer to filter and dry.

Dealing with Oils and Colloidal Suspensions. When the solid being recrystallized is quite impure or has a low melting point, it may separate as an *oil* (a second liquid phase) upon cooling. Oils are undesirable because, even if they solidify on cooling, the solid retains most of the original impurities.

If the solid to be purified has a melting point below the boiling point of the recrystallization solvent, it may be possible to prevent oiling by substituting a lower boiling solvent with similar properties. For example, methanol (bp 65°C) might be substituted for 95% ethanol (bp 78°C), or acetone (bp 57°C) for 2-butanone (bp 80°C). Oiling may also be prevented by using more recrystallization solvent, adding seed crystals, or both. Seed crystals can sometimes be obtained by dissolving a small amount of the oil in an equal volume of a volatile solvent in a small, open test tube and letting the solvent evaporate slowly.

Using a lower boiling solvent usually results in a lower percent recovery.

If oiling occurs, try the following remedies, in order:

1. Heat the solution until the oil dissolves completely, adding more solvent if necessary, then cool it slowly while rubbing the inside of the container (see "Inducing Crystallization").
2. Add an amount of pure recrystallization solvent equal to about 25% of the total solvent volume and repeat the process described in Step **1**.
3. Follow the procedure in Step **1**, but add a seed crystal or two at the approximate temperature where oiling occurred previously.
4. Try to crystallize the oil by either (a) cooling the solution in an ice–salt bath, rubbing the oil with a stirring rod and adding seed crystals if necessary, or (b) removing all of the oil with a Pasteur pipet, dissolving it in an equal volume of a volatile solvent, and letting the solvent evaporate slowly in an open test tube. Then collect the solid by vacuum filtration and recrystallize it from the same solvent or a more suitable one. If necessary, use one or more of the previous methods to prevent further oiling.

A *colloid* is a suspension of very small particles dispersed in a liquid or another phase. Colloids generally have a cloudy appearance and cannot be filtered through ordinary filtering media because the particles pass right through. If a solid separates from a cooled solution as a colloidal suspension, the colloid can often be coagulated to form normal crystals by extended heating in a hot water bath, or (if the solvent is polar) by adding an electrolyte such as sodium sulfate. Colloid formation can sometimes be prevented by treating a recrystallization solution with Norit, as described previously, or by cooling the solution very slowly.

General Directions for Single-Solvent Recrystallization

Safety Notes

Standard Scale

Use this method (illustrated in Figure E1) when you have approximately 1 g or more of crude solid.

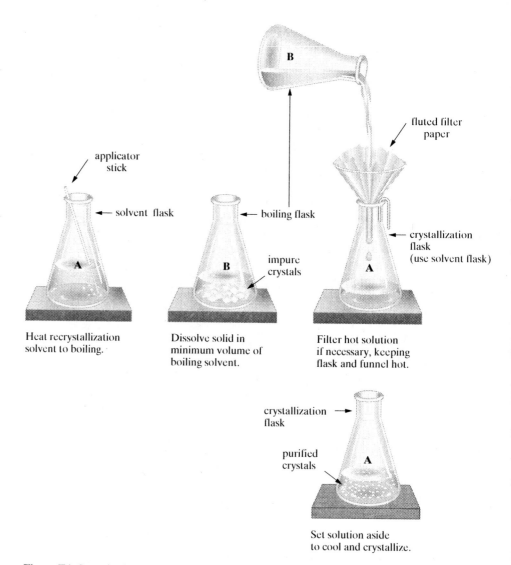

Figure E1 Steps in the recrystallization of a solid

Equipment and Supplies

2 Erlenmeyer flasks
recrystallization solvent
graduated cylinder
heat source
boiling (applicator) stick or boiling chips
flat-bottomed stirring rod
small watch glass
Buchner funnel with filter paper
filter flask
cold washing solvent
watch glass or beaker (to cover crystallization flask)

powder funnel (optional)
fluted filter paper (optional)
Norit (optional)

If you know the solubility of the solid in the boiling recrystallization solvent, calculate the approximate volume of solvent that you will need to recrystallize it and measure out that amount plus 10–20% extra. Otherwise, start with about 10 mL per gram of solid and use more if needed. Measure the solvent into an Erlenmeyer flask (the *solvent flask,* **A**). Add a boiling stick or a few boiling chips, insert a powder funnel in the flask mouth, and heat the solvent to boiling with an appropriate heat source [OP-7a]. A steam bath is preferred for organic solvents that boil below 100°C; a hot plate can be used for water and higher boiling organic solvents. Place the solid to be purified in a second Erlenmeyer flask (the *boiling flask,* **B**) and add about one-quarter of the hot liquid in the solvent flask to the boiling flask (use about three-quarters of the liquid if you have calculated the approximate volume needed). Heat the mixture *at the boiling point* with continuous swirling or stirring, breaking up any large particles with a spatula or flat-bottomed stirring rod, until it appears that no more solid will go into solution. If undissolved solid remains, add more portions of hot solvent, about 10% of the total each time, heating the solution *at the boiling point* with swirling or stirring after each addition. Continue this process until (1) the solid is completely dissolved *or* (2) no more solid dissolves when a fresh portion of solvent is added and it appears that only solid impurities remain.

If the solution has an intense color but the pure product should not be colored, decolorize it as directed in the section *Removing Colored Impurities.* If the boiling solution contains no solid impurities, use the boiling flask as a crystallization flask and go to the next paragraph. If it does contain solid impurities (including Norit for decolorizing), add about one-tenth as much recrystallization solvent as you have used so far and heat the mixture back to boiling. Put a preheated powder funnel on the neck of the emptied solvent flask **A** (with a bent wire or paper clip between them) and set this flask (which is now the *crystallization flask*) on the heat source. Insert a coarse fluted filter paper and rapidly filter the hot solution while it is still near the boiling point, keeping any unfiltered solution hot throughout the filtration. If any solid crystallizes on the filter paper or inside the funnel, redissolve it as described in the section *Filtering the Hot Solution.*

Set the crystallization flask on the benchtop, cover it with a watch glass or inverted beaker, and let the solution cool slowly to room temperature. If no crystals form by the time the solution reaches room temperature, see *Inducing Crystallization.* If an oil separates or the solution becomes cloudy but no solid precipitates, see *Dealing with Oils and Colloidal Suspensions.* Once crystals have begun to form, allow at least 15 minutes (sometimes much longer) for complete crystallization. If desired, cool the flask further in an ice bath for five minutes or more to improve the yield.

Collect the crystals by vacuum filtration [OP-16] on a Buchner funnel or Hirsch funnel of appropriate size. Transfer any crystals remaining in the crystallization flask to the funnel with a small amount of ice-cold recrystallization solvent (or another appropriate solvent) and use more of the cold solvent to wash the solid on the filter [OP-26a]. Air-dry the crystals by leaving the vacuum on for a few minutes after the last washing, then dry [OP-26b] them further as necessary.

See the section Filtering the Hot Solution *for additional information.*

Summary

1. Measure recrystallization solvent into solvent flask; heat to boiling.
2. Add some hot solvent to solid in boiling flask; boil with stirring.
3. Add more hot solvent in portions (as necessary) until solid dissolves.
   IF solution contains colored impurities, GO TO 4.
   IF solution contains undissolved impurities, add more hot solvent, GO TO 5.
   IF not, GO TO 6.
4. Cool below boiling point, stir in Norit, heat to boiling.
5. Filter hot solution by gravity.
6. Cover flask and set aside to cool until crystallization is complete.
7. Collect crystals by vacuum filtration; wash and air-dry on filter.
8. Clean up, dispose of solvent as directed.

 ## Microscale (Test-Tube Method)

Use this method (illustrated in Figure E2) when you have approximately 0.1 g to 1 g of crude solid.

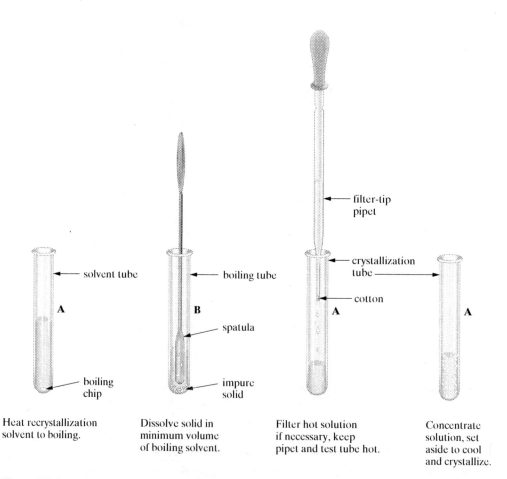

solvent tube

boiling chip

A

Heat recrystallization solvent to boiling.

boiling tube

spatula

impure solid

B

Dissolve solid in minimum volume of boiling solvent.

filter-tip pipet

crystallization tube

cotton

A

Filter hot solution if necessary, keep pipet and test tube hot.

A

Concentrate solution, set aside to cool and crystallize.

Figure E2 Steps in microscale recrystallization using a test tube

Equipment and Supplies

2 test tubes, 13 × 100 mm or 15 × 125 mm

recrystallization solvent

calibrated Pasteur pipet (preferably filter-tipped)

heat source

boiling stick or boiling chip

flat-bladed microspatula

Hirsch funnel with filter paper

small filter flask

cold washing solvent

small Erlenmeyer flask (to hold test tube)

beaker (to cover flask and test tube)

shortened filter-tip pipet or filtering pipet (optional)

pelletized Norit (optional)

If you know the solubility of the solid in the boiling recrystallization solvent, calculate the approximate volume of solvent you will need to recrystallize it and measure out that amount plus 20–50% extra (use the larger amount if you think you will need to filter the hot solution). Otherwise, start with about 1 mL per 0.1 g of solid and use more if needed. Obtain a test tube large enough so that the solvent will fill it no more than half full. Measure the recrystallization solvent into this test tube (the *solvent tube*, **A**) with a calibrated Pasteur pipet, add a boiling chip or boiling stick, and heat it to boiling using a heating block or sand bath. Place the solid to be purified in an identical test tube (the *boiling tube*, **B**) and clamp this test tube *above* the heat source (not in it, as the solid may melt) so that you can control the heating rate quickly by raising or lowering it. Use the calibrated Pasteur pipet to transfer about one-quarter of the liquid in the solvent tube to the boiling tube (use about three-quarters of the liquid if you have calculated the approximate volume needed). Lower the boiling tube and heat the mixture *at the boiling point* while continuously stirring it by twirling the rounded end of a flat-bladed microspatula in the test tube, until it appears that no more solid will go into solution. If undissolved solid remains, add more portions of hot solvent, about 10% of the total each time, heating the solution *at the boiling point* with swirling or stirring after each addition. Continue this process until (1) the solid is completely dissolved *or* (2) no more solid dissolves when a fresh portion of solvent is added and it appears that only solid impurities remain.

If the solution has an intense color but the pure product should not be colored, see the section *Removing Colored Impurities*. If the boiling solution contains no solid impurities, use the boiling tube as a crystallization tube and go to the next paragraph. If it does contain solid impurities (including Norit used for decolorizing), add 30–50% as much recrystallization solvent as you used to dissolve the solid and heat the mixture back to boiling. Use a preheated filter-tip pipet to transfer the hot solution to a *crystallization tube* (use the emptied solvent tube), leaving the solid impurities behind. (Alternatively, you can filter the solution into the crystallization tube through a preheated filtering pipet.) If crystals begin to form in the pipet, redissolve them as described in the section *Filtering the Hot Solution*.

If you will need less than ~2 mL of recrystallization solvent, it isn't necessary to preheat the solvent.

See the section Filtering the Hot Solution *for additional information.*

Concentrate the filtered solution by boiling it while stirring with a flat-bladed microspatula until traces of solid begin to form on the microspatula just above the solvent level.

Set the crystallization tube in a small Erlenmeyer flask, cover it with an inverted beaker, and let the solution cool slowly to room temperature. If no crystals form by the time the solution reaches room temperature, see *Inducing Crystallization*. If an oil separates or the solution becomes cloudy but no solid precipitates, see *Dealing with Oils and Colloidal Suspensions*. Once crystals have begun to form, allow at least 10 minutes (sometimes much longer) for complete crystallization. If desired, cool the test tube further in an ice bath for 5 minutes or more to increase the yield.

Collect the product by vacuum filtration [OP-16] on a Hirsch funnel. Transfer any crystals remaining in the crystallization flask to the funnel with a small amount of ice-cold recrystallization solvent (or another appropriate solvent) and use more cold solvent to wash the solid on the filter [OP-26a]. Air-dry the crystals by leaving the vacuum on for a few minutes after the last washing, then dry [OP-26b] them further as necessary.

Summary

1. Measure recrystallization solvent into solvent tube; heat to boiling.
2. Add some hot solvent to solid in boiling tube; boil with stirring.
3. Add more hot solvent in portions (as necessary) until solid dissolves.
   IF solution contains colored impurities, GO TO 4.
   IF solution contains undissolved impurities, add more hot solvent, GO TO 5.
   IF not, GO TO 6.
4. Cool below boiling point, stir in pelletized Norit, heat to boiling.
5. Filter and concentrate hot solution.
6. Cover hot solution and set aside to cool until crystallization is complete.
7. Collect crystals by vacuum filtration; wash and air-dry on filter.
8. Clean up, dispose of solvent as directed.

 ## Microscale (Craig-Tube Method)

You can use this method (illustrated in Figure E3) when you have approximately 0.1 g or less of crude solid or know that the recrystallization will require no more than 2 mL of recrystallization solvent.

Equipment and Supplies

Craig tube with plug
recrystallization solvent
calibrated Pasteur pipet (preferably filter tipped)
heat source
flat-bladed microspatula
25-mL Erlenmeyer flask (to hold Craig tube)
100-mL beaker
copper wire
centrifuge tube (preferably plastic)

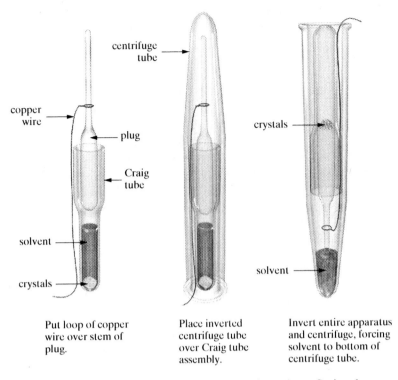

centrifuge
tube

copper
wire

plug

Craig
tube

crystals

solvent

solvent

crystals

Put loop of copper
wire over stem of
plug.

Place inverted
centrifuge tube
over Craig tube
assembly.

Invert entire apparatus
and centrifuge, forcing
solvent to bottom of
centrifuge tube.

Figure E3 Steps in microscale recrystallization using a Craig tube

centrifuge
13 × 100-mm test tube (optional)
shortened filter-tip pipet (optional)
pelletized Norit (optional)

If it may be necessary to filter the hot solution because of the likelihood of insoluble or colored impurities, put the impure solid in a 13 × 100-mm test tube and dissolve it in the boiling recrystallization solvent (adding 30–50% more solvent than is needed to dissolve it), as described in the previous directions. If necessary, decolorize the solution with pelletized Norit as described in *Removing Colored Impurities.* Then transfer the hot solution to a Craig tube using a preheated filter-tip pipet. Concentrate the solution by boiling it while stirring with a flat-bladed microspatula until traces of solid begin to form on the microspatula just above the solvent level. Skip the next paragraph and go on to the following one.

The experimental procedure or your instructor may indicate whether the hot solution will need to be filtered.

If it will not be necessary to filter the hot solution, place the impure solid in a Craig tube and add just enough room-temperature recrystallization solvent to cover it. Heat the mixture *at the boiling point* with a heating block or sand bath while continuously stirring it by twirling the rounded end of a flat-bladed microspatula in the Craig tube, until it appears that no more solid will go into solution. If undissolved solid remains, add more solvent drop by drop with stirring, keeping the solution at the boiling point, until the solid is completely dissolved.

Don't add solvent to the level where the Craig tube widens; if it takes that much solvent to dissolve the solid, transfer the hot solution to a test tube for recrystallization as described in the previous directions.

Insert the plug into the Craig tube, support it in a 25-mL Erlenmeyer flask, and set the flask inside a 100-mL beaker. If desired, cover the apparatus with a large inverted beaker. Let the solution cool slowly to room temperature. If no crystals form by the time the solution reaches room temperature, see *Inducing Crystallization*. If an oil separates or the solution becomes cloudy but no solid precipitates, see *Dealing with Oils and Colloidal Suspensions*. Once crystals have begun to form, allow at least 10 minutes (sometimes much longer) for complete crystallization. If desired, cool the Craig tube further in an ice bath for 5 minutes or more to increase the yield.

Obtain a length of thin copper wire that is about as long as the combined length of the assembled Craig tube and its plug. Make a loop in the copper wire that is large enough to fit over the narrow stem of the plug and bend it so that it is at right angles to the straight part of the wire (see Figure E3). Holding the Craig tube vertically, slip the loop over the plug stem and slide it down as far as it will go, with the straight part of the wire parallel to the Craig tube. Hold the wire against the bottom of the Craig tube between your thumb and forefinger, then invert a conical centrifuge tube over the Craig tube assembly until the narrow end of the plug is nested in (or nearly touching) the V-shaped end of the tube. Pressing the wire firmly against the bottom of the Craig tube to keep the plug in place, take the centrifuge tube in your other other hand and carefully invert the entire apparatus so that the centrifuge tube is upright. Bend the wire over the lip of the centrifuge tube and down the outside so that it doesn't stick out. Place the centrifuge tube in a centrifuge [OP-17] opposite another tube containing enough water to balance it *or* a centrifuge tube containing another student's Craig tube assembly. If the wire scrapes against any part of the centrifuge, cut it shorter, but not so short that you won't be able to grasp it to lift out the Craig tube. Run the centrifuge for 3 minutes or more. When the centrifuge has stopped turning, remove the centrifuge tubes. All of the solvent should be at the bottom of the Craig tube's centrifuge tube, and most of the solid should be packed near the ground joint where the plug fits into the Craig tube. If there still is solvent in the Craig tube, replace the assembly in the centrifuge tube and centrifuge it again. Grasp the copper wire and lift the Craig tube assembly out of its centrifuge tube. Carefully remove the plug and use a microspatula to transfer the solid to another container. Dry [OP-26b] it further if necessary.

Summary

1. IF filtering the hot solution will be necessary, dissolve solid in excess hot recrystallization solvent in test tube, Go To 2.
   IF filtering the hot solution should not be necessary, dissolve solid in boiling recrystallization solvent in Craig tube, GO TO 3.
2. Transfer hot solution to Craig tube with filter-tip pipet and boil to concentrate.
3. Insert plug, cover hot solution, cool until crystallization is complete.
4. Collect crystals by centrifugation.
5. Clean up, dispose of solvent as directed.

When Things Go Wrong

If you are attempting to dissolve a solid product in a hot recrystallization solvent but no more solid appears to dissolve with each addition of fresh solvent, ask yourself the following questions: (1) Do the crystals that remain in solution look different than those that already dissolved? If so, they probably belong to an impurity. Follow the directions in *Filtering the Hot Solution*. (2) Does the volume of the solution stay about the same each time you add fresh solvent and bring the mixture back to the boiling point? If the volume isn't increasing significantly, you must be boiling away solvent about as fast as you are adding it. You should be using an Erlenmeyer flask, test tube, Craig tube, or another narrow-mouthed vessel for recrystallization, never a beaker or similar container that promotes rapid evaporation. If you are using an appropriate container, try turning down the heat to the lowest setting that keeps the solvent boiling. It may also help to reduce the time you allow the solution to boil after each addition of fresh solvent.

Suppose you cool the recrystallization solution by putting it in a container of ice and water that is too big for it, so that as the ice melts, the flask tips over and fills with water. Since most organic solids are even less soluble in water than in the common recrystallization solvents, you may be able to recover your product by waiting until the ice has completely melted and then putting all of the liquid through a vacuum filtration apparatus [OP-16]. The solid that collects on the filter will be wet and impure, so it will have to be dried [OP-26b] and recrystallized again, but at least you won't have to start from square one.

Suppose you carry out the recrystallization of a solid product and let the solution cool to room temperature but no solid crystallizes. This usually implies one of three things: (1) the solution is supersaturated; (2) you used too much recrystallization solvent; (3) you used an inappropriate recrystallization solvent. The section *Inducing Crystallization* tells you how to deal with each of these situations. In case you used too much solvent, refer to *Dissolving the Impure Solid* before you attempt another recrystallization.

If you filter a hot recrystallization solution by vacuum filtration [OP-16] and your product crystallizes in the filtrate (which is very likely), transfer the mixture in the filter flask to a suitable recrystallization vessel (don't use the filter flask), heat it with enough additional recrystallization solvent to get the solid back into solution, and then filter the hot solution by gravity [OP-15]. Vacuum filtration should *not* be used to filter recrystallization solutions because the hot solvent vaporizes rapidly under vacuum.

If, while you are centrifuging a Craig tube assembly, you hear a loud scraping noise, the copper wire is probably scraping against the centrifuge sides or lid. Remove the Craig tube assembly and shorten the wire or bend it out of the way.

If, after cooling the recrystallization solution, you obtain considerably less solid product than you started with, you may have used too much recrystallization solvent or an inappropriate solvent. Collect the product that has already crystallized, using vacuum filtration [OP-16], and then follow the appropriate directions in the second part of *Inducing Crystallization* to obtain another crop of crystals. The crystals you collected first should be purer than the second crop, so they should be used for any melting-point measurement or spectral analysis.

If you are recrystallizing a solid product from water and obtain a low yield of crystals on cooling, you can try dissolving some sodium chloride in the cold solution to salt out (see OP-18b) more of the product, Don't use more than about 3 g of salt per 10 mL of solution. Make sure it all dissolves or your product will be contaminated with salt, which you will have to wash out [OP-26a] with fresh water.

If, after your impure solid has dissolved, a liquid separates or a suspension of fine particles forms on cooling, see *Dealing with Oils and Colloidal Suspensions*.

If the solid that crystallizes is colored but your product is supposed to be colorless or a different color, see *Removing Colored Impurities*.

If the crystals that form on cooling are very fine or very coarse, you may have cooled the recrystallization solution too rapidly or too slowly. See *Cooling the Hot Solution*.

b. Recrystallization from Mixed Solvents

Some compatible solvent pairs

ethanol–water
methanol–water
acetic acid–water
acetone–water
ethyl ether–methanol
ethanol–acetone
absolute ethanol–petroleum ether
ethyl acetate–cyclohexane

Solids that cannot be recrystallized readily from any single solvent can usually be purified by recrystallization from a mixture of two compatible solvents such as those listed in the margin. The solvents must be miscible in one another, and the solid compound should be quite soluble in one solvent and relatively insoluble in the other. If the composition of a suitable solvent mixture is known beforehand (such as 40% ethanol in water, for example), the recrystallization can be carried out with the premixed solvent in the same way as for a single-solvent recrystallization. But a mixed-solvent recrystallization is usually performed by heating the compound in the solvent in which it is most soluble (which we will call solvent A) until it dissolves, and then adding enough of the second solvent (solvent B) to bring the solution to the saturation point.

If the compound is *very* soluble in solvent A, the total volume of solvent may be quite small compared to that of the crystals, which may then separate as a dense slurry. In such a case, you should use more of solvent A than is needed to just dissolve the compound, and add correspondingly more of solvent B to bring about saturation. Be careful to avoid adding so much of solvent A that *no* amount of solvent B will result in saturation. If that occurs, you will have to concentrate the solution (see OP-19) or remove all of the solvent and start over.

General Directions for Mixed-Solvent Recrystallization

Most of the steps in a mixed-solvent recrystallization are the same as for single-solvent recrystallization; refer to the previous General Directions for experimental details.

Place the crude solid in a boiling flask or tube and add solvent A (previously heated to boiling) in portions, while boiling and stirring, until the solid dissolves or only undissolved impurities remain. If necessary, the solution can be decolorized with Norit at this point. If the hot solution contains undissolved impurities (including Norit), add enough solvent A to prevent premature crystallization, heat it to boiling, and filter it. For microscale work, concentrate the solution after filtration to remove most of the excess solvent. Add hot solvent B in small portions, with stirring, keeping the mixture boiling after each addition, until a persistent cloudiness appears or a precipitate starts to form. Then add just enough hot solvent A drop by drop to the boiling mixture, with stirring, to clear it up or dissolve the precipitate. Set the mixture aside to cool to room temperature, allow sufficient time for crystallization, and cool it further in an ice/water bath if desired. When crystallization is complete, collect the product by vacuum filtration, or by centrifugation if you are using a Craig tube. After vacuum filtration, wash the product on the filter [OP-26a] with cold solvent B or another suitable solvent.

When Things Go Wrong

Most of the things that can go wrong during a single-solvent recrystallization can also go wrong during a mixed-solvent recrystallization, so you can refer to the previous *When Things Go Wrong* section for help. The following case applies only to mixed-solvent recrystallization.

Suppose you have dissolved the impure solid in solvent A, added enough B to produce cloudiness or a precipitate, and are now adding more A drop by drop at the boiling point of the mixture to clear it up, but it doesn't clear up. You may be boiling away solvent A (which is often the more volatile solvent) faster than solvent B, preventing the solid material from dissolving. Turn down the heat to the lowest setting that will keep the solvent boiling. Continue adding solvent A, perhaps at a faster rate than before, until the solution is clear; then remove it from the heat source without delay.

c. Choosing a Recrystallization Solvent

A solvent suitable for recrystallization of a given solid should meet the following criteria, or as many of them as is practical:

- Its boiling point should be in the 60–100°C range.
- Its freezing point should be well below room temperature.

Table E1 Properties of common recrystallization solvents

Solvent	bp	fp	Comments
water	100	0	solvent of choice for polar compounds; crystals dry slowly
methanol	65	−94	good solvent for relatively polar compounds; easy to remove
95% ethanol	78	−116	excellent general solvent; usually preferred over methanol because of higher boiling point
2-butanone	80	−86	good general solvent; acetone is similar but its boiling point is lower
ethyl acetate	77	−84	good general solvent
toluene	111	−95	good solvent for aromatic compounds; high boiling point makes it difficult to remove
petroleum ether (high-boiling)	~60–90	low	a mixture of hydrocarbons; good solvent for less polar compounds
hexane	69	−94	good solvent for less polar compounds; easy to remove
cyclohexane	81	6.5	good solvent for less polar compounds; freezes in some cold baths

Note: bp and fp (freezing point) are in °C. Solvents are listed in approximate order of decreasing polarity.

- It must not react with the solid.
- It should not be excessively hazardous to work with.
- It should dissolve between 5 g and 25 g of the solid per 100 mL at the boiling point, and less than 2 g per 100 mL at room temperature, with at least a 5:1 ratio between the two values.

As a rule, the recrystallization solvent should be either somewhat more or somewhat less polar than the solid, since a solvent of very similar polarity will dissolve too much of it. Solubility information from reference books listed in Category A of the Bibliography may help you choose a suitable solvent. For example, a solvent in which the compound is designated as *sparingly soluble* or *insoluble* when cold and *very soluble* or *soluble* when hot may be suitable for recrystallization. If solubility data for the compound are not available, a solvent may have to be chosen by trial and error from a selection such as that in Table E1. Once you have identified some possible solvents, test them as described next. You can test several solvents at the same time and choose the best one.

 ## General Directions for Testing Recrystallization Solvents

Standard Scale and Microscale

J. Chem. Educ. **1989**, *66*, *88 describes another microscale method for testing recrystallization solvents.*

For microscale (μS) work, use the quantities in parentheses. Drops should be added using a medicine dropper, not a Pasteur pipet.

Weigh about 0.1 g (μS, 10 mg) of the finely divided solid into a test tube or Craig tube and add 1 mL (μS, 3 drops) of the solvent. Stir the mixture by twirling the round end of a flat-bladed microspatula in it and carefully observe what happens. If the solid dissolves in the cold solvent, the solvent is unsuitable. If the solid doesn't dissolve, heat the mixture to boiling, with stirring. If it dissolves after heating, try to induce crystallization as described next. If it doesn't dissolve completely, add more solvent in 0.5-mL (μS, 1 drop) portions, gently boiling and stirring after each addition, until it dissolves *or* until the total volume of added solvent is about 3 mL (μS, 10 drops). If the solid *does not* dissolve in that amount of hot solvent, the solvent is probably not suitable. If it *does* dissolve at any point, record the volume of boiling solvent required to dissolve it. Let the solution cool while rubbing the inside of the tube with a glass rod to see whether crystallization occurs. If crystals separate, examine them for apparent yield and evidence of purity (absence of extraneous color and good crystal structure).

If no single solvent is satisfactory, choose one solvent in which the compound is quite soluble and another in which it is comparatively insoluble (the two solvents must be miscible). Dissolve the specified amount of solid in the first solvent with stirring and boiling, and record the amount of solvent required. Add the second hot solvent drop by drop while boiling and stirring until saturation occurs. Then cool the mixture and try to induce crystallization.

Once a suitable solvent or solvent pair has been identified, estimate the volume of solvent needed to dissolve all of the solid to be purified, based on the amounts of solid and solvent you used in the test.

Sublimation

Principles and Applications

Sublimation is a phase change in which a solid passes directly into the vapor phase without going through an intermediate liquid phase. Many solids that have appreciable vapor pressures below their melting points can be purified by (1) heating the solid to sublime it (convert it to a vapor), (2) condensing the vapor on a cold surface, and (3) scraping off the condensed solid. This method works best if impurities in the crude solid do not sublime appreciably. Sublimation isn't as selective as recrystallization or chromatography, but it has some advantages in that no solvent is required and losses in transfer can be kept low.

Experimental Considerations

Sublimation is usually carried out by heating the *sublimand* (the solid before it has sublimed) with a suitable heat source, and collecting the *sublimate* (the solid after it has sublimed and condensed) on a cool surface. For best results, the sublimand should be dry and finely divided, and the distance between the sublimand and the condensing surface should be minimized. A simple but effective sublimator consists of two nested beakers of appropriate sizes. For example, a 100-mL beaker can be nested inside a 150-mL beaker, with the inner beaker rotated so as to leave a gap of about 1 cm at the bottom. In some cases, it may be necessary to place separators made of folded-over strips of filter paper or paper toweling between the beakers to get the right spacing. Good beaker combinations are 100 mL/150 mL (for microscale work), 250 mL/400 mL, and 400 mL/600 mL. The sublimand is spread out at the bottom of the outer beaker, and the condensing (inner) beaker is partially filled with crushed ice or ice water. As the outer beaker is heated, crystals of sublimate collect on the bottom of the condensing beaker. Figure E4 illustrates a sublimator operating on the same principle, except that an Erlenmeyer flask is used as a condenser and the temperature is controlled by flowing water.

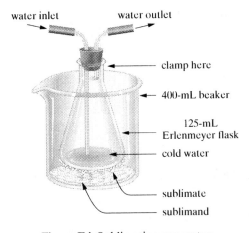

water inlet   water outlet

clamp here

400-mL beaker

125-mL
Erlenmeyer flask

cold water

sublimate

sublimand

Figure E4 Sublimation apparatus

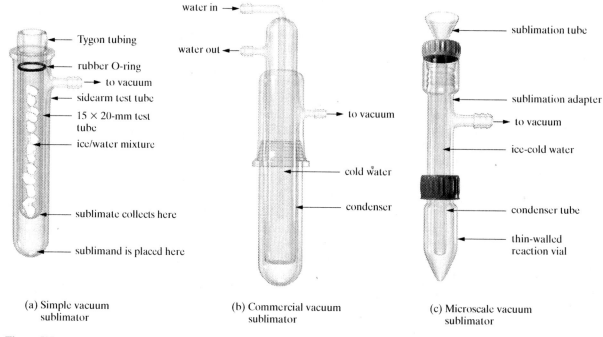

Figure E5 Apparatus for vacuum sublimation

J. Chem. Educ. **1991**, *68*, A63

Solids that do not sublime rapidly at atmospheric pressure may do so under vacuum. Figure E5a illustrates a vacuum sublimator that can be assembled by fitting a 15 × 125-mm test tube snugly inside an 18 × 150-mm sidearm test tube, using a rubber O-ring from a microscale lab kit to act as a vacuum seal. A short section of 15-mm i.d. Tygon tubing (or several layers of masking tape) is placed around the lip of the inner tube to keep it from slipping inside the sidearm test tube. A commercial vacuum sublimator, such as the one in Figure E5b, is more efficient because the condenser has a flat, wide bottom that is close to the sublimate. The microscale vacuum sublimation apparatus pictured in Figure E5c can be constructed using a special sublimation tube and adapter.

Depending on the temperature required and the nature of the sublimation apparatus, heat sources such as hot water baths, oil baths, steam baths, hot plates (for nested beakers), heating blocks, and sand baths can be used for sublimation. An oil bath provides the most uniform heating, but oil baths are messy and somewhat hazardous to work with. With some of these heat sources, crystals tend to collect on the sides of the sublimation container as well as on the condenser. Wrapping the base of the sublimation container with aluminum foil or other insulation will help prevent this.

General Directions for Sublimation

Safety Notes

Because of the possibility of an implosion, safety glasses must be worn during a vacuum sublimation. It is best to work behind a hood sash or safety shield while the apparatus is under vacuum.

Standard Scale and Microscale

Equipment and Supplies

> sublimation apparatus
> heat source
> cooling fluid
> flat-bladed spatula

Assemble one of the sublimation setups described or one suggested by your instructor. Powder the dry sublimand finely and spread it in a thin, uniform layer over the bottom of the sublimation container. If you are using a vacuum sublimator, attach the apparatus to a trap (if necessary) and vacuum source and then turn on the vacuum. Turn on the cooling water *or* partly fill the condensing tube or beaker with crushed ice or ice water. Heat the sublimation container with an appropriate heat source until some sublimate begins to collect on the condenser, then adjust the temperature to attain a suitable rate of sublimation without melting or charring the sublimand. As needed, add small pieces of ice to the condenser to replace melted ice. When the condenser is well covered with sublimate (but before sublimate crystals begin to drop into the sublimand), stop the sublimation and remove the sublimate as described in the next paragraph, and then resume sublimation. If the sublimand hardens or becomes encrusted with impurities, stop the sublimation, grind it to a fine powder, and then resume sublimation.

When all of the compound has sublimed or only a nonvolatile residue remains, remove the apparatus from the heat source, break the vacuum if you are using a vacuum sublimator, and let the apparatus cool. Carefully remove the condenser (avoid dislodging any sublimate) and scrape the crystals into a suitable tared container using a flat-bladed spatula.

Summary

1. Assemble sublimation apparatus; add sublimand.
2. Turn on vacuum if necessary; add cooling mixture or turn on cooling water.
3. Heat until sublimation begins; adjust heat to maintain good sublimation rate.
4. When sublimation is complete, stop heating, break vacuum if necessary, let cool.
5. Scrape sublimate into tared container.
6. Clean sublimation apparatus.

When Things Go Wrong

Suppose you carried out a vacuum sublimation using a sublimator like the one in Figure E5a, but the sublimate was wet when you removed it. You may have added ice to the inner tube before you turned on the vacuum, causing moisture to condense inside the apparatus. If your product seems pure except for the moisture, you may be able to dry it [OP-26b] without repeating the sublimation.

If you are carrying out a sublimation and the sublimand begins to melt, turn dark, or char, you are overheating it. Disassemble the apparatus, let the liquid (if any) in the sublimation container solidify, and break up the solid (which may have a hard crust) with a stirring rod. Then resume sublimation using a lower heat setting or a low-intensity heat source such as a steam bath.

Simple Distillation

a. Distillation of Liquids

A pure liquid in a container open to the atmosphere boils when its vapor pressure equals the external pressure, which is usually about 1 atm (760 torr, 101.3 kPa). The vapor contains the same molecules as the liquid, so its composition is identical to that of the pure liquid. A mixture of two (or more) liquids with different vapor pressures will boil when the total vapor pressure over the mixture equals the external pressure, but the composition of the vapor will be different than that of the liquid itself, being richer in the more volatile component (the one with the higher vapor pressure). If this vapor is condensed into a separate receiving vessel, the condensed liquid will have the same composition as the vapor; that is, it will also be richer in the more volatile component.

The process of purifying a liquid by vaporizing a liquid mixture in one vessel and condensing the vapors into another is called *distillation*. The liquid mixture being distilled, the *distilland*, can be heated in a boiling vessel— sometimes called the *pot*—using an apparatus such as the one shown in Figure E7 on a following page. The vapors are condensed on a cool surface such as the inside of a water-cooled condenser and the resulting liquid, the *distillate*, is collected in a suitable *receiver*. If the components of the distilland have sufficiently different vapor pressures, most of the more volatile component will end up in the receiver and most of the less volatile component(s) will remain in the pot.

The purity of the distillate increases with the number of vaporization–condensation cycles it experiences—the number of times it is vaporized and condensed—on its way to the receiver. *Simple distillation* involves only a single vaporization–condensation cycle. It is most useful for purifying a liquid that contains either nonvolatile impurities or small amounts of higher or lower boiling impurities. *Fractional distillation* [OP-32] allows for several vaporization–condensation cycles in a single operation. It can be used to separate liquids with comparable volatilities and to purify liquids that contain relatively large amounts of volatile impurities. *Vacuum distillation* [OP-31] is carried out under reduced pressure, which reduces the temperature of the distillation. It is used to purify high-boiling liquids and liquids that decompose when distilled at atmospheric pressure.

The distilland is usually a solution of two or more miscible liquids, but it may also be a liquid–solid solution. The distillation of immiscible liquids is discussed in OP-20.

Principles and Applications

To understand how distillation works, consider a mixture of two ideal liquids with ideal vapors, which obey both Raoult's law (Equation **1**) and Dalton's law (Equation **2**).

$$\text{Raoult's law: } P_A = X_A \cdot P_A^o \qquad \textbf{(1)}$$

$$\text{Dalton's law: } P_A = Y_A \cdot P \qquad \textbf{(2)}$$

In these expressions, P_A is the partial pressure of component A over the mixture, P_A^o is the equilibrium vapor pressure of pure A at the same temperature, X_A is the mole fraction of A in the liquid, and Y_A is the mole fraction of

A in the vapor. No real liquids obey these laws perfectly, so we shall consider the behavior of two imaginary hydrocarbons, *entane* (bp = 50°C) and *orctane* (bp = 100°C), that obey them both. If a mixture of entane and orctane is heated at normal atmospheric pressure, it will begin to boil at a temperature that is determined by the composition of the liquid mixture, producing vapor of a different composition. For example, an equimolar mixture of entane and orctane will start to boil at a temperature just above 66°C, and the vapor will contain more than 4 moles of entane for every mole of orctane. The liquid and vapor composition of an entane–orctane mixture at any temperature can be calculated using Equations **3** and **4**, which are derived from Dalton's law and Raoult's law.

Presumably, entane and orctane exist only in J. R. R. Tolkien's Middle-Earth, where they are used for fuel by Ents and Orcs, respectively.

$$X_A = \frac{P - P_B^o}{P_A^o - P_B^o} \qquad (3)$$

$$Y_A = \frac{P_A^o}{P} X_A \qquad (4)$$

P is the total pressure over the mixture, assumed here to be 1 atm (760 torr).

For example, at 70°C the vapor pressure of orctane is 315 torr and that of entane is 1370 torr (see Table E2), so the mole fraction of entane in a distilland that boils at 70°C will be (from Equation **3**),

$$X_{entane} = \frac{760 - 315}{1370 - 315} = 0.422$$

Its mole fraction in the vapor will be (from Equation **4**),

$$Y_{entane} = \frac{1370}{760} \times 0.422 = 0.761$$

showing that the vapor (and thus the distillate) is considerably richer in entane than is the liquid. The vapor pressures and approximate liquid–vapor compositions for entane and orctane at this and other temperatures are given in Table E2.

The key to an understanding of distillation is this: *The vapor over any mixture of volatile liquids contains more of the lower boiling component than does the liquid mixture itself.* So at any time during a distillation, the

Table E2 Equilibrium vapor pressures and mole fractions of entane and orctane at different temperatures

T, °C	Entane			Orctane		
	P°, torr	X	Y	P°, torr	X	Y
50°	760	1.00	1.00	160	0.00	0.00
60°	1030	0.67	0.90	227	0.33	0.10
70°	1370	0.42	0.76	315	0.58	0.24
80°	1790	0.24	0.57	430	0.76	0.43
90°	2300	0.11	0.32	576	0.89	0.68
100°	2930	0.00	0.00	760	1.00	1.00

Note: P° = equilibrium vapor pressure of the pure liquid; X = mole fraction in liquid mixture; Y = mole fraction in vapor.

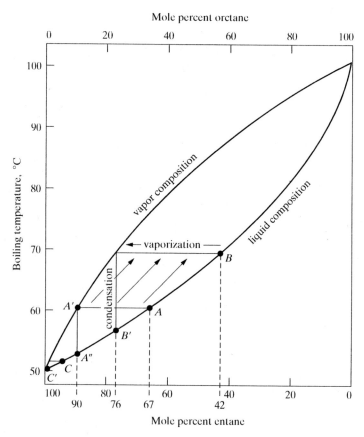

Figure E6 Temperature-composition diagram for entane–orctane mixtures

liquid condensing into the receiver contains more of the lower boiling component than does the liquid in the pot. As more of the lower boiling component distills away, the pot liquid becomes richer in the higher boiling liquid, so by the end of the distillation, most of the lower boiling liquid is in the receiver and most of the higher boiling liquid is in the pot.

The purification process is illustrated by Figure E6, in which the liquid and vapor compositions are plotted against the boiling temperatures of entane–orctane mixtures. Suppose we distill a mixture containing 2 moles of entane for every mole of orctane (67 mole percent entane). From the graph and Table E2, you can see that such a mixture will boil at 60°C (point A) and that its vapor will contain 90 mole percent (mol%) entane (point A′). Thus the distillate that is condensed from this vapor (point A″) will be much richer in entane than was the original mixture in the pot. As the distillation continues, however, the more volatile component will boil away faster and the pot will contain progressively less entane. Therefore, the vapor will also contain less entane and the boiling temperature will rise. When the percentage of entane in the pot has fallen to 42 mol% (point B), the boiling temperature will have risen to 70°C and the distillate will contain only 76 mol% entane (point B′). Only if the distillation were to be continued after nearly all of the entane had distilled would the distillate contain more of the less volatile component; at 90°C, for example, more than two-thirds of the molecules in the distillate would be orctane molecules.

This example shows that the purification effected by simple distillation of a mixture of volatile liquids may be very imperfect. In the example, the distillate never contains more than 90 mol% entane, and it may be considerably less pure than that, depending on the temperature range over which it is collected. If we start with a mixture containing only 5 mol% orctane (point C), however, considerably better purification can be accomplished. The initial distillate will be 99 mol% entane (point C′) at 51°C, and if the distillation is continued until the temperature rises to 55°C, the final distillate will be 95 mol% entane. The average composition of the distillate will lie somewhere between these values, so most of the orctane will remain in the boiling flask along with some undistilled entane, and the distillate will be relatively pure entane.

Thus simple distillation can be used to purify a liquid containing *small* amounts of volatile impurities if (1) the impurities have boiling points appreciably higher or lower than that of the liquid, and (2) the distillate is collected over a narrow temperature range (usually 4–6°C), starting at a temperature that is within a few degrees of the liquid's normal boiling point.

Experimental Considerations

Apparatus. The size of the glassware used for distillation should be consistent with the volume of the distilland; otherwise, excessive losses will occur. For example, suppose you recovered 20 mL of crude isopentyl acetate from Experiment 5 and decided to distill it from a 250-mL boiling flask over a 137–143°C boiling range. At the end of the distillation, the flask would be filled with 250 mL of undistilled vapor at a temperature of 143°C (416 K). An ideal gas law calculation shows that this is about 0.0073 mol of vapor.

$$n = \frac{PV}{RT} = \frac{(1.00 \text{ atm})(0.25 \text{ mL})}{(0.0821 \text{ L atm mol}^{-1} \text{ K}^{-1})(416 \text{ K})} = 0.0073 \text{ mol}$$

Isopentyl acetate has a molecular weight of 130 and a liquid density of 0.876 g/mol, so this is about 1.1 mL of the liquid.

$$0.0073 \text{ mol} \times \frac{130 \text{ g}}{1 \text{ mol}} \times \frac{1 \text{ mL}}{0.876 \text{ g}} = 1.1 \text{ mL}$$

Thus the vapor would condense to about 1.1 mL of liquid isopentyl acetate, which would not be recovered. By comparison, a 50-mL boiling flask will retain only one-fifth as much condensed vapor. Liquid can also be lost by surface adsorption on glass, so the lower surface area of the smaller boiling container will reduce losses as well. Additional liquid losses occur in the still head, condenser, and any other parts of the apparatus that can trap vapors or adsorb liquid.

Figure E7 illustrates a typical setup for standard scale simple distillation. The boiling flask should be no more than half full and should contain several boiling chips or a magnetic stir bar to prevent bumping [OP-7b]. A thermometer is used to monitor the temperature of the distillate vapors during a distillation, both to ensure that the right substance is being collected and to provide an estimate of its boiling point. The thermometer is inserted into the *still head* (a connecting adapter) through a stopper or thermometer

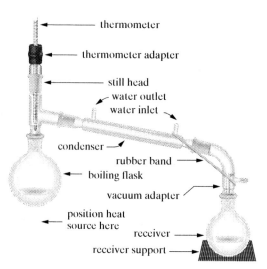

Figure E7 Conventional apparatus for standard scale simple distillation

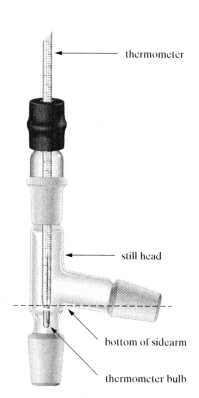

Figure E8 Thermometer placement in still head

Both flasks must be clamped or otherwise supported.

adapter. It should be well centered in the still head (not closer to one wall than another) with the entire bulb below an imaginary line extending from the bottom of the sidearm. In other words, the *top* of the bulb should be aligned with the *bottom* of the sidearm, as illustrated in Figure E8. In this way, the entire bulb will become moistened by the condensing vapors of the distillate. The placement of the thermometer is extremely important. If the thermometer is too high the temperature reading will be too low and much of the desired product may be discarded. If the thermometer is only a few millimeters below the sidearm its reading should not be affected, but if it is quite close to the boiling liquid its reading will be too high.

The condenser should have a straight inner section of comparatively small diameter. A West condenser of the type used for heating under reflux can also be used for distillation, but a jacketed distillation column should not be used for that purpose; its larger inner diameter affords less efficient condensation. The vacuum adapter (and sometimes the still head as well) should be secured to the condenser by a joint clip or a rubber band. An open container such as a tared vial can be used as a receiver unless the distillate is quite volatile or has vapors that might present a health hazard. In that case, a ground-joint flask that fits snugly on the vacuum adapter should be used. When a ground-joint receiver is used, the vacuum adapter sidearm must never be plugged up—otherwise, heating the system will build up pressure that could result in an explosion and severe injury from flying glass.

If you distill a comparatively small quantity of a liquid (~1–10 mL) in a conventional distillation apparatus, a large fraction of your product can be lost due to trapping of vapors and adsorption of liquid in the apparatus. You can reduce such losses by using the smallest available boiling flasks from a standard scale lab kit and assembling the compact distillation apparatus shown in Figure E9, in which the distillate is condensed by a cold bath surrounding the receiver rather than a West condenser. The vacuum adapter should be secured by a wire or joint clip, because a rubber band exposed to the heat of the vapors may break. If a bulky heat source is used, it may be

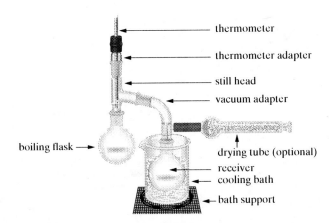

thermometer
thermometer adapter
still head
vacuum adapter
boiling flask
drying tube (optional)
receiver
cooling bath
bath support

Figure E9 Compact apparatus for standard scale simple distillation

necessary to twist the vacuum adapter and receiver at an angle to the still head to allow room for the heat source.

For the cooling bath to work properly, the ground-glass joint between the receiver (which cannot be an open container) and vacuum adapter must be tight, so the receiving flask should be clamped in place or secured by a joint clip. The type of coolant needed depends on the boiling point of the distillate. The following coolants should be suitable for the distillate boiling-point ranges indicated:

$<50°C$ ice–salt
$50–100°C$ ice/water
$101–150°C$ cold tap water
$>150°C$ no coolant if reveiver is well insulated from heat source

When an ice/water or ice–salt bath is used, the vacuum adapter sidearm should be connected to a drying tube [OP-27] so that moisture will not condense inside the receiver.

With the following qualifications, distillation with the compact apparatus is carried out as described in the general directions for standard scale simple distillation. (1) If vapors begin to come out of the vacuum adapter outlet, use a colder bath or add ice (or salt) to the one you are already using. (2) If you need to change receivers, stop the distillation and remove the heat source first.

In a distillation apparatus such as the one in Figure E7, the vapor has a long way to travel between pot and receiver. This increases the likelihood that distillate will be lost by sticking to the glass surfaces it encounters along the way or being diverted into dead ends such as the top of the still head. For a microscale distillation, the path between the pot and the receiver should be kept as short as is practicable. The *Hickman still* (also called a *Hickman head*) illustrated in Figure E10 is the most commonly used short-path receiver for microscale distillation. Vapors from the pot (usually a conical vial or small round-bottom flask) rise through the narrow neck of the still and collect in a *well* (also called a *reservoir*), which functions as the receiver. The well resembles a hollow doughnut sliced in half and has a capacity of 1–2 mL. Vapors of high-boiling liquids condense on the upper walls of the Hickman still, from which the condensed liquid drains into the well.

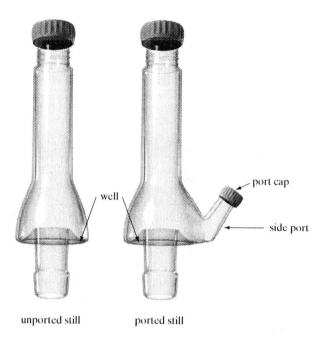

unported still   ported still

Figure E10 Hickman stills

For lower boiling liquids, a condenser must be attached to the top of the still (see Figure E11) to keep the vapors from escaping. The following rough guidelines tell you what kind of condenser you should use for a given distillate boiling point range *if* the distillation is conducted slowly:

<100°C   water-cooled condenser
100–150°C   air condenser
>150°C   no external condenser

Of course you can *always* use a water-cooled condenser, just to be on the safe side.

If, during a distillation, you smell any escaping distillate or see distillate droplets collecting near the top of the condenser (or the top of the still, with no external condenser), either slow the distillation rate or attach a more efficient condenser, or both. If you are distilling a high-boiling liquid, it is advisable to insulate the narrow neck of the still by wrapping it with glass wool, which can be held in place with aluminum foil.

Two types of Hickman still are illustrated in Figure E10. The ported still has a capped *side port*, allowing distillate to be withdrawn from the well with a Pasteur pipet and transferred to a tared screw-cap vial or another collecting vessel. To remove distillate from a ported still during the course of a distillation, carefully remove the port cap, transfer the liquid without delay, and quickly recap the port. It is important to avoid leaving the port uncapped for long, because distillate vapors will escape through it.

With an unported still, the distillate is collected by inserting a 9-inch Pasteur pipet through the top of the still. To remove distillate from an unported still during a distillation, you may need to remove the condenser and thermometer, if you are using them, and replace them immediately after you withdraw the liquid. Before you remove a condenser, raise the apparatus above the heat source so that the liquid in the pot stops boiling. If the Pasteur pipet doesn't reach all of the distillate, try attaching a short length of fine plastic tubing

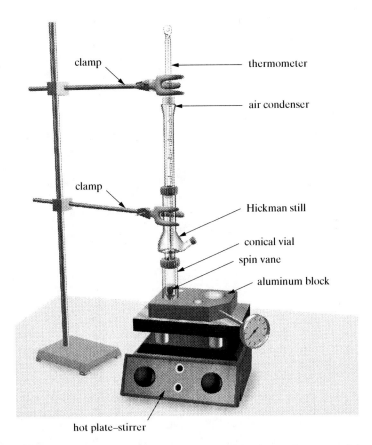

clamp — thermometer

clamp — air condenser

clamp — Hickman still

— conical vial
— spin vane
— aluminum block

hot plate–stirrer

Figure E11 Microscale distillation apparatus with air condenser and internal thermometer

(of the type used for microscale gas collection) to its capillary tip or bending the tip in a microburner flame. With a longer plastic tube attached to the pipet, you may be able to withdraw liquid without removing the condenser.

To measure the vapor temperature, a thermometer is inserted through the condenser (if there is one) into the Hickman still and held in place with a small clamp, as shown in Figure E11. It should *never* be secured by a thermometer adapter (unless the adapter's cap is removed), because that makes the distillation apparatus a closed system, which could shatter when heated and cause severe injury due to flying glass. The thermometer should be placed so that the *top* of its bulb is even with the *bottom* of the well, or slightly below it, as illustrated in Figure E12. The bulb should not touch the sides of the well.

Take Care! Never heat a closed system!

Heating. Almost any of the heat sources described in OP-7a can be used for simple distillation. Heating mantles and oil baths are preferred for standard scale distillation, because they provide reasonably constant, even heating over a wide temperature range. Such a heat source should be positioned so that it can be easily lowered and removed at the end of the distillation. For microscale simple distillation, the heat source is usually an aluminum block or a sand bath.

The heating rate should be adjusted to maintain gentle boiling in the pot and a suitable distillation rate. For most standard scale distillations, a distillation rate of 1–3 drops per second (about 3–10 mL of distillate per minute) is recommended. It is hard to count drops if you are using a Hickman still, but

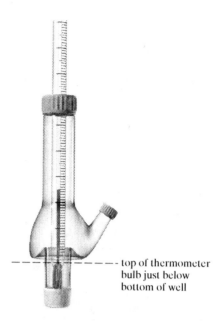

- - - - - - - - - - top of thermometer
bulb just below
bottom of well

Figure E12 Thermometer placement in Hickman still

a distillation rate sufficient to fill its well in 5–15 minutes or more should be suitable. Distilling at higher rates will decrease the purity of the product and, in microscale work, may make it difficult to record the boiling temperature accurately.

If you are distilling a liquid over a wide boiling range, it may be necessary to increase the heating rate gradually to maintain the same distillation rate. Otherwise the temperature at the heat source should be kept relatively constant throughout a simple distillation. If you are using an aluminum block or a sand bath as a heat source, you can monitor its temperature with an external thermometer to help you maintain a suitable heating rate. For example, if the temperature at the heat source is 50°C or so higher than the expected boiling point of the distilland, the heating rate is probably too high and you should reduce it. An excessive heating rate, in addition to reducing distillation efficiency, may cause mechanical carryover or decomposition of the distilland. On the other hand, an insufficient heating rate can cause the still-head temperature to fluctuate and distillation to slow down or stop altogether. Keep in mind that the temperature an external thermometer records will always be higher than the still-head temperature, so it is no substitute for an internal thermometer, which records the actual vapor temperature of the liquid being distilled.

Boiling Range.   An approximate boiling range for a distillation may be specified in an experimental procedure. For example, when isopentyl acetate is prepared by the reaction of isopentyl alcohol and acetic acid, the product may be collected over a boiling range of about 137°C to 143°C.

$$\underset{\substack{\text{acetic acid}}}{CH_3\overset{\displaystyle O}{\overset{\|}{C}}OH} + \underset{\substack{\text{isopentyl alcohol}\\\text{b.p. 130°C}}}{HOCH_2CH_2\overset{\displaystyle CH_3}{\overset{|}{C}HCH_3}} \longrightarrow \underset{\substack{\text{isopentyl acetate}\\\text{b.p. 142°C}}}{CH_3\overset{\displaystyle O}{\overset{\|}{C}}OCH_2CH_2\overset{\displaystyle CH_3}{\overset{|}{C}HCH_3}} + H_2O$$

Because the major impurity in this case is the more volatile isopentyl alcohol, the isopentyl acetate should distill below its normal boiling point of 142°C; using 143°C as the end of the range allows for experimental error. If the boiling range for a distillation isn't specified, you should collect the *main fraction* (the distilled liquid containing the desired component) over a relatively narrow boiling range, usually 4–6 °C, that brackets the boiling point of the desired component.

A liquid fraction that distills below the expected boiling range for the main fraction is called a *forerun*. It should be collected in a different container than the main fraction and saved until distillation is complete, then disposed of as directed. Sometimes, because of improper thermometer placement or an excessive heating rate, the internal thermometer will record a temperature lower than the actual vapor temperature. In that case, some or all of the liquid collected as forerun will actually be part of the expected main fraction, and should *not* be discarded. So if you collect more "forerun" than expected, save it and redistill it later after readjusting the heating rate or thermometer position. It may be advisable to redistill *all* of the liquid that distilled previously, since the rest of the distillate may also have been collected over the wrong temperature range.

General Directions for Simple Distillation

 Standard Scale

Equipment and Supplies

heat source
supports for heat source and receiver
clamps, ring stands
round-bottom flask
boiling chips *or* stir bar and magnetic stirrer
connecting adapter (still head)
thermometer adapter
thermometer
condenser
condenser tubing
vacuum adapter
receiver(s)
joint clip(s) or rubber band

For distilling ~10 mL of liquid or more, assemble [OP-2] the conventional distillation apparatus pictured in Figure E7. For distilling ~1–10 mL of liquid, assemble the compact apparatus pictured in Figure E9. Use the smallest round-bottom flask for which the liquid volume will be roughly half or less of the container's capacity; for example, use a 50–mL flask if the liquid volume is 13–25 mL and a 25 mL flask if it is less than 13 mL. Be sure to position the thermometer correctly in the still head (see Figure E8). If the liquid being distilled is quite volatile or hazardous, use a round-bottom flask as the receiver; otherwise you can use an open container (usually tared), such as an Erlenmeyer flask, graduated cylinder, or large vial. Remove the thermometer assembly and add the distilland to the boiling flask through a stemmed funnel, then drop in a few boiling chips or a stir bar. Replace the thermometer assembly and, if you are using the conventional apparatus, turn on the water to provide a slow but steady stream of cooling water. Note that the water should flow *in* the lower end of the condenser and *out* the upper end. If you are using the compact apparatus, immerse the receiver in a suitable cooling bath.

Start the stirrer (if you are using one) and turn on the heat source, adjusting the heating rate so that the liquid boils gently and a reflux ring of condensing vapors rises slowly into the still head. Shortly after the reflux ring reaches the thermometer bulb, the temperature reading should rise rapidly and vapors should begin passing through the sidearm into the condenser or vacuum adapter, coalescing into droplets that run into the receiving flask. As the first few droplets come over, the thermometer reading should rise to an equilibrium value and stabilize at that value. At this time, the entire thermometer bulb should be covered by a thin film of condensing liquid, which will drip off the end of the bulb into the pot. Record the temperature at which the thermometer reading stabilizes; if it is lower than expected, check the thermometer placement and adjust it if necessary. Distill the liquid at a rate of about 1–3 drops per second for the conventional apparatus or about 1 drop per second or less for the compact apparatus, monitoring the temperature frequently throughout the distillation.

If the initial thermometer reading is below the expected boiling range, carry out the distillation until the lower end of the range is reached, collecting the forerun in the receiver; then replace the receiver by another one. Try to make the switch quickly enough so that no distillate is lost. (Stop the distillation before changing receivers if you are using the compact apparatus.) If the initial thermometer reading is within the expected boiling range, there is no need to change receivers. Continue distilling until the upper end of the expected boiling range is reached or until only a small volume of liquid remains—enough to just moisten the bottom of the boiling flask.

Turn off and remove the heat source before the boiling flask is completely dry; heating a dry flask might cause tar formation or even an explosion. Disassemble and clean the apparatus as soon as possible after the distillation is completed.

Summary

1. Assemble distillation apparatus.
2. Add distilland and boiling chips or stir bar.
3. Turn on condenser water and stirrer (if used).
4. Start heating; adjust heat so that vapors rise slowly into still head.
5. Record temperature after distillation begins and thermometer reading stabilizes.
   IF initial temperature is below expected boiling range, GO TO 6.
   IF initial temperature is within expected boiling range, GO TO 7.
6. Collect forerun, change receivers at low end of boiling range.
7. Distill until temperature reaches high end of boiling range or until pot is nearly dry.
8. Turn off and remove heat source.
9. Disassemble and clean apparatus, dispose of forerun (if any) and residue in pot.

Microscale (Hickman-Still Method)

Equipment and Supplies

heat source
microclamps
conical vial or round-bottom flask
boiling chip *or* stirring device and magnetic stirrer
Hickman still
thermometer(s)
condenser (omit for high-boiling liquids)
collecting container(s)

Transfer the distilland to a clean, dry conical vial or round-bottom flask of appropriate size (the pot) and add a boiling chip, stir bar, or spin vane. Use the smallest container for which the liquid volume will be roughly half or less of the container's capacity; for example, use a 3-mL conical vial if the liquid volume is 1.5 mL or less. Assemble [OP-2] a distillation setup like the one shown in Figure E11. Depending on the boiling point of the liquid to be distilled, you can use a water-cooled condenser, an air condenser, or no condenser (see the *Apparatus* section). Clamp the apparatus securely to a ring stand with the pot positioned correctly in the heat source (usually a sand bath or heating block). Insert the thermometer through the condenser (if you are using one) and into the Hickman still so that its bulb is positioned correctly in the neck of the still (see Figure E12), and clamp it in place. If desired, place a thermometer in the heat source to monitor its temperature as well. If you are distilling a high-boiling liquid, it's a good idea to insulate the narrow neck of the Hickman still with glass wool, which can be held in place with aluminum foil. If you are using a water condenser, turn on the condenser water to provide a slow but steady stream of coolant. Note that the condenser water should flow *in* the lower end of the condenser and *out* the upper end.

Start the stirrer (if you are using one) and adjust the heat control so that the liquid boils *gently* and condensing vapors rise *slowly* into the neck of the Hickman still. If vapors begin rising rapidly into the still shortly after boiling begins, raise the apparatus above the heat source to reduce the heating rate; then carefully reposition the apparatus and readjust the heating rate for slow distillation (be careful not to bump the thermometer when you move the apparatus). Shortly after the vapors reach the thermometer bulb, the temperature reading should rise rapidly and liquid droplets should begin to appear on the sides of the still and drain into the well. During this time, the thermometer reading should rise to an equilibrium value and stabilize at that value, and the entire thermometer bulb should be covered by a thin film of condensing liquid that drips off the end of the bulb into the pot. Record the temperature when the reading stabilizes and continue to monitor it during the distillation.

If the initial thermometer reading is below the expected boiling range, carry out the distillation until the lower end of the range is reached, and then transfer any liquid in the well to a small beaker or another container (save this forerun, if any, for later disposal or possible redistillation). The distillate should collect in the well slowly, taking 5–15 minutes (or even more) to fill the well completely, if there is enough liquid to fill it. If it distills more rapidly than this, especially if the temperature reading never stabilizes or stabilizes at the wrong temperature, return the distillate to the pot and redistill it more slowly. If the well becomes filled (or nearly so) with distillate, use a Pasteur pipet to transfer it to a collecting container, such as a tared screw-cap vial. (If you don't collect the distillate when the well is full, the excess will simply overflow and drip down into the pot.) Continue distilling and transferring the distillate as needed until the upper end of the expected boiling range is reached *or* until only a drop or so of liquid remains in the pot.

Turn off the heat and raise the apparatus away from the heat source before the pot is completely dry. Let the apparatus cool down, then transfer any remaining distillate to the collecting container. You should be able to recover more distillate by tilting the Hickman still so that the liquid flows to one side of the well as you collect it. Gently tapping the sides of the still or cooling it with air or cold water may cause more distillate to collect in the well. You can also rinse the still with a volatile solvent in which the distillate is soluble, and then evaporate [OP-19] the solvent. Disassemble and clean the apparatus as soon as possible after the distillation is completed.

Summary

1. Add distilland and boiling chips or stirring device to pot.
2. Assemble distillation apparatus.
3. Turn on condenser water and stirrer (if used).
4. Start heating; adjust heating rate so that vapors rise slowly into Hickman still.
5. Record temperature after distillation begins and thermometer reading stabilizes.
   IF initial temperature is below expected boiling range, GO TO 6.
   IF initial temperature is within expected boiling range, GO TO 7.
6. Remove forerun when temperature reaches low end of boiling range.
7. Distill until temperature reaches high end of boiling range or until pot is nearly dry, transferring distillate to collector as well fills up.
8. Stop heating and let apparatus cool.
9. Transfer any remaining distillate to collector.
10. Disassemble and clean apparatus, dispose of forerun (if any) and residue in pot.

When Things Go Wrong

Note that the substance being distilled can be any liquid you're trying to purify, not just a liquid reaction product

If the liquid in the flask is boiling but the thermometer records a temperature well below your product's boiling point, don't worry. The temperature is low because the product's vapors haven't reached the still head yet. When they do, you'll see the temperature rise rapidly.

If a substantial amount of liquid has distilled but the thermometer at the still head records a temperature below the expected boiling range, the thermometer is probably positioned incorrectly. Refer to Figure E8 (standard scale) or Figure E12 (microscale) and move the thermometer farther down (or rarely, farther up) if it isn't positioned as shown in the illustration.

Then return all of the distilled liquid to the pot and begin the distillation again. If the thermometer was originally positioned correctly, your product must contain a substantial amount of low-boiling impurity (forerun). Continue distilling until the temperature nears the expected boiling range, transfer the forerun to a separate container or change receivers, and begin collecting the main fraction when the temperature reaches the lower end of its boiling range.

If, after the main fraction begins distilling, the temperature drops below it's boiling range, either (1) the heating rate isn't high enough to maintain distillation, or (2) all of the main fraction that can distill has distilled. Try increasing the heat setting to see whether the temperature will rise again (vapors should be rising into the still head in this case). If raising the heat over a period of several minutes has no effect on the still head temperature and vapors are no longer rising toward the still head, the distillation is over; continue as described in the General Directions.

Suppose your standard-scale distillation is going well and product is dripping into the receiver at a gratifying rate when suddenly the vacuum adapter slips off the condenser and goes crashing onto the desk top—along with the receiver and the product you distilled up to then. Because you failed to secure your vacuum adapter to the apparatus, you have just lost a battle with gravity. There isn't much you can do to get out of this situation gracefully; trying to recover a spilled liquid from a flat (and probably highly contaminated) surface is like trying to shovel water out of a flooded basement. You will probably have to rely on whatever liquid is left in the pot, or start over.

Suppose you carried out a distillation until all (or nearly all) of the liquid in the pot was gone, but later noticed some liquid in the pot that wasn't there before. That liquid condensed from the vapors that didn't make it into the receiver. Trying to get the undistilled liquid into the receiver by boiling it again is pointless, because its vapors will only condense into the pot again when the apparatus cools. You might recover more product by using a *chaser* as described in OP-32, but that's usually not worth the trouble for a simple distillation.

b. Distillation of Solids

A Hickman still distillation setup can be used for the distillation of small amounts of low-melting solids. The apparatus is constructed as illustrated in Figure E11, except that no condenser is attached, and the solid—which liquefies when the pot is heated—is distilled as described in the General Directions for distilling liquids. Most low-melting solids will not crystallize during or immediately after distillation and can be transferred to a suitable container with a Pasteur pipet while the distillation apparatus is still warm.

If the solid does crystallize in a Pasteur pipet, it can be melted with a heat gun or dissolved in a volatile solvent. If the solid crystallizes in the well of a Hickman still, it can be dissolved by pipetting a small amount of a volatile solvent, such as diethyl ether or dichloromethane, into the well. The sides of the well should be rinsed down with the solvent to dissolve any adherent solid. The solution is then removed with a Pasteur pipet and the solvent is evaporated to recover the solid. Alternatively, the solid in the well can be melted with a heat gun or heat lamp and then transferred.

Larger quantities of low-melting solids can be distilled using the apparatus pictured in Figure E9 or with a special apparatus designed for that purpose. Unless the solid has a boiling point below 150°C, a cooling bath should not be necessary. Use a vacuum adapter with the drip tip (outlet tube) removed, if any are available; otherwise, have a heat gun or heat lamp handy to melt any solid that forms in the drip tip. When distillation is complete, melt the solid by heating the receiving flask and transfer it to a tared vial or another container. The last traces of solid can be transferred with a small amount of diethyl ether or another volatile solvent, which is then evaporated under a hood.

A "cool" burner flame can be used if no flammable liquids are in the vicinity.

OPERATION **31**   Vacuum Distillation

Principles and Applications

The boiling point of a liquid decreases when the external pressure is reduced, so under reduced pressure a liquid distills at a temperature that is lower than its normal boiling point. For example, a liquid that boils at 200°C at a pressure of 1 atm (760 torr) will boil near 100°C at 25 torr, as shown in Table E3.

Reducing the boiling point of a liquid reduces the likelihood that thermal decomposition and other high-temperature reactions will occur during distillation. As a rule, most liquids that are heat sensitive or have boiling points of 200°C or higher at atmospheric pressure should be purified by distillation under reduced pressure, which is also called *vacuum distillation*.

Vacuum distillation has certain inherent features that can potentially cause hazards and experimental difficulties. According to Boyle's law ($V \propto 1/P$), the volume of vapor generated by boiling a given amount of liquid increases when the pressure is reduced. For example, a vapor bubble that occupies a volume of 1 μL (0.001 mL) at 760 torr will expand to 0.3 mL at 25 torr. This can cause excessive bumping in the boiling flask and mechanical carryover of liquid to the receiver. Special devices such as capillary bubblers and microporous boiling chips are used to maintain smooth boiling. Carryover can be reduced by interposing a Claisen adapter between the pot and still head for a standard scale vacuum distillation, or by using the largest available boiling flask for a microscale vacuum distillation.

Reducing the pressure in a distillation apparatus also increases the vapor velocity because there are fewer molecules around to bump into. This can cause superheating of the vapor at the still head or in the neck of a Hickman still and produce a pressure differential throughout the system, making the observed distillation temperature too high and the pressure reading too low.

These problems can be circumvented by using the right kind of apparatus and by carrying out the distillation slowly. Even then, the separation attainable under vacuum distillation doesn't equal that possible at

Table E3 Approximate boiling points of liquids at 25 torr

Normal bp	bp at 25 torr
150°C	60°C
200°C	100°C
250°C	140°C
300°C	180°C

atmospheric pressure, and there is always the danger that the apparatus may implode because of the unbalanced external pressure on the system. Accordingly, a vacuum distillation must be performed with great care and attention to detail to obtain satisfactory results and prevent accidents.

Experimental Considerations

Apparatus. Standard scale and microscale setups for vacuum distillation are shown in Figures E15 and E16 in the General Directions sections. The glassware used to assemble them must be free of cracks, star fractures, and other imperfections that might cause the apparatus to shatter under reduced pressure. All rubber connecting tubing should be thick-walled to prevent collapse under vacuum and as short as possible to reduce the pressure differential between the vacuum source and the system. The rubber tubes should be stretched or bent to see that they are pliable and free of cracks. Connections between rubber and glass, as well as ground-joint connections, must be secure and airtight. If the distilland may contain a volatile impurity, at least two receivers should be available to collect it and the main fraction. It is usually necessary to remove the heat source and break the vacuum before changing receivers. Although commercially available rotating receivers (sometimes called "cows" because of their udderlike appearance) make it possible to switch receivers without breaking the vacuum, they are seldom available in undergraduate laboratories.

Before assembly of a standard scale apparatus such as the one shown in Figure E15, the ground glass joints should be lubricated with vacuum grease to prevent leaks (be sure to clean the joints afterward as described in OP-2). The compact apparatus for simple distillation pictured in Figure E9 of OP-30 can be used under reduced pressure if the vacuum adapter sidearm is connected to a vacuum line through a trap, as illustrated in Figure E15. Except that a cooling bath is used instead of condenser water, the procedure is essentially the same as that used for a conventional vacuum distillation. It may be necessary to insert a Claisen adapter between the boiling flask and the still head to prevent mechanical carryover of the distilland or to allow for insertion of a bubbler.

Although vacuum distillation can be used successfully in some microscale experiments, it isn't practicable for liquid volumes much less than 0.5 mL. The apparatus for microscale vacuum distillation is similar to that for simple distillation except that a Claisen adapter or multipurpose adapter is needed to connect it to the vacuum source. Vacuum grease should ordinarily *not* be used on ground joints held together by compression caps, because the O-ring should provide a tight seal. Ungreased joints that are under vacuum for some time may freeze, however, so your instructor may suggest that you grease them lightly (see OP-2) to prevent this (be sure to clean the joints after use). All O-rings should be inspected to see that they are flexible and free of cracks or nicks that might allow air to enter. Microporous boiling chips are generally used to reduce bumping, but a stir bar may work for relatively large quantities of distilland. To help prevent mechanical carryover of liquid, it is advisable to use the largest distilling vessel available; thus a 10-mL round-bottom flask is preferable to a conical vial. If you are distilling a high-boiling liquid, it's a good idea to insulate the narrow neck of the Hickman still by wrapping it with glass wool, which can be held in place with aluminum foil.

Vacuum Sources.   Most organic chemistry laboratories are provided with either water aspirators or vacuum lines that are connected to a remote vacuum pump. In principle, an aspirator should be able to attain a reduced pressure equal to the vapor pressure of the water flowing through it, which is a function of the water temperature, as shown in Table E4. In practice, the pressure is often 5–10 torr higher because of insufficient water pressure, leaks in the system, or deficiencies of the aspirator itself. Thus with cold tap water at a temperature of 14°C running through it, a typical aspirator may provide a vacuum of around 20 torr. Because the water flow rate in your laboratory probably varies with the number of people using aspirators, the pressure will also vary as aspirators are turned on and off. An aspirator must be provided with a trap to prevent backup of water into the receiving flask during changes of water pressure and to reduce pressure fluctuations throughout the system. A thick-walled filter flask of the largest convenient size makes a good trap, although a Pyrex glass bottle wrapped with plastic tape (see OP-16) may also be adequate. The trap should be provided with a pressure release valve and hooked up to the aspirator and distillation apparatus as illustrated in Figure E15.

As a rule, a vacuum pump can attain a much lower pressure than an aspirator, and the pressure should not vary significantly. However, lab systems with vacuum lines connected to a central vacuum pump usually provide pressures that are comparable to those attained by water aspirators. All vacuum pumps must be protected from condensable vapors that might damage them. A central vacuum system should have its own purge assembly to remove such vapors, but it's not a bad idea to interpose a cold trap between the vacuum line and your apparatus, just in case the purge assembly is inadequate or poorly maintained. A cold trap can be assembled by immersing an ordinary aspirator trap (or a commercial cold trap) in an ice–salt bath, dry ice in acetone, or some other efficient coolant, and clamping the trap to a ring stand. A cold trap should always be used with any free-standing vacuum pump that isn't equipped with a purge system.

Pressure Measurement.   To know with any certainty when the desired substance is distilling, you need to know (or be able to estimate) the pressure inside the vacuum distillation system so that you can estimate the boiling point of the desired component at that pressure. Devices called *manostats* can be used to maintain a desired pressure in a system,

Take Care! Never use a thin-walled container, such as an Erlenmeyer flask, as a trap; it may shatter under vacuum.

Table E4 Vapor pressure of water below 30°C

T, °C	P, torr	T, °C	P, torr
30	31.8	18	15.5
28	28.3	16	13.6
26	25.2	14	12.0
24	22.4	12	10.5
22	19.8	10	9.2
20	17.5	8	8.0

but they are seldom available in undergraduate laboratories. More often you must work with the pressure provided by your aspirator or other vacuum system and use a *manometer* to measure that pressure. For example, if the closed-end manometer shown in Figure E13 is attached to an evacuated system, the pressure (in torr) inside the system will be equal to the vertical distance (Δh, in millimeters) between the mercury levels in the inner tube and the cylinder. Another kind of closed-end manometer is shown in Figure E15; its pressure reading (in torr) is the difference in height (in millimeters) between the mercury levels in the two arms of the bent tube. The mercury in a manometer can present a hazard if the vacuum is broken suddenly—air rushing in can push the mercury column forcefully to the closed end of the tube, breaking it and releasing toxic mercury into the laboratory. For this and other reasons, the vacuum must always be released *slowly*. It is advisable to open the valve connecting a manometer to an evacuated system only while a pressure reading is being made.

When the boiling points of any impurities are quite different from that of the main fraction, you may get by without a manometer if you can make a rough estimate of the pressure. For example, if you are using a water aspirator, you can measure the temperature of the water, estimate its vapor pressure from Table E4, and estimate the minimum boiling temperature of the liquid as described in the next paragraph. The product will probably distill somewhat *above* that temperature; its actual boiling range will depend on the efficiency of the aspirator and the air-tightness of your apparatus. You can also use a vacuum gauge to measure the pressure at the outlet of an aspirator or vacuum line, keeping in mind that the pressure in your distillation apparatus will be somewhat higher than the measured pressure.

Boiling Points under Reduced Pressure.
If the boiling point of a substance at a given pressure is not known, it can be estimated using the vapor pressure–temperature nomograph shown in Figure E14. More precise estimates can be made using tables such as those in R. R. Dreisbach, *Pressure–Volume–Temperature Relationships of Organic Compounds* (New York: McGraw-Hill, 1952), or by using various empirical relationships. The main fraction should be collected over a range that brackets the expected boiling point, keeping in mind that the distillation temperature of a liquid may vary by 10°C or more under vacuum because of pressure fluctuations and other factors.

Heat Sources.
The heat source should be capable of providing constant, uniform heating to prevent bumping and superheating and to maintain a constant distillation rate. For a typical standard scale vacuum distillation, a good heating mantle or an oil bath should be suitable. A microscale vacuum distillation can be carried out using a heating block or sand bath. (See OP-7a for additional information about the use of these heat sources.)

Smooth-Boiling Devices.
You can reduce bumping and foaming during a vacuum distillation by using microporous boiling chips, a magnetic stirring device, or (for standard scale vacuum distillation) a bubbler. A standard

A manometer measures the pressure inside a system; a manostat controls it.

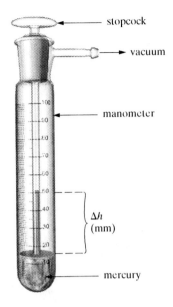

Figure E13 A closed-end manometer

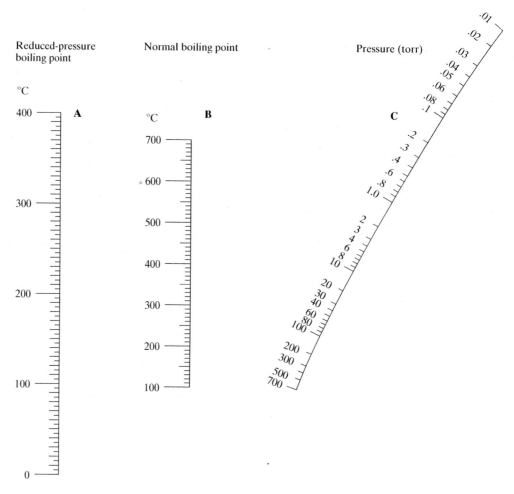

To estimate the boiling point at pressure P given the boiling point at another pressure P': (a) connect pressure P' in **C** with the boiling point at that pressure in **A** using a ruler, and place a sharp pencil point (or the point of a pin) where the ruler intersects line **B**; (b) pivot the ruler around the point until it reaches the desired pressure (P) in **C**, and then read the boiling point at that pressure from **A**.

Example: To estimate the boiling point of dibutyl phthalate at 10 torr from its reported boiling point of 236°C at 40 torr, place a ruler at 40 torr in **C** and 236° in **A**, causing it to intersect line **B** at about 345° . Then hold a pencil point at 345° on **B**, pivot the ruler about that point to 10 torr in **C**, and read the boiling point from **A**. This yields an estimated boiling point of 197°C at 10 torr.

Figure E14 Reduced-pressure boiling-point nomograph

scale lab kit may contain a narrow-tipped tube with a very fine hole at the outlet, which can be used as a bubbler. Otherwise, you can construct a flexible capillary bubbler, about as fine as a cat's whisker, by drawing it from a length of *thick-walled* capillary tubing (capillary bubblers drawn from thin-walled tubing break easily). See your instructor for help in constructing a capillary bubbler. A short rubber tube with a screw clamp should be placed at the top of either kind of bubbler to control the rate of bubbling. The bubbler is inserted through a thermometer adapter (or a rubber stopper, if necessary) so that its tip extends to within a millimeter or two of the bottom of the boiling flask. Under vacuum, this device should deliver a very fine stream of air bubbles, preventing development of the large bubbles that cause bumping. However, a bubbler has several drawbacks. Air entering the system raises the pressure slightly and may oxidize the product at high temperatures, and a capillary bubbler may plug up when the vacuum is broken. Nevertheless, a well-constructed capillary bubbler works better than any other smooth-boiling device.

General Directions for Vacuum Distillation

Safety Notes

> **Because of the possibility of an implosion, safety glasses *must* be worn during a vacuum distillation. Work behind a hood sash or safety shield while the apparatus is under vacuum. A rapid pressure increase, accompanied by a thick fog in the distilling flask, indicates decomposition of the distilland. If this occurs, remove the heat source *immediately* and get away (warning others to do so also) until the flask cools. Report the incident to your instructor.**

 Standard Scale

Before beginning a vacuum distillation you should make a rough estimate of the boiling temperature of your sample under vacuum, or be ready to estimate it using the nomograph in Figure E14 once you obtain a manometer reading. Italicized instructions apply only if you are using a manometer; otherwise, disregard them.

Equipment and Supplies

heat source
rings, clamps, supports
round-bottom flask
bubbler or other smooth-boiling device
Claisen adapter
connecting adapter (still head)
stopper (if bubbler is not used)
thermometer
thermometer adapter
condenser
vacuum adapter
receiving flask(s)
joint clips (or wire, etc.)
condenser tubing
thick-walled rubber tubing
screw clamps

trap with pressure-release valve
manometer (optional)
glass tee (optional)

Inspect all glassware and rubber tubing and replace any damaged items; if you have any doubt about their condition, see your instructor. For distilling ~10 mL of liquid or more, assemble [OP-2] the conventional vacuum-distillation apparatus as pictured in Figure E15. Be sure that the thermometer is placed correctly as shown in Figure E8 of OP-30. For distilling ~1–10 mL of liquid, assemble the compact apparatus pictured in Figure E9 but connect it to the rest of the system as shown in Figure E15; interpose a Claisen head between the boiling flask and still head if you will be using a capillary bubbler or if carryover may be a problem. Apply vacuum grease at all joints as described in OP-2. If you are not using a manometer, omit the 3-way tubing connector shown. Connect the vacuum adapter to a trap having a pressure-release valve, and connect the trap to the vacuum source. If you are using a capillary bubbler, provide it with a screw clamp on a short length of rubber tubing to control the bubbling rate. If you are using microporous boiling chips or a stir bar to prevent bumping, drop them into the reaction flask and insert a ground-glass stopper into the short arm of the Claisen adapter (if your apparatus has one). If you are using a vacuum line rather than an aspirator, cool the trap if directed to do so by your instructor.

Make sure that all joints and connections are tight; then add the liquid to be distilled through a funnel. (If the liquid contains a volatile solvent, remove it first by distilling at atmospheric pressure, then let the apparatus cool before proceeding.) If you are using a bubbler, open its screw clamp a turn or two. Raise the heat source into position, but do not

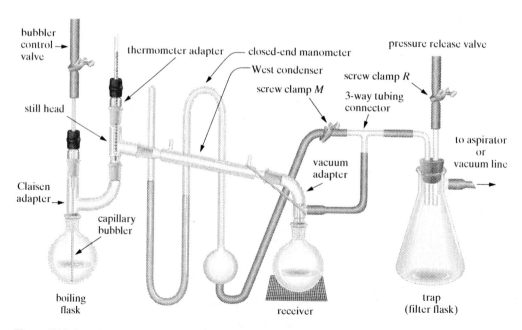

Figure E15 Standard scale apparatus for vacuum distillation

begin heating yet. Turn on the condenser's cooling water (if you are using one), open screw clamp R, and turn on the vacuum fully. Slowly close clamp R and adjust the screw clamp on the bubbler (if you are using one) so that it emits a fine stream of bubbles. (If bumping and foaming occur, there may be some residual solvent in the distilland; open clamp R slowly and then close it down to a point at which the solvent will evaporate without excessive bumping. If a rubber tube collapses under vacuum, open clamp R slowly to break the vacuum and replace it with sturdier tubing.) *Open clamp M, wait a minute or two until the pressure equilibrates, then read the manometer and close clamp M. If the observed pressure is more than ~10 torr above the estimated pressure, check the system for leaks caused by loose joints, cracked tubing, and so on. If you find any leaks, release the vacuum by slowly opening clamp R and fix them. If the pressure is satisfactory, use the nomograph shown in Figure E14 to estimate the boiling range at that pressure.*

If you are using a magnetic stirrer, turn it on. Begin heating to bring the mixture to the boiling point. If you are using a bubbler, adjust the bubbler clamp as necessary to maintain a very fine stream of bubbles as the temperature rises. If excessive foaming occurs on boiling, reduce the heating rate. If bubbles form inside any joint during the distillation, that joint is leaking air into the system; remove the heat source, release the vacuum, and regrease the joint or replace the glassware part by another one. When liquid begins to condense into the receiver, record the temperature reading. *Open clamp M, let the pressure equilibrate, and record the pressure reading; then close clamp M and leave it closed except when you need to make another pressure reading.* Adjust the heat source so that a distillation rate of about one drop per second or less is attained. If the temperature at the still head jumps up or down while the liquid is distilling, the pressure may be fluctuating because of changes in the aspirator flow rate; adjust the heating rate as necessary to maintain a suitable distillation rate. (To minimize such fluctuations, only a few students should use aspirators at the same time.)

If the initial distillation temperature is markedly lower than the estimated boiling range for the product, you are probably distilling a volatile forerun. Continue distilling until the temperature reaches the low end of the expected boiling range. Then change receivers without turning off the vacuum by the following procedure:

1. Lower the heat source and let the system cool down.
2. Open the bubbler clamp (if you are using one), then open clamp R slowly until the system is at atmospheric pressure.
3. Replace the receiver with another one. If you are using microporous boiling chips, add another chip or two.
4. Slowly close clamp R and adjust the bubbler.
5. Heat until distillation resumes.
6. Record the boiling temperature.

A capillary bubbler will sometimes plug up when the vacuum is broken; if that happens, you will have to replace it with a new one. Continue distilling until the upper end of the estimated boiling range is reached or until a significant drop in temperature indicates that the product is completely distilled. Stop the distillation before the boiling flask is completely dry.

When the distillation is complete, follow steps 1 and 2 for bringing the system back to atmospheric pressure, then turn off the vacuum. Disassemble the apparatus and clean the glassware promptly. Dispose of any residue and forerun as directed by your instructor.

Summary

1. Inspect glassware and tubing, assemble apparatus, check connections and joints.
2. Add distilland and smooth-boiling device.
3. Position heat source, start cooling water, open clamp R, turn on vacuum.
4. Close R, let pressure equilibrate.
5. *Open M, read pressure, close M, estimate boiling range.*
6. Heat until distillation begins, adjust bubbler (if used).
7. Record temperature *and pressure.*
   IF temperature is below estimated range, GO TO 8.
   IF temperature is within estimated range, GO TO 12.
8. Adjust heat and distill until temperature reaches lower end of expected boiling range.
9. Lower heat source, let cool, open bubbler clamp and clamp R.
10. Change receiver, close R.
11. Adjust bubbler (if used), heat until distillation resumes, record temperature.
12. Distill until upper end of temperature range is attained or only a little distilland remains.
13. Lower heat source, let cool, open R.
14. Turn off vacuum, remove distillate.
15. Disassemble and clean apparatus, dispose of residue and forerun.

 Microscale

Before beginning a vacuum distillation, you should have made a rough estimate of the boiling temperature of your sample under vacuum, or be ready to estimate it using the nomograph in Figure E14 once you obtain a manometer reading. Italicized instructions apply only if you are using a manometer; otherwise, disregard them.

Equipment and Supplies

heat source
clamps, supports
boiling flask
smooth-boiling device
Hickman still
*Claisen adapter
thermometer
*2 thermometer adapters (or 1 adapter and a 1-hole rubber stopper)
*bent glass tube (8 mm o.d.)
thick-walled rubber tubing
screw clamps
trap with pressure-release valve

manometer (optional)

3-way tubing connector (optional)

*These parts can be replaced by a multipurpose adapter, which is provided in some microscale lab kits.

Inspect all glassware, O-rings, and rubber tubing and replace any damaged items; if you have any doubt about their condition, see your instructor. Add the liquid to be distilled and a few microporous boiling chips (or a stir bar) to the boiling flask, and then assemble the apparatus illustrated in Figure E16. If you are using a multipurpose adapter, connect it directly to the Hickman still. If you are using a Claisen adapter and have only one thermometer adapter, use it for the thermometer and insert the bent glass tube into a

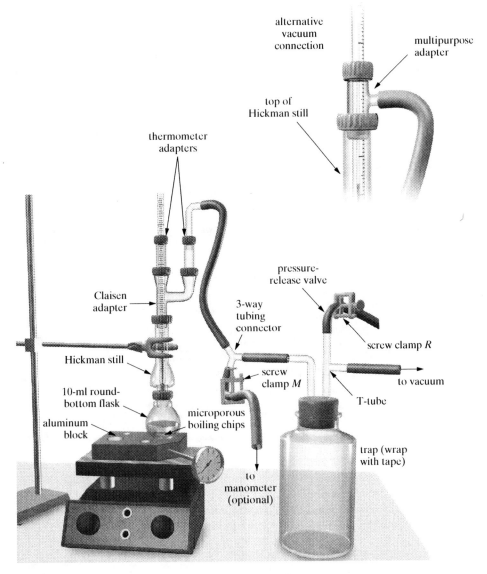

Figure E16 Microscale apparatus for vacuum distillation

one-hole #0 rubber stopper (it must fit snugly). Make sure that the top of the thermometer bulb is just below the well of the Hickman still, as shown in Figure E12 of OP-30. If a manometer is *not* used, omit the 3-way tubing connector shown. Connect the apparatus to a trap having a pressure-release valve, and connect the trap to the vacuum source. If you are using a vacuum line rather than an aspirator, cool the trap if directed to do so by your instructor.

Make sure that all joints and connections are tight. (If the liquid contains a volatile solvent, remove it by distilling at atmospheric pressure and let the apparatus cool before proceeding.) Open screw clamp R and turn on the vacuum fully. Slowly close clamp R. (If a rubber tube collapses, replace it with heavy-walled tubing after releasing the vacuum. If bumping and foaming occur, there may be some residual solvent in the distilland; open clamp R and then close it down to a point at which the solvent will evaporate without excessive bumping.) *Open clamp M, wait a minute or two until the pressure equilibrates, then read the manometer and close clamp M. If the observed pressure is more than ~10 torr above the estimated pressure, check the system for leaks caused by loose joints, cracked tubing, and so on. If you find any, release the vacuum by opening clamp R and fix them. If the pressure is satisfactory, use the nomograph shown in Figure E14 to estimate the boiling range at that pressure.*

If you are using a magnetic stirrer, turn it on. Begin heating to bring the mixture to the boiling point. If excessive foaming occurs upon boiling, reduce the heating rate. If bubbles form inside any joint during the distillation, that joint is leaking air into the system; remove the heat source, release the vacuum by opening clamp R, and replace the O-ring or (if necessary) the glassware part containing the joint. When liquid begins to appear in the well of the Hickman still, record the temperature reading. *Open clamp M, let the pressure equilibrate, and record the pressure reading; then close clamp M and leave it closed except when you need to make another pressure reading.* Adjust the heat control to maintain a slow distillation rate without excessive bumping. If the temperature jumps up or down while the liquid is distilling, the pressure may be fluctuating owing to changes in the aspirator flow rate; adjust the heating rate as necessary to maintain a suitable distillation rate. (To minimize such fluctuations, only a few students should use aspirators at the same time.)

If the initial distillation temperature is markedly lower than the estimated boiling range for the product, you are probably distilling a volatile forerun. Continue distilling until the temperature reaches the low end of the expected boiling range. Then remove the forerun by the following procedure:

1. Raise the apparatus away from the heat source and let the system cool down.
2. Open clamp R slowly until the system is at atmospheric pressure, then turn off the vacuum.
3. Remove the liquid in the well of the Hickman still with a Pasteur pipet. If you are using microporous boiling chips, add another chip.
4. Turn on the vacuum and slowly close clamp R.
5. Heat until distillation resumes.
6. Record the boiling temperature.

Continue distilling until the upper end of the estimated boiling range is reached or until a significant drop in temperature indicates that the product is completely distilled. If the well of the Hickman still becomes full of liquid (or if you will be collecting a higher-boiling fraction), remove it by again following Steps 1–6. Stop the distillation before the boiling flask is completely dry.

When the distillation is completed, follow steps 1 and 2 for bringing the system back to atmospheric pressure; then turn off the vacuum and remove any distillate. Disassemble the apparatus, clean the glassware promptly, and dispose of any residue and forerun as directed by your instructor.

Summary

1. Inspect glassware and tubing, add distilland and smooth-boiling device.
2. Assemble apparatus, check connections and joints.
3. Open clamp R, turn on vacuum.
4. Close R, let pressure equilibrate.
5. *Open M, read pressure, close M, estimate boiling range.*
6. Heat until distillation begins.
7. Record temperature *and pressure.*
   IF temperature is below estimated range, GO TO 8.
   IF temperature is within estimated range, GO TO 12.
8. Adjust heat and distill until temperature reaches lower end of expected boiling range.
9. Raise apparatus above heat source, let cool, open clamp R.
10. Remove distillate, close R.
11. Heat until distillation resumes, record temperature.
12. Distill until upper end of temperature range is attained or only a little distilland remains.
13. Raise apparatus above heat source, let cool, open R.
14. Turn off vacuum and remove distillate.
15. Disassemble and clean apparatus, dispose of residue and forerun.

When Things Go Wrong

Most of the things that go wrong during a simple distillation can also go wrong during a vacuum distillation, so you can refer to *When Things Go Wrong* in OP-30 for help. The following cases apply only to vacuum distillation.

Suppose you are performing a vacuum distillation using an aspirator (no manometers are available), but when liquid begins to distill the still-head temperature is well above the estimated boiling point at 20 torr. Probably the temperature is higher than expected because the aspirator isn't operating efficiently or because there are leaks in your system. Check for leaks by looking for bubbles that form around the inside of joints under vacuum, and repair any leaks as described in the General Directions. If there are no leaks, continue collecting the liquid until the temperature either rises or drops markedly. You can check to make sure your product is the right one by obtaining an IR spectrum or analyzing it in some other way.

If severe bumping occurs in the pot during a vacuum distillation, your smooth-boiling device isn't working properly. (See *Smooth Boiling Devices.*) Depending on the device you are using, either prepare a new bubbler, replace boiling chips with fresh ones, or stir faster, using a larger stir bar if necessary. The stir bar should rotate smoothly, without hopping or wobbling.

OPERATION 32

Fractional Distillation

Principles and Applications

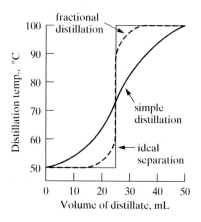

Figure E17 Separation efficiency of simple and fractional distillations

Although simple distillation can purify organic liquids that contain small amounts of volatile impurities, it isn't a very effective means of separating the components of a mixture when each component makes up a substantial fraction of the mixture, unless the boiling points of the components are far apart. With mixtures of closer boiling liquids, the distillate composition and boiling point will change continually throughout the distillation as illustrated in Figure E17, and most of the distillate will be a relatively impure mixture of the components.

The separation could be improved by redistilling portions of the initial distillate and subsequent distillates. Such a process is diagrammed in Figure E18, in which the initial distillate is delivered directly to a second distilling flask, which redistills the condensed vapors and delivers its distillate to a third distilling flask, which distills that liquid into a receiver. (Note that the apparatus is purely hypothetical; there are more practical ways of accomplishing the same result.) This process can be understood by referring to the temperature-composition diagram in Figure E19 for the imaginary entane–orctane mixture discussed in OP-30.

When the mixture containing 50.0 mol% of both entane (bp = 50°C) and orctane (bp = 100°C) is boiled in flask **A** (point *A* on the diagram in

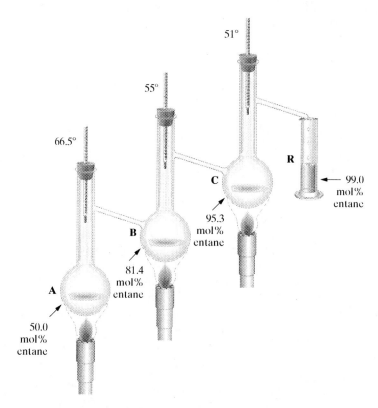

Figure E18 Hypothetical multistage distilling apparatus

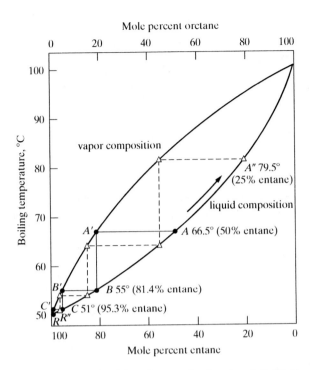

Figure E19 Temperature-composition diagram for fractional distillation of entane–orctane mixture

Figure E19), the vapor over the mixture has a composition of 81.4 mol% entane and only 18.6 mol% orctane (point A'). This is because entane has a considerably higher vapor pressure at the boiling point of the mixture (66.5°C) than does orctane. The vapor is condensed into flask **B** (line $A'-B$), where it boils at 55°C to yield a vapor containing 95.3 mol% entane (line $B-B'$), which is condensed (line $B'-C$) into flask **C**. The condensed liquid in **C** boils at 51°C to yield a vapor that is 99.0 mol% entane (line $C-C'$), which is delivered to receiver **R** as a liquid of the same composition (line $C'-R$).

Only the first few drops of distillate will attain this degree of purity, because as the more volatile entane is removed, the less volatile orctane will accumulate in the distilling flasks, reducing the proportion of entane in the vapor. The efficiency of the hypothetical apparatus would be improved considerably if some of the orctane-enriched liquid in each flask were drained back into the preceding flask through an overflow tube so that orctane did not accumulate as rapidly in the upper stages. Even then, the purity of the distillate would decrease with time. For example, when the amount of entane in flask **A** had decreased to 25 mol% (point A''), the vapor condensing into the receiver (point R'') would be only 96.6 mol% entane, as shown by the broken lines in Figure E19. In other words, during a distillation, the distilland climbs inexorably up the temperature-composition graph, and the distillate composition changes accordingly. Nevertheless, the separation effected by several distillation stages is considerably better than by only one, as can be seen by comparing the curves for simple distillation and fractional distillation shown in Figure E17.

The boiling point decreases with each subsequent distillation because the distilland has become richer in the more volatile component.

Fractional distillation refers to a distillation process involving several concurrent vaporization–condensation cycles. During a fractional distillation, the distillate is collected in several separate *fraction collectors* (receivers), the contents of each being a different *fraction*. Each fraction is collected over a different temperature range, with the lower boiling fractions containing a greater percentage of the more volatile component and the higher boiling fractions containing more of the less volatile component. For a fractional distillation having the distillation curve illustrated by the broken line in Figure E17, the first 30% (15 mL) or so of distillate would be nearly pure entane and the last 30 percent would be nearly pure orctane. The middle fraction would be a mixture of the two, which could be redistilled if desired.

Like the hypothetical distillation diagrammed in Figure E18, fractional distillation is a multistage distillation process performed in a single operation. A vertical *distilling column* performs the function of boiling flasks **B** and **C** in the hypothetical apparatus—in effect redistilling the original distillate from flask **A**. The column is filled with some kind of *column packing*, which provides a large surface area from which repeated vaporization and condensation cycles can take place.

Suppose our 50.0 mol% mixture of entane and orctane is distilled through such a column, as diagrammed in Figure E20. The vapor leaving the pot will have the same composition (81.4 mol% entane) as it did in the hypothetical apparatus, but as it passes onto the column it will cool, condense onto the packing surface, and begin to trickle down the column on its way back to the pot. Since the temperature is higher near the pot, part of the condensate will vaporize on the way down, yielding a vapor richer in entane. This vapor will rise up the column until, at a higher level than before (since its boiling point is lower), it cools enough to recondense. This process of vaporization and condensation may be repeated a number of times on the way to the top of the column, so that, when the vapor finally arrives at that point, it is nearly pure entane. Although this is a continuous process involving the

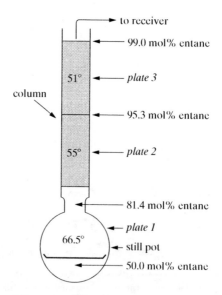

Figure E20 Operation of a fractionating column

simultaneous upward flow of vapor and downward flow of liquid, the net result is the same as that produced by successive discrete distillations.

Each section of the distillation apparatus that provides a separation equivalent to one cycle of vaporization and condensation (one "step" on the temperature-composition graph) is called a *theoretical plate*. The first cycle occurs in the pot, which provides one theoretical plate; the column illustrated in Figure E20 contributes two more. Note that there is a continuous variation in both temperature and vapor composition as one proceeds up the column, and that the temperature is fixed by the liquid–vapor composition. For instance, the entane-rich liquid near the top of the column boils at a lower temperature than the original mixture, so the average temperature of plate 3 is lower than that of the plates below it.

In certain multistage columns, the "theoretical" plates are real. A Bruun column consists of a series of horizontal plates stacked at intervals inside a vertical tube, and one vaporization–condensation cycle occurs on each plate.

Column Efficiency.   The efficiency of a fractional-distillation apparatus can be determined using the *Fenske equation,*

$$ n = \frac{\log = \dfrac{Z_A}{X_A} - \log \dfrac{Z_B}{X_B}}{\log \alpha} $$

where n is the total number of theoretical plates, X_A and X_B are the mole fractions of liquids A and B in the distilland, and Z_A and Z_B are the mole fractions of the same components in the vapor that emerges from the top of the column. The term α is the *relative volatility* of the two liquids, where the volatility of a liquid in a mixture is the ratio of its mole fraction in the vapor to its mole fraction in the liquid. The pot provides one theoretical plate, so the number of theoretical plates provided by the column is $n - 1$. The efficiency of a particular kind of column or column packing is given by its *HETP* (height equivalent to a theoretical plate), which is equal to the height of the column divided by the number of theoretical plates it provides. For example, a 24-cm column that provides four theoretical plates has an HETP of 6 cm. The lower its HETP, the more efficient is the column.

In practice, a number of factors limit the efficiency of a given column. Efficiency is highest under the equilibrium condition of *total reflux*, in which all of the vapors are returned to the pot. In practice, some of the vapors are continuously distilling into the receiver, which disturbs the equilibrium and reduces the column's efficiency. For good separation the *reflux ratio (R)* — the ratio of liquid returned to liquid distilled measured over the same time interval — should be kept reasonably high by keeping the distillation rate low.

$$ R = \frac{\text{liquid volume returning to pot}}{\text{liquid volume distilled}} $$

According to one rule of thumb, the reflux ratio should at least equal the number of theoretical plates for efficient operation. Reflux ratios of 5 to 10 are common for routine separations. To a lesser extent, the *holdup* of a column (the amount of liquid that adheres to a column's surface and packing) can also affect the efficiency of a distillation.

Azeotropes.   In discussing the principals of simple distillation and fractional distillation, we considered only the imaginary ideal liquids, entane and orctane. No real liquid displays ideal behavior, although some may come very close. Non-ideal behavior arises from intermolecular interactions, with the greatest deviations from ideality occurring in liquids having strong intermolecular forces. Thus water, whose molecules exhibit hydrogen bonding, behaves much less ideally than cyclohexane, whose molecules are held together by weak dispersion forces.

Large deviations from ideal behavior in liquid mixtures may result in the formation of azeotropes, where an *azeotrope* is a solution of two or more liquids whose composition doesn't change during distillation. Azeotrope formation can make it difficult or impossible to purify certain liquids by distillation. For example, ethanol and water form a *minimum-boiling azeotrope* called (inaccurately) 95% ethanol, which contains 95.6% ethanol and 4.4% water by mass and boils at 78.15°C—a temperature lower than the boiling point of either ethanol (78.5°C) or water. Distilling ethanol–water mixtures that contain a higher percentage of water eventually yields 95.6% ethanol, but it is impossible to obtain pure ethyl alcohol by distilling 95.6% ethanol, as illustrated by the phase diagram in Figure E21. Suppose we start with a mixture that contains X% ethanol (by mass), with X being less than 95.6%. The liquid will initially boil at the temperature shown by point A and its vapor (point A') can be condensed and revaporized in a column until it eventually reaches point C' on the diagram. At that point the liquid and vapor composition are the same, so the liquid 95.6% ethanol will yield a vapor that is also 95.6% ethanol, and no further purification will occur.

In this case, the problem caused by an azeotrope can be solved by a different azeotrope. Benzene and water form an azeotrope with ethanol that boils at a lower temperature (64.9°C) than the ethanol–water azeotrope, so by distilling 95% ethanol with some benzene, the benzene–water–ethanol

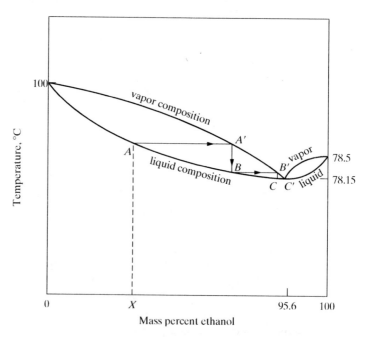

Figure E21 Temperature-composition diagram for ethanol–water mixtures

azeotrope can be distilled off until all of the water is removed and absolute (100%) ethanol is obtained. This is an example of *azeotropic drying*, the removal of water from an organic liquid by distillation with another liquid that forms a low-boiling azeotrope with water.

Experimental Considerations

Apparatus. The column is the most important part of any fractional-distillation apparatus. One of the simplest is the *Vigreux column*, which has a series of indentations in the column (see Figure E22a). The surface area for condensation inside the column is comparatively small, so Vigreux columns are not very efficient, having HETP values in the 8–12-cm range. At the other extreme, a *spinning-band column*, which contains a spiral of Teflon or some other material that is rotated inside the column by a magnetic stirrer or another device, can have HETP values around 0.5 cm or lower. The Hickman–Hinkle still shown in Figure E22d is a modified Hickman still with a long neck that functions as a column and a Teflon spinning band that is rotated by a magnetic stirrer.

The simplest type of column for fractional distillation is a straight tube (sometimes jacketed) filled with a suitable packing material. Packed columns are more efficient than Vigreux columns, but their holdup is higher and distillation rates are lower. The packing is introduced into a wide-bore jacketed column (not the narrower West condenser) from a standard scale lab kit or an air condenser from a microscale lab kit. Packing materials include metal turnings and glass or porcelain beads, rings, helices, and saddles. Highly

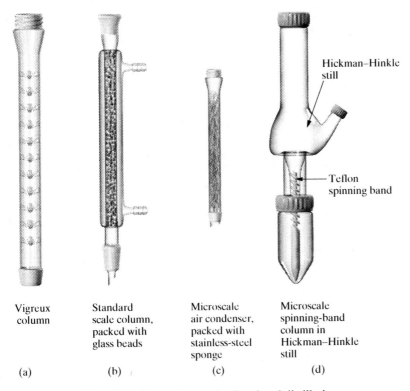

Vigreux column

(a)

Standard scale column, packed with glass beads

(b)

Microscale air condenser, packed with stainless-steel sponge

(c)

Hickman–Hinkle still

Teflon spinning band

Microscale spinning-band column in Hickman–Hinkle still

(d)

Figure E22 Some columns for fractional distillation

efficient packing materials such as glass helices may provide HETP ratings down to about 1 cm, but they are quite expensive. Glass beads and stainless-steel sponge are more practical for use in most undergraduate laboratories. The jacketed standard scale column shown in Figure E22b is packed with glass beads, while the microscale column in Figure E22c is constructed of an air condenser packed with stainless-steel sponge. Stainless-steel sponge pads can be stretched and cut into 6–8 inch lengths for use in a typical standard scale distilling column. Approximately 1.5 g of stainless-steel sponge can be stretched to pack a microscale column. With stainless-steel packing, the column is sometimes deliberately flooded by strong heating to wet the packing (see "Flooding"). Then the heat is reduced to drain the column before distillation is begun. Stainless-steel packing should not be used to distill halogen compounds, which corrode it.

The apparatus for standard scale fractional distillation is like that for simple distillation except that a distilling column is interposed between the pot and the still head, as shown in Figure E23. For high-boiling liquids, the column should be insulated to prevent heat losses that may reduce efficiency or prevent distillation entirely. It can be covered by one or more layers of crumpled aluminum foil (shiny side in), or glass wool can be wrapped around it and held in place by aluminum foil. The insulation should provide "windows" that can be opened for observation of the vapors in the still head and the packing

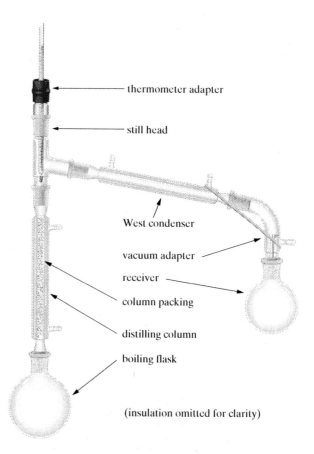

thermometer adapter

still head

West condenser

vacuum adapter

receiver

column packing

distilling column

boiling flask

(insulation omitted for clarity)

Figure E23 Apparatus for standard scale fractional distillation

near the bottom of the column. The still head can be insulated with the same materials as the column.

Typical microscale parts can be used to construct fractional-distillation setups like the one illustrated in Figure E24. Unless you are using a spinning-band column or another high-throughput column, it is seldom practical to distill less than 3–5 mL of liquid through a fractionating column, so the pot should ordinarily be a round-bottomed flask with a capacity of 10 mL or more. A microscale column made from an air condenser should be insulated. The insulating material can be glass wool or aluminum foil, as described for a standard scale column, or two concentric layers of clear PVC tubing as shown in Figure E24. To insulate an air condenser with PVC tubing, obtain two split 20-cm lengths of tubing of different diameters; then wrap the smaller (in diameter) tube around the column and the larger tube around the smaller one. For an air condenser having an outer diameter (o.d.) at its narrowest of $\frac{1}{2}$ inch, use $\frac{1}{2}$-inch i.d. (inner diameter) by $\frac{5}{8}$-inch o.d. and $\frac{5}{8}$-inch i.d. by $\frac{7}{8}$-inch o.d. tubing.

A Hickman still is set atop the column to collect the condensing vapors. The narrow neck of the still should be insulated by wrapping it with glass wool, which can be held in place with aluminum foil, if desired. If the distilland contains components that boil much below 100°C, it is a good idea to attach a condenser to the Hickman still. For most fractional distillations,

The throughput *of a column is the volume of distillate collected per unit of time.*

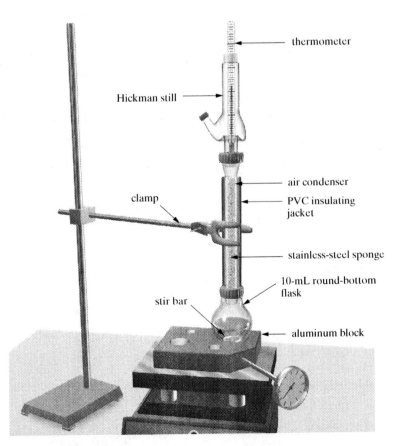

Figure E24 Apparatus for microscale fractional distillation

- thermometer
- Hickman still
- air condenser
- PVC insulating jacket
- clamp
- stainless-steel sponge
- 10-mL round-bottom flask
- stir bar
- aluminum block

Take Care! Capping the thermometer adapter will create a closed system, which could shatter and cause serious injury.

however, no condenser is needed if the heating rate is kept low enough to prevent the escape of distillate vapors from the top of the still. If desired, an *uncapped* (important!) thermometer adapter or a multipurpose adapter with a sidearm can be inserted on top of the Hickman still to support the thermometer and reduce the likelihood of vapor escape.

Heat Sources. For good results, the heat source should provide constant, uniform heating. An oil bath works best, but a heating mantle will suffice if its heat output is adjusted carefully to bring about the desired distillation rate. A heating block or sand bath is suitable for microscale work. (See OP-7a for additional information about the use of these heat sources.)

Flooding. One problem often encountered during a fractional distillation is *flooding*, in which the column becomes partly or entirely filled with liquid. Flooding is usually caused by an excessive heating rate, but it may also be caused by poor insulation, an unsuitable packing support, or improper packing. For example, sponge packing that is too tightly compressed or a glass-wool plug used as a packing support may hold up enough liquid to cause flooding. If flooding occurs, the apparatus should be separated from the heat source until all of the excess liquid has returned to the pot. Heating can then be resumed at a slower rate. Flooding decreases the efficiency of a separation because it reduces the surface area of packing available for the separation.

Chasers. A column with a high holdup will retain a relatively large amount of distillate. To improve the distillate recovery, the held-up liquid can be driven off the column into the receiving container or the well of a Hickman still by a suitable high-boiling substance, called a *chaser*. For example, *p*-xylene (bp 138°C) is a good chaser for liquids boiling around 100°C or below. After the last fraction has distilled, the chaser is added to the pot and heated to boiling, and the temperature is monitored as it climbs up the column. When the temperature begins to rise sharply, indicating that the chaser has reached the still head, distillation is immediately stopped and the recovered distillate is collected.

General Directions for Fractional Distillation

 Standard Scale

Equipment and Supplies

heat source
clamps, rings, supports
boiling flask
boiling chips *or* stir bar and stirrer
distilling column
column packing
insulating material (optional)
connecting adapter (still head)
thermometer

thermometer adapter
condenser
condenser tubing
vacuum adapter
fraction collectors
joint clips or rubber bands

If you are using column packing, pack the column to within a centimeter or so of the upper ground joint. To pack a column with stainless-steel sponge, *pull* it into the column using a copper wire bent into a hook at one end, making sure that it is as uniform as possible. Wear gloves or you might cut yourself on the sharp edges of the packing. To pack a column with glass beads, hold the column nearly horizontal with your hand over the bottom opening and place a few large beads in the top of the column. Then quickly pivot the column to an upright position so that (with a little luck) the beads will jam together at the constriction and support the remainder of the packing. (A small plug of stainless-steel sponge can also serve as a support.) Slowly pour in the rest of the packing from a beaker, with continuous shaking, so that it is as uniform as possible. Other packing can be added similarly, except that glass helices should be dropped in one at a time, with shaking.

Assemble the apparatus illustrated in Figure E23, using a boiling flask large enough so that it will be no more than half full of distilland. The fraction collectors can be screw-cap vials or other containers of an appropriate size; they should be numbered and tared. For very volatile distillates, small ground-joint flasks can be used as fraction collectors. The vacuum adapter drip tube should extend into the collector to reduce losses by evaporation. Make sure that all joints are tight, the column is perpendicular to the bench top, the boiling flask and fraction collector are supported properly, and none of the joints are under excessive strain. If the boiling temperature will exceed 100°C or so during the distillation, it is advisable to insulate the apparatus from the bottom of the column to the top of the still head. Position the thermometer correctly, as illustrated in Figure E8 of OP-30. Add the distilland to the distilling flask and drop in a few boiling chips or a stir bar. Start cooling water flowing through the condenser, but *not* through the column jacket.

Position the heat source, start the stirrer if you are using one, and begin heating to bring the mixture to the boiling point. When it boils, adjust the heating rate so that the reflux ring of condensing vapor passes up the column at a slow, even rate—it should take 5–10 minutes or more to reach the top of the column. Watch the packing at the bottom of the column closely for evidence of flooding. If flooding occurs, remove the heat source immediately and let the liquid drain into the boiling flask; then resume heating at a lower rate. If flooding is still a problem, you may need to reinsulate or repack the column. When the vapors rise above the column packing, adjust the heat to keep the reflux ring between the packing and the sidearm for a minute or so, giving the column time to equilibrate. After vapors begin to condense into the collector, read the thermometer when the temperature reading stabilizes. Distill at a rate of about 1 drop every 1–3 seconds, or at a rate that gives the desired reflux ratio. (Estimate the reflux ratio by counting the drops that drip into the pot and those that drip into the receiver during a short time period.) If the initial distillate is cloudy (due to dissolved water), change receivers when it becomes clear.

If the column isn't very efficient or the components' boiling points are close together, the temperature may rise only gradually throughout the distillation. In that case, it is best to collect fractions continuously at regular temperature intervals and redistill them. Otherwise, continue distilling until the still-head temperature begins to rise sharply or until a predetermined target temperature is reached, change fraction collectors, and record the temperature. Change collectors again when the temperature begins to stabilize at a higher value, or when the next target temperature is reached, and record the temperature. Increase the heating rate as necessary to maintain a suitable distillation rate. Continue to collect fractions over the appropriate temperature ranges until the pot is nearly dry, the temperature drops sharply, or the final target temperature is reached. (Note that the temperature may drop if the heating rate is too low, so that the hot vapors no longer reach the thermometer bulb.) Then lower and turn off the heat source and let the column drain. Any fractions collected while the temperature was rising rapidly are impure; unless you wish to redistill them, they should be placed in a solvent recovery container. Disassemble the apparatus and clean it promptly.

Summary

1. Pack column and assemble apparatus.
2. Add distilland and boiling chips or stir bar.
3. Turn on cooling water, stirrer (if used), and heat source.
4. Adjust heat so that reflux ring passes slowly up the column.
5. Record temperature after distillation begins when thermometer reading stabilizes.
6. Distill until temperature rises sharply or target temperature is reached, record temperature.
   IF more fractions are to be collected, change collector, REPEAT 6.
   IF you are distilling the last (or only) fraction, CONTINUE.
7. Distill until temperature drops sharply or final target temperature is reached.
8. Remove heat source, drain column.
9. Disassemble and clean apparatus, dispose of residue and any impure fractions.

 Microscale

Note: The italicized instructions apply only if you are using a Hickman-Hinkle still or another apparatus using a spinning band.

Equipment and Supplies

    heat source
    clamps, support
    boiling flask or vial

boiling chips *or* stir bar and stirrer
air condenser (for distilling column)
column packing *or spinning band*
insulating material
Hickman still
thermometer
condenser (optional)
condenser tubing (optional)
fraction collectors

If you are using column packing, pack the column to within a centimeter or less of the upper ground joint. To pack a column with stainless-steel sponge, pull it into the column using a copper wire bent into a hook at one end, making sure that it is as uniform as possible. Wear gloves or you might cut yourself on the sharp edges of the packing. *If you are using a Hickman-Hinkle still or a similar apparatus having a magnetically rotated spinning band, insert the spinning band in the bottom of the still (pointed end down) and lower it into the conical vial when you assemble the apparatus.*

Add the distilland to the boiling flask or vial and add one or two boiling chips or a stir bar. Assemble the apparatus illustrated in Figure E24, using a conical vial or a 10-mL round-bottomed flask as the pot and small screw-cap vials, numbered and tared, as fraction collectors. *If you are using a Hickman-Hinkle still, do not insert a separate column between the still and the pot; its column is an integral part of the still.* Make sure that all joints are tight, the column is perpendicular to the bench top, and the apparatus and thermometer are supported securely. Position the thermometer correctly, as illustrated in Figure E12 of OP-30. Insulate the column and the narrow neck of the Hickman still.

Position the boiling flask or vial in the heat source, start the stirrer if you are using one, and begin heating to bring the mixture to the boiling point. *A spinning band should be rotated at a low spin rate when heat is applied, changed to an intermediate spin rate when reflux begins, and then adjusted to the highest practicable spin rate when liquid begins to enter the column. It should rotate freely, without significant vibration.* When the liquid boils, adjust the heating rate so that the reflux ring of condensing vapor passes up the column at a slow, even rate—it should take 5 minutes or more to reach the top of the column. Watch the packing at the bottom of the column closely for evidence of flooding. If flooding occurs, raise the apparatus away from the heat source immediately and let the liquid drain into the pot; then resume heating at a lower rate. If flooding is still a problem, you may need to reinsulate or repack the column. After distillate begins to collect in the well of the Hickman still, read the thermometer when the temperature reading stabilizes. Distill the liquid so that the liquid level in the well rises very slowly; it should take 10 minutes or more to fill the well.

If the column isn't very efficient or the components' boiling points are close together, the temperature may rise only gradually throughout the distillation. In that case, it is best to collect fractions continuously at regular temperature intervals and redistill them. Otherwise, continue distilling until the still-head temperature begins to rise sharply or until a predetermined target temperature is reached, then transfer the liquid from the well to the first fraction collector. Close the port (of a ported still) immediately after the liquid has been withdrawn. If you are collecting more than one fraction, transfer each fraction to a different collector and record the temperature range over which it was collected. If necessary, increase the heating rate to maintain a suitable distillation rate. Continue to collect fractions over the appropriate temperature ranges until the pot is nearly dry, the temperature drops sharply, or the final target temperature is reached. (Note that the temperature may drop if the heating rate is too low, so that the hot vapors no longer reach the thermometer bulb.) Then remove the apparatus from the heat source and let the column drain. Any fractions collected while the temperature is rising rapidly are impure; unless you wish to redistill them, they should be placed in a solvent recovery container. Disassemble the apparatus and clean it promptly.

Summary

1. Pack column, add distilland and boiling chips or stir bar.
2. Assemble apparatus.
3. Turn on stirrer (if used) and heat source.
4. Adjust heat so that reflux ring passes slowly up the column.
5. Record temperature after distillation begins when thermometer reading stabilizes.
6. Distill until temperature rises sharply or target temperature is reached, transfer distillate to collector, record temperature.
    IF more fractions are to be collected, change collector, REPEAT 6.
    IF you are distilling the last (or only) fraction, CONTINUE.
7. Distill until temperature drops sharply or final target temperature is reached.
8. Remove apparatus from heat source, drain column.
9. Disassemble and clean apparatus, dispose of residue and any impure fractions.

When Things Go Wrong

Most of the things that go wrong during a simple distillation can also go wrong during a fractional distillation, so you can refer to *When Things Go Wrong* in OP-30 for help. The following cases apply only to fractional distillation.

If, while you're collecting a fraction, the temperature drops below that fraction's boiling range while there is still liquid in the pot, it's likely that all of the liquid making up that fraction has distilled but the vapors of the next fraction have yet not reached the still head. Keep heating, increasing the heating rate as necessary, so that the next fraction will start distilling. It's also possible that the heat setting was not high enough to maintain distillation of the first fraction, in which case it will resume distilling after you turn up the heat.

Suppose when you assembled your standard-scale apparatus you were careful to connect the cooling water tap to all of the available hose connections. Now the liquid in the pot is boiling, but none of its vapors are reaching the still head, where the temperature is still below 30°C. By connecting your distilling column to the cooling water, you have turned it into a reflux condenser; any vapors that enter it will condense and drip back into the pot. Yes, a distilling column does look a lot like a condenser, but it's not to be used as one during a fractional distillation!

F. Measuring Physical Constants

After a reaction product has been purified it is usually *analyzed* to find out whether it is, in fact, the desired product, and how pure it is. Another function of analysis is to determine the identity of an unknown compound. Analysis may involve measurement of a substance's physical properties, as described in this section, or recording a spectrum or chromatograph of the substance, as described in section G. A solid substance is usually analyzed by measuring its melting point [OP-33] and a liquid by measuring its boiling point [OP-34]. Measuring a liquid's refractive index [OP-35] provides a good indication of its purity and can be used to help identify it. An optically active compound, either liquid or solid, can be characterized by measuring its optical rotation [OP-36].

OPERATION 33

Melting Point

Principles and Applications

The *melting point (mp)* of a pure substance is defined as the temperature at which the solid and liquid phases of the substance are in equilibrium at a pressure of 1 atm. At a temperature slightly below the melting point, a mixture of the two phases solidifies; at a temperature slightly above the melting point, the mixture liquefies. Melting points can be used to characterize organic compounds and to assess their purity. The melting point of a pure compound is a unique property of that compound, which is essentially independent of its source and method of purification. This is not to say that no two compounds will have the same melting point; many compounds have melting points that differ by no more than a fraction of a degree. If two pure samples have *different* melting points, however, they are almost certainly different compounds.

The melting point of an organic solid is usually measured by grinding the solid to a powder and packing the powder inside a *melting-point tube*, a capillary tube that is closed at one end. The melting-point tube is then placed in an appropriate heating device and the *melting-point range* of the sample—the range of temperatures over which the solid is converted to a liquid—is observed and recorded. A pure substance usually melts within a range of no more than 1–2°C; that is, the transition from a crystalline solid to a clear, mobile liquid occurs within a degree or two if the rate of heating is sufficiently slow and the sample is properly prepared.

The presence of impurities in a substance *lowers* its melting point and *broadens* its melting-point range. To better understand the effects of impurities on melting points, consider the phase diagram for phenol (P) and diphenylamine (D) in Figure F1. Pure phenol melts at 41°C and pure diphenylamine at 53°C. If a sample of phenol contains diphenylamine as an impurity, its melting point will be lower than 41°C by an amount that depends on the mole percent of diphenylamine present. Similarly, the melting point of diphenylamine will be lower than 53°C if it contains phenol as an impurity. For example, the melting point of phenol containing 10 mol%

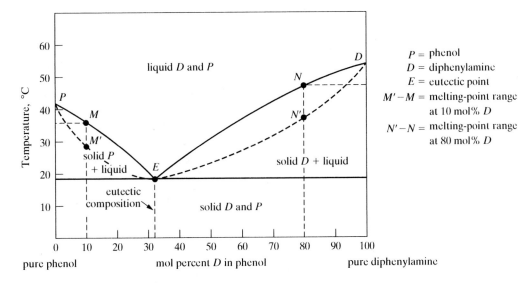

Figure F1 Phase diagram for the phenol–diphenylamine system

diphenylamine is given by point *M* on the phase diagram, and that of diphenylamine containing 20 mol% phenol is given by point *N*.

Pure phenol and pure diphenylamine both have sharp melting points, meaning that the transition from solid to liquid occurs over a narrow temperature range. Mixtures of the two (except the *eutectic mixture* at point *E*) exhibit broader melting-point ranges that depend on their composition. The approximate melting-point range for a mixture is given by the distance between the broken lines connecting points *P* and *E* or points *E* and *D* and the solid lines connecting the same points on the phase diagram. For example, the melting-point range of phenol containing 10 mol% diphenylamine is given by the distance between *M′* and *M*, and melting-point range of diphenylamine containing 20 mol% phenol is given by the distance between *N′* and *N*.

Since the melting point decreases in a nearly linear fashion as the amount of impurity increases (up to a point, at least), the difference between the observed and expected values can make it possible to estimate a compound's purity. The melting-point lowering effect can also be used to confirm the identity of a substance, such as the product of a reaction, when it is thought to be a certain known compound. The compound in question is mixed with a sample of the known compound and the melting point of the mixture is measured. If the two compounds are identical, the mixture melting point will be essentially the same as that of the known compound. If they are not identical, the known compound will act as an impurity in the unknown, so the melting point of the mixture will be lower and its range broader than for the known compound.

Experimental Considerations

Melting Behavior. The melting-point range of a sample is reported as the range between (1) the temperature at which the sample first begins to liquefy and (2) the temperature at which it is completely liquid, called the

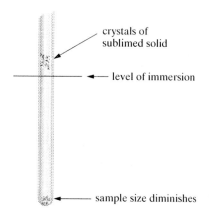

Figure F2 Sublimation of a sample in a melting-point tube

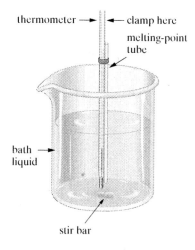

Figure F3 Stirred heating bath

Newer Mel-Temp models are provided with mercury-free precision digital thermometers.

liquefaction point. When a single melting point is to be reported, the liquefaction point is generally used. A compound may also be characterized by its *meniscus point*, the temperature at which the liquid meniscus is just clear of the solid below it. Some automatic melting-point devices report melting points that are closer to the meniscus point than to the liquefaction point.

If traces of solvent remain in a sample because of insufficient drying or other causes, you may observe "sweating" of solvent from the sample or bubbles in the molten sample, which may resolidify when all of the solvent is driven off. If a sample does show this behavior, it should be dried and the melting point remeasured. Even a dry sample will tend to soften and shrink before it begins to liquefy; this process begins at the *eutectic temperature* (point *E* in Figure F1). In any case, softening, shrinking, and sweating should not be mistaken for melting behavior. The melting-point range doesn't begin until the first free liquid is clearly visible.

Some compounds sublime—change directly from the solid to the vapor state—when they are heated in an open container. Sublimation can be detected during a melting-point determination by a pronounced shrinking of the sample, accompanied by the appearance of crystals higher up inside the melting-point tube (Figure F2). The melting point of a sample that sublimes at or below its melting temperature can be measured using a sealed melting-point tube. An ordinary melting-point tube should be cut (see OP-3) short enough so that it doesn't project above the block of a melting-point apparatus such as the one in Figure F4, or so that it can be entirely immersed in a heating-bath liquid. The sample is introduced and the open end is sealed in a burner flame (see OP-3). Then the melting point is measured by one of the methods described in the directions section.

If a sample becomes discolored and liquefies over an unusually broad range during a melting-point determination, it is probably undergoing thermal decomposition. Compounds that decompose on heating usually melt at temperatures that vary with the rate of heating. The approximate melting point (also called the *decomposition point*) of such a compound should be measured by heating the melting-point apparatus to within a few degrees of the expected melting temperature before inserting the sample, and then raising the temperature at a rate of about 3–6°C per minute.

Apparatus for Measuring Melting Points. Melting points can be determined with good accuracy using a Thiele tube or Thiele–Dennis tube filled with a heating-bath liquid such as mineral oil, as illustrated in Figure F5 for a Thiele–Dennis tube. The design of such tubes promotes good circulation of the heating liquid without stirring. A melting-point tube containing the solid is secured to a thermometer, which is then immersed in the bath liquid. The solid is observed carefully for evidence of melting as the apparatus is heated. Melting points can also be measured in an ordinary beaker if the bath liquid is stirred constantly (Figure F3), either manually or with a magnetic stirrer [OP-10].

Melting-point instruments that use capillary melting-point tubes, such as the Mel-Temp illustrated in Figure F4, are available commercially. Such melting-point devices are quite accurate if operated properly, and they can be used to make several measurements at once. This feature is useful when a mixture melting point is being determined, since the melting

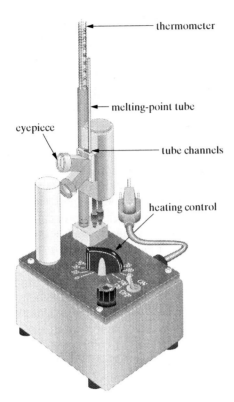

thermometer

melting-point tube

eyepiece

tube channels

heating control

Figure F4 Apparatus for measuring melting points (Mel-Temp method)

points of the unknown compound, the known, and the mixture can be measured and compared at the same time. The heating rate is adjusted by a dial controlling the voltage input to a heating coil. The dial reading required to attain the desired heating rate at the sample's melting point can be estimated from a heating-rate chart furnished with the instrument. The dial is initially set higher than this to bring the temperature to within 20°C or so of the expected melting point, and then reset to the value estimated from the chart.

Either kind of apparatus can be used for both standard scale and microscale experiments, but commercial melting-point instruments are easier to use and provide more uniform heating than melting-point baths.

Thermometer Corrections. The observed melting point of a compound may be inaccurate because of defects in the thermometer used or because of the *emergent stem error* that results when a thermometer is not immersed to its intended depth in a heating bath. Many thermometers are designed to be immersed to the depth indicated by an engraved line on the stem, usually 76 mm from the bottom of the bulb; slight deviations from this depth will not result in serious error. Other thermometers are designed for total immersion; if they are used under other circumstances, the temperature readings will be in error. An emergent stem error can be corrected by placing the bulb of a second thermometer opposite the middle of the exposed part of the thermometer's mercury column when the sample is melting, recording the temperature readings of both thermometers, and calculating an *emergent stem correction* with the following equation:

$$\text{emergent stem correction} = 0.00017 \cdot N(t_1 - t_2)$$

N = length in degrees of exposed mercury column

t_1 = observed melting temperature

t_2 = temperature at middle of exposed column

The emergent stem correction is added to the observed melting temperature, t_1. Melting points that have been corrected in this way should be reported as, for example, "mp 123–124° (corr.)."

Errors arising from thermometer defects as well as an emergent stem can be corrected by *calibrating* the thermometer under the conditions in which it is to be used. A thermometer used for melting-point determinations, for example, is calibrated by measuring the melting points of a series of known compounds, subtracting the observed melting point from the true melting point of each compound, and plotting this difference—the *melting-point correction*—as a function of temperature. The correction at the melting point of any other compound can then be read from the graph and added to its observed melting point, taking into account the sign (+ or −) of the correction. A list of pure compounds that can be used for melting-point calibrations is given in Table F1. Sets of pure calibration substances are available from chemical supply houses.

Table F1 Substances used for melting-point calibrations

Substance	mp, °C
ice	0
diphenylamine	54
m-dinitrobenzene	90
benzoic acid	122.5
salicylic acid	159
3,5-dinitrobenzoic acid	205

Mixture Melting Points. A mixture melting point is obtained by grinding together approximately equal quantities of two solids (a few milligrams of each) until they are thoroughly intermixed, then measuring the melting point of the mixture by the usual method. Usually one of the compounds, X, is an "unknown" that is believed to be identical to a known compound, Y. A sample of pure Y is mixed with X and the melting points of this mixture and of pure Y (and sometimes of X as well) are measured. If the melting points of pure Y and of the mixture are the same, to within a degree or so, then X is probably identical to Y. If the mixture melts at a lower temperature and over a broader range than pure Y, then X and Y are different compounds.

General Directions for Melting-Point Measurement

Safety Notes

> Mineral oil begins to smoke and discolor below 200°C and can burst into flames at higher temperatures. Oil fires can be extinguished with solid-chemical fire extinguishers or powdered sodium bicarbonate.

Standard Scale and Microscale

Equipment and Supplies   (Starred items are for method **B** only):

Mel-Temp (method **A** only)

watch glass

flat-bottomed stirring rod or flat-bladed spatula

1-meter length of glass tubing

thermometer

capillary melting-point tube

*Thiele tube or Thiele–Dennis tube

*clamp, ring stand

*heating oil

*burner

*cut-away cork

*3-mm rubber ring

Put a few milligrams of the dry solid on a small watch glass and grind it to a fine powder with a flat-bottomed stirring rod or a flat-bladed spatula. Use a spatula to make a small pile of powder near the middle of the watch glass. Press the open end of a capillary melting-point tube into the pile until enough has entered the tube to form a column 1–2 mm high. Using too much sample can result in a melting-point range that is too broad and a melting-point value that is too high. Tap the closed end of the tube gently on the bench top (or rub its sides with a small file); then drop it (open end up) through a 1-meter length of small-diameter glass tubing onto a hard surface, such as a bench top. Repeat this process several times to pack the sample firmly into the bottom of the tube. Use one of the methods that follow to measure the melting-point range of the sample. If you don't know the sample's expected melting point, determine its approximate value with rapid heating (6°C per minute or more); then carry out a more accurate melting-point measurement with a second sample as described here.

A melting-point tube can be constructed by sealing one end of a 10-cm length of 1-mm (i.d.) capillary tubing.

A. Mel-Temp Method.   Place the melting-point tube (sealed side down) in one of the channels on the Mel-Temp's heating block. Use the heating-rate chart to estimate the heat-control dial setting that will cause the temperature to rise at a rate of 1–2°C per minute at the expected melting point of the sample. Set the dial to a higher value at first, so that the temperature rises quite rapidly to about 20°C below the expected melting point, and then reduce the dial setting to the value that you estimated. As the temperature nears the expected melting point, adjust the dial as necessary so that the temperature rises at a rate of 1–2°C per minute while the sample is melting. Observe the sample through the eyepiece and record (as the limits of the melting-point range) the temperatures (1) when the first free liquid appears in the melting-point tube and (2) when the sample is completely liquid. For best results, do at least two measurements on each compound. Let the block cool to 15–20°C below the melting point before you do another measurement. Cooling can be accelerated by passing an air stream over the block.

B. Thiele-Tube Method.   (You can use a similar method with other melting-point baths, such as the one shown in Figure F3, but stirring and a hot plate or another flameless heat source will be required in that case.) Clamp the Thiele tube or Thiele–Dennis tube securely to a ring stand and add enough mineral oil (or another appropriate bath liquid) to just cover the top of the sidearm outlet, as shown in Figure F5a. (Refer to OP-7a for

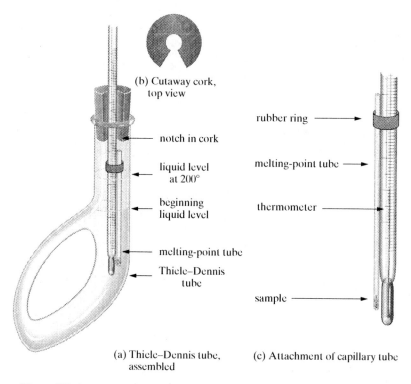

(b) Cutaway cork, top view

notch in cork

liquid level at 200°

beginning liquid level

rubber ring

melting-point tube

thermometer

melting-point tube

Thiele–Dennis tube

sample

(a) Thiele–Dennis tube, assembled

(c) Attachment of capillary tube

Figure F5 Apparatus for measuring melting points (Thiele-tube method)

precautions to be followed when using oil baths.) Secure the melting-point tube to a broad-range thermometer by the method that follows:

1. Cut a 3-mm-thick rubber ring from $\frac{1}{4}$-inch i.d. thin-walled rubber tubing (rubber rings may be provided).
2. Place the rubber ring around the thermometer about 9 cm from its bulb end.
3. Pinch the rubber ring between your fingers to create a gap and insert the open end of the melting-point tube through the gap.
4. Move the melting-point tube until the sample is adjacent to the middle of the thermometer bulb (see Figure F5c).

The cork is bored to accommodate the thermometer and then cut with a single-edged razor blade or a sharp knife so that the thermometer can be snapped into place from the side rather than inserted from the top.

Snap the thermometer into the center of a cutaway cork (Figure F5b) at a point above the rubber ring, with the melting-point tube and the degree markings on the same side as the opening in the cork. Insert this assembly into the bath liquid and move the thermometer, if necessary, so that its bulb is centered in the tube about 3 cm below the sidearm junction and the temperature can be read through the opening in the cork (Figure F5a). The rubber ring should be 2–3 cm above the liquid level so that the bath liquid, as it expands on heating, will not come in contact with it. If the hot oil meets the rubber, it may soften and allow the melting-point tube to drop out.

Heat the bottom of the Thiele tube with a burner flame (or use a microburner for more precise heat control) until the temperature is about 15°C below the expected melting point. Reduce the heating rate by turning down the flame and applying it at the sidearm, so that the temperature will rise at a rate of 1–2°C per minute at the melting point. Continue heating at that rate as you observe the sample closely, and record (as the limits of the melting-point range) the temperatures (1) when the first free liquid appears in the melting-point tube and (2) when the sample is completely liquid. For best results, do at least two measurements on each compound. Let the heating bath cool to 15–20°C below the melting point before you do another measurement. Cooling can be accelerated by passing an air stream over the tube.

Clean up the apparatus and either place the oil in a designated container or store it in the Thiele tube. Mineral oil can be removed from glassware by rinsing the glassware with petroleum ether followed by acetone, and then washing it with a detergent and water.

Summary

1. Assemble apparatus for melting-point determination.
2. Grind solid to powder, fill melting-point tube(s) to depth of 1–2 mm. IF you are using Method **B**, GO TO 4.
3. Insert sample in Mel-Temp heating block. GO TO 5.
4. Secure melting-point tube to thermometer; insert assembly in heating bath.
5. Heat rapidly to ~15–20°C below mp; then reduce heating rate to 1–2°C/min.
6. Observe and record melting-point range.
7. Disassemble apparatus as needed, clean up.

When Things Go Wrong

If, during a melting-point determination, the solid sample begins to shrink before any liquid appears, don't assume that it's melting. The melting point range doesn't begin until there is some liquid with the solid. But if the sample shrinks until there is little if any solid left at the bottom of the melting point tube, your sample is subliming and may completely disappear before it starts to melt. In that case you should prepare a new sample tube and seal it as described in *Melting Behavior*.

If your sample discolors significantly during a melting-point determination, it is probably decomposing. Prepare a new sample tube and proceed as described under *Melting Behavior*.

If the melting point you measure for a substance is substantially higher than its literature value, you may have heated the sample too rapidly. If so, carry out another measurement with the temperature rising at a rate of no more than 2°C per minute at the melting point.

Small melting-point errors may result from reading the thermometer incorrectly. Make sure that your eyes are at the same level as the top of the mercury column and that the thermometer's graduation marks extend to the mercury column when you take your readings. If repeating the melting point measurement more carefully with a fresh sample (don't remelt the original sample) still gives a low value, your product may be wet or impure.

If your sample appears to melt over an excessively wide range, you may have used too much sample or the sample may not have been packed tightly enough; the height of the sample in the melting-point tube should be no more than 2 mm. If you used the right amount of sample, you may have heated the sample too rapidly; the temperature should rise at a rate of no more than 2°C per minute at the melting point. Otherwise, you may have recorded the initial temperature before the sample began to liquefy, or your product may have decomposed near its melting point. (See *Melting Behavior.*) If repeating the melting point measurement more carefully with a fresh sample (don't remelt the original sample) still gives a broad range, your product may be wet or impure.

OPERATION **34**

Boiling Point

The *boiling point (bp)* of a liquid is defined as the temperature at which the vapor pressure of the liquid is equal to the external pressure at the surface of the liquid, and also as the temperature at which the liquid is in equilibrium with its vapor phase at that pressure. These definitions are the basis for various standard scale and small-scale methods for measuring boiling points. For example, the boiling point of a liquid can be determined by distilling a small quantity of the liquid and observing the temperature at the still head, where the liquid and its vapors are assumed to be in equilibrium. It can also be determined by measuring the temperature at which the liquid's vapor, trapped inside a capillary tube immersed in the liquid, exerts a pressure equal to the external pressure. Like the melting point of a solid, the boiling point of a liquid can be used to help identify it and assess its purity.

Boiling-Point Corrections

The *normal boiling point* of a liquid is its boiling point at an external pressure of 1 atmosphere (760 torr, 101.3 kPa). Since the atmospheric pressure at the time of a boiling-point determination is seldom exactly 760 torr, observed boiling points may differ somewhat from values reported in the literature and should be corrected. If a laboratory boiling-point determination is carried out at a location reasonably close to sea level, atmospheric pressure will rarely vary by more than 30 torr from 760 torr. For deviations of this magnitude, a *boiling-point correction, t,* can be estimated using Equation 1.

$$\Delta t \approx y(760 - P)(273.1 + t) \tag{1}$$

Δt = temperature correction, to be added to the observed boiling point
P = barometric pressure, in torr
t = observed boiling point, in °C

The value 1.0×10^{-4} is used for the constant y if the liquid is water, an alcohol, a carboxylic acid, or another associated liquid; otherwise y is assigned the value 1.2×10^{-4}. For example, the boiling point of water at 730 torr is 98.9°C. Use of Equation **1** leads to a correction factor of $[(1.0 \times 10^{-4})(760 - 730)(273.1 + 98.9)]°C = 1.1°C$, which yields the correct normal boiling point of 100.0°C.

At high altitudes, the atmospheric pressure may be considerably lower than 1 atmosphere, resulting in observed boiling points substantially lower than the normal values. For example, water boils at 93°C on the campus of the University of Wyoming at Laramie which has an elevation of 2290 m (7520 ft). For major deviations from atmospheric pressure, Equation **2** can be used in conjunction with approximate entropy of vaporization values obtained from the section "Correction of Boiling Points to Standard Pressure" in older editions of the *CRC Handbook of Chemistry and Physics.*

$$\Delta t \approx \frac{(273 + t)}{\varphi}\log\frac{760}{P} \qquad (2)$$

$\varphi = $ (entropy of vaporization at normal boiling point)$/2.303R$

Equation **2** *is a simplified form of the Hass–Newton equation found in the* CRC Handbook, *64th Edition, p. D-189.*

For example, hexane boils at 49.6°C at 400 torr. The value of φ for alkanes is, according to the *CRC Handbook,* about 4.65 at that temperature. Substituting these values into Equation **2** yields a correction factor of

$$\Delta t \approx \frac{(273°C + 49.6°C)}{4.65}\log\frac{760 \text{ torr}}{400 \text{ torr}} = 19.3°C$$

which gives a normal boiling point of 68.9°C. A second approximation using the value of φ at the corrected boiling point gives 68.8°C, which compares very favorably to the reported normal boiling point of 68.7°C. As for melting-point determinations, it may be necessary to correct a boiling point for thermometer error, especially when working with high-boiling liquids. See *Thermometer Corrections* in OP-33 for instructions.

a. Distillation Boiling Point

During a carefully performed distillation of a pure liquid, the vapors surrounding the thermometer bulb are in equilibrium (or nearly so) with the liquid condensing on the bulb. Thus the vapor temperature recorded during the distillation of a pure liquid should equal its boiling point. If the liquid is contaminated by impurities, the distillation boiling point may be either too high or too low, depending on the nature of the impurity. Volatile impurities in a liquid lower its boiling point, whereas nonvolatile impurities raise it; in either case, its boiling point will vary throughout its distillation. Therefore, if there is any doubt about the purity of a liquid, it should be distilled or otherwise purified prior to a boiling-point determination.

General Directions for Distillation Boiling-Point Measurement

 Standard Scale and Microscale

For standard scale work, assemble the compact distillation apparatus pictured in Figure E9 [OP-30] using a 25-mL round-bottom flask and placing the thermometer as shown in Figure E8. For microscale work, assemble the apparatus pictured in Figure E11 [OP-30] using a conical vial or a 10-mL round-bottom flask and placing the thermometer as shown in Figure E12. If the liquid boils below 100°C, use a water-cooled condenser; otherwise you can use an air condenser. For high-boiling liquids, be sure that the still head or Hickman still is well insulated from the heat source to prevent superheating of the vapors. Add 5–10 mL (standard scale) or 2–5 mL (microscale) of the pure liquid to the flask or conical vial and drop in a boiling chip or two, or a stirring device. Distill the liquid slowly, recording the temperature readings when (1) the liquid has begun to collect in the well or receiver and the temperature has equilibrated; (2) about half of the liquid has distilled; (3) the pot is nearly empty but the temperature has not begun to drop.

The first and last readings represent the boiling-point range and the middle reading is the *median boiling point*. The median value should provide a good estimate of the actual boiling point. A boiling-point range greater than 2°C suggests that the liquid may need to be purified before further work is done. Record the barometric pressure so that you can make a pressure correction, if necessary.

When Things Go Wrong

Most of the things that go wrong during a simple distillation can also go wrong during a distillation boiling-point measurement, so refer to *When Things Go Wrong* in OP-30 for help. The following case applies only to distillation boiling points.

If your liquid distills at too low or too high a temperature, if it distills over a broad range, or if the still-head temperature rises or drops abruptly after only part of it has distilled, it is probably impure. Purify it by an appropriate distillation technique and then repeat the measurement.

b. Capillary-Tube Boiling Point

When a liquid is heated to its boiling point, the pressure exerted by its vapor becomes just equal to the external pressure at the liquid's surface. If a tube that is closed at one end is filled with a liquid and immersed (open end down) in a reservoir containing the same liquid, the tube will begin to fill with vapor as the liquid is heated to its boiling point. At the boiling point, the vapor pressure inside the tube will balance the pressure exerted on the liquid surface by the surrounding atmosphere, so that the liquid levels inside and outside the tube will be equal (see Figure F6). If the temperature is raised above the boiling point, vapor will escape in the form of bubbles; if the temperature is lowered below the boiling point, the tube will begin to fill with liquid.

This behavior is the basis for semimicroscale and microscale methods for measuring boiling points. The semimicroscale method, which uses 2–4 drops of liquid, is suitable for standard scale and some microscale work. It requires a *bell*—a capillary tube sealed at one end—that is inverted inside a *boiling tube* containing the liquid. A bell for the semimicroscale method can be made by cutting [OP-3] an ordinary capillary melting-point tube in half and saving the sealed half. The boiling tube is constructed by sealing one end of a piece of 4–5-mm o.d. soft-glass tubing, cutting it to a length of 8–10 cm, and fire polishing the open end (see OP-3). The bell is placed in the boiling tube containing the liquid and the assembly is secured to a thermometer and heated in a Thiele or Thiele-Dennis tube. Alternatively, you can use a stirred oil bath as illustrated in Figure F3 [OP-33].

The microscale method, which requires only about 5 μL of liquid, is similar in practice except that a melting-point tube is used as the boiling tube and the bell is constructed from a very fine capillary tube. A bell can be made by cutting an 8–10 mm length from a 10 μL micropipet (such as a Drummond Microcap) and sealing one end of it (see OP-3), leaving a little glob of glass at the sealed end to help weigh it down. The sealed tube should be 7–8 mm long and should fit comfortably inside the melting point tube used as the boiling tube. If micropipets aren't available, you can make a pair

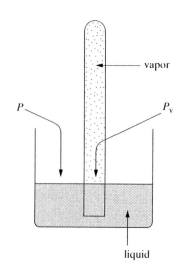

vapor

P P_v

liquid

Figure F6 Boiling-point principle. At the boiling point, $P_v = P$, where P_v = pressure exerted by the vapor on the liquid surface and P = pressure exerted by the atmosphere on the liquid surface

of bells by heating an open-ended melting-point tube in the middle (see OP-3), drawing it out to an appropriate diameter, breaking it in two, sealing the narrow ends, and then breaking off a length of fine capillary tube from each half.

It is important to avoid overheating during a boiling-point measurement; otherwise the liquid will boil away. It may be necessary to add more liquid if the sample is very volatile. For best results, you should repeat the boiling-point measurement at least once. If repeated determinations on the same sample give appreciably different (usually higher) values for the boiling point, the sample is probably impure and should be distilled.

General Directions for Capillary-Tube Boiling-Point Measurement

  **Semimicroscale**

Equipment and Supplies

boiling tube
semimicroscale bell
thermometer
rubber ring
Thiele tube or Thiele-Dennis tube
mineral oil or other bath liquid
burner

Construct a semimicroscale bell and a boiling tube as described previously. Add 2–4 drops of the liquid to the boiling tube and insert the bell into the boiling tube with its open end down. Secure the assembly to a thermometer by means of a rubber ring cut from thin-walled rubber tubing, as illustrated in Figure F7. Insert this assembly into a Thiele tube or Thiele–Dennis tube as for a melting-point determination (see Figure F5, OP-33), with the rubber ring 2 cm or more above the liquid level. (Refer to OP-7a for precautions to be followed when using oil baths.) Heat the tube with a burner flame (or use a microburner, for more precise heat control) until the temperature is within a few degrees of the expected boiling point of the liquid (if it is known). Then continue heating more slowly until you see a *rapid, continuous* stream of bubbles emerging from the bell. Immediately remove the heat source and let the bath liquid cool slowly until the bubbling stops, and then record the temperature when liquid *just* begins to enter the bell. Let the temperature drop a few degrees so that the liquid partly fills the bell. Then heat very slowly until the first bubble of vapor emerges from the mouth of the bell. Record the temperature at this point also. The two temperatures represent the boiling-point range; they should be within a degree or two of each other. Record the barometric pressure so that you can make a pressure correction, if necessary. Clean and dry the boiling tube, and save it for future boiling-point measurements. (See the Summary following the next section.)

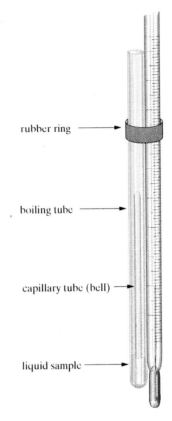

rubber ring →

boiling tube →

capillary tube (bell) →

liquid sample →

Figure F7 Semimicroscale boiling-point assembly

The expansion of heated air in the bell will cause a slow *evolution of bubbles; if you stop heating before rapid bubbling occurs, your reported boiling point will be too low.*

Sometimes the capillary tube will stick to the bottom of the boiling tube. This can be prevented by cutting a small nick in the open end of the capillary tube with a triangular file.

Microscale

Equipment and Supplies

boiling tube (capillary melting-point tube)
microscale bell
syringe
Mel-Temp (or other capillary melting-point apparatus)

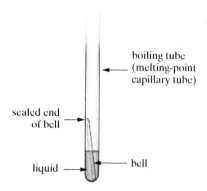

Figure F8 Microscale boiling-point assembly

The expansion of heated air in the bell will cause a slow *evolution of bubbles; if you stop heating before rapid bubbling occurs, your reported boiling point will be too low.*

Construct one or more microscale bells by one of the methods described previously. Use a syringe with a thin needle to introduce enough liquid into the boiling tube to fill it to a depth of approximately 5 mm. Holding the open end of the boiling tube securely, carefully "shake down" the liquid, as you would shake down the mercury column in a clinical thermometer, until it is all at the bottom of the tube and there are no air bubbles. (Alternatively, you can secure the boiling tube in a centrifuge tube with glass-wool packing and centrifuge it.) Insert a bell, open end down, into the boiling tube, and push it all the way to the bottom as shown in Figure F8 (use a thin copper wire if necessary). Insert the boiling tube into a Mel-Temp or another capillary melting-point apparatus. Heat until the temperature is within a few degrees of the expected boiling point of the liquid (if it is known). Then continue heating more slowly (preferably at a rate of no more than 2°C per minute) until you see a *rapid, continuous* stream of bubbles emerging from the bell. Immediately lower the heat setting so that the temperature drops slowly. When the bubbling stops, record the temperature as the liquid *just* begins to enter the bell. Then heat very slowly (don't wait for the bell to fill with liquid) until the first bubble of vapor emerges from the mouth of the bell. Record the temperature at that point also. These two temperatures represent the boiling-point range; they should be within a degree or two of each other. Record the barometric pressure so that you can make a pressure correction, if necessary.

Summary

1. Construct boiling-point bell.
2. Add liquid to boiling tube, insert inverted bell.
   IF you are using the microscale method, GO TO 4.
3. Secure boiling tube assembly to thermometer.
4. Place assembly in Mel-Temp or heating bath.
5. Heat until continuous stream of bubbles emerges from bell; lower heat setting or stop heating.
6. Record temperature when liquid just enters bell.
7. Heat slowly until first vapor bubble emerges from bell; record temperature.
8. Let liquid cool below boiling point.
   IF another measurement is needed, add more liquid if necessary, GO TO 5. IF not, CONTINUE.
9. Record barometric pressure, correct observed boiling point.
10. Disassemble apparatus, clean up.

When Things Go Wrong

If, after seeing bubbles in the boiling tube, you remove the heat source but the liquid level in the bell doesn't change, you probably mistook air bubbles for vapor bubbles. Resume heating until you see a *rapid* stream of bubbles emerging from the bell (which should contain no liquid at that point), and then proceed as described in the General Directions.

If after a while you can't see any liquid in the boiling tube, it has probably boiled away. Add more liquid and continue with a lower rate of heating, making sure that you don't heat it too long and miss the actual boiling point.

If the bell for a microscale determination moves up the boiling tube when the liquid boils, try shaking it down and heating at a lower rate. If that doesn't work, make another bell using a somewhat longer length of micro-pipet and leave a bigger glob of glass on the sealed end (not too big or it won't fit in the boiling tube). The total length of the tube shouldn't be longer than 8 mm.

If you just can't get the capillary-tube method to work for you and you have enough liquid, try this instead. Transfer 0.3 mL of your liquid to a Craig tube, add a boiling chip, and clamp a thermometer inside the tube so that its bulb is about 5 mm above the liquid surface. Heat the sample to boiling and continue heating (don't overheat) until the ring of condensing vapors is 1–2 cm above the *top* of the thermometer bulb. Keep the vapors at that level as you monitor the temperature. When it has stabilized at one value for at least a minute, record that value as your boiling point. Unless your liquid decomposes near its boiling point, the remaining liquid should be pure enough to put back with the rest of your product.

Refractive Index

The *refractive index* (index of refraction) of a substance is defined as the ratio of the speed of light in a vacuum to its speed in the substance in question. When a beam of light passes into a liquid its velocity is reduced, causing it to bend downward. The refractive index (n) is related to the angles that the incident and refracted (bent) beams make with a line perpendicular to the liquid surface, as shown in Figure F9 and Equation **1**.

$$n^t_\lambda = \frac{c_{vac}}{c_{liq}} = \frac{\sin \theta_{vac}}{\sin \theta_{liq}} \qquad (1)$$

n^t_λ = refractive index at temperature t using light of wavelength λ

c = speed of light

The refractive index is a unique physical property that can be measured with great accuracy (up to eight decimal places), so it is very useful for characterizing pure organic compounds. Refractive-index measurements are also used to assess the purity of known liquids and to determine the composition of solutions.

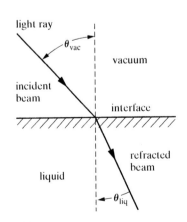

Figure F9 Refraction of light in a liquid

Experimental Considerations

It is much easier to make measurements in air than in a vacuum, so most refractive-index measurements are made in air, which has a refractive index of 1.0003, and the small difference is corrected by the instrument. The refractive-index value for a liquid depends on both the wavelength of the light used for its measurement and the density of the liquid, which varies with its temperature. Most values are reported with reference to light from a yellow line in the sodium emission spectrum, called the D line, which has a wavelength of 589 nm. Thus a refractive index value may be reported as $n_D^{20} = 1.3330$, for example, where the superscript is the temperature in °C and "D" refers to the sodium D line. Since most refractive index readings are made at or corrected to this wavelength, the "D" is sometimes omitted. If a different light source is used, its wavelength is specified in nanometers.

The Abbe refractometer (see Figure F10) is a widely used instrument for measuring refractive indexes of liquids. It measures the *critical angle* (the smallest angle at which a light beam is completely reflected from a surface) at the boundary between the liquid and a glass prism and converts it to a refractive-index value. The instrument employs a set of compensating prisms so that white light can be used to give refractive-index values corresponding to the sodium D line. A properly calibrated Abbe-3L refractometer should be accurate to within ±0.0002. The calibration of the instrument can be checked by measuring the refractive index of distilled water, which should be 1.3330 at 20°C and 1.3325 at 25°C.

The sample block of an Abbe refractometer can be kept at a constant temperature of 20.0°C by water pumped through it from a thermostatted water bath. If a refractive index is measured at a temperature other than 20.0°C, the temperature should be read from the thermometer on the instrument and the refractive index corrected by using the following equation,

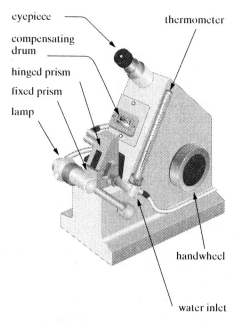

Figure F10 Abbe-3L refractometer

$$\Delta n = 0.00045 \times (t - 20.0)$$

where t is the temperature of the measurement in °C. The correction factor (including its sign) is added to the observed refractive index. For example, if the refractive index of an unknown liquid is found to be 1.3874 at 25.1°C, its refractive index at 20.0°C should be about $1.3874 + [0.00045 \times (25.1 - 20.0)] = 1.3897$. Alternatively, the refractive index of a reference liquid similar in structure and properties to the unknown liquid can be measured at the same temperature (t) as the unknown and a correction factor calculated from the equation

$$\Delta n = n^{20} - n^t$$

where n^t is the measured refractive index of the reference liquid at temperature t and n^{20} is the reported value of its refractive index at 20°C. The correction factor is then added to the measured refractive index of the unknown liquid at temperature t. This method can correct for experimental errors (improper calibration of the instrument, etc.) as well as temperature differences.

Small amounts of impurities can cause substantial errors in the refractive index. For example, the presence of just 1% (by mass) of acetone in chloroform reduces the refractive index of the latter by 0.0015. Such errors can be critical when refractive-index values are being used for qualitative analysis, so it is essential that an unknown liquid be pure when its refractive index is measured.

General Directions for Refractive-Index Measurements

 Standard Scale and Microscale

These directions apply to an Abbe-3L refractometer; if you are using a different instrument, your instructor will demonstrate its operation.

Equipment and Supplies

Abbe-3L refractometer
dropper
washing liquid
soft tissues

Raise the *hinged prism* of the refractometer and, using an eyedropper, place 2–3 drops of the sample in the middle of the *fixed prism* below it. Never allow the tip of a dropper or another hard object to touch the prisms—they are easily damaged. (If the liquid is volatile and free flowing, it may be introduced into the channel alongside the closed prisms.) With the prism assembly closed, switch on the *lamp* and move it toward the prisms to illuminate the visual field as viewed through the eyepiece. Rotate the *handwheel* until two distinct fields (light and dark) are visible in the eyepiece, and reposition the lamp for the best contrast and definition at the borderline between the fields. Rotate the *compensating drum* on the front

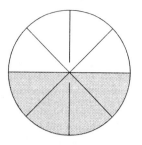

Figure F11 Visual field for a properly adjusted refractometer

of the instrument until the borderline is sharp and achromatic (black and white) where it intersects an inscribed vertical line. If the borderline cannot be made sharp and achromatic, the sample may have evaporated. Rotate the handwheel (or the fine adjustment knob, if there is one) to center the borderline exactly on the crosshairs (Figure F11).

Depress and hold down the *display switch* on the left side of the instrument to display an optical scale in the eyepiece. Read the refractive index from this scale (estimate the fourth decimal place). Record the temperature, if it is different from 20.0°C. Open the prism assembly and remove the sample by gently *blotting* it with a soft tissue (do not rub!). Wash the prisms by moistening a tissue or cotton ball with a suitable solvent (acetone, methanol, etc.) and blotting them gently. When the residual solvent has evaporated, close the prism assembly and turn off the instrument.

Summary

1. Add sample to fixed prism, close hinged prism.
2. Switch on and position lamp.
3. Rotate handwheel until two fields are visible, reposition lamp for best contrast.
4. Rotate compensating drum until borderline is sharp and achromatic.
5. Rotate handwheel or fine adjustment knob until line is centered on crosshairs.
6. Depress display switch, read refractive index, record temperature.
7. Clean and dry prisms, close prism assembly.
8. Make temperature correction if necessary.

When Things Go Wrong

If, while you are adjusting the refractometer for a reading, the borderline between the fields becomes fuzzy and colored or the fields seem to fade away, your sample is evaporating. Make sure the lamp is properly positioned, then add more sample (use an extra drop or two), adjust the instrument as quickly as you can, and take the reading. Repeat if necessary.

If the refractive index you record is significantly different from the literature value, read the thermometer and, if the temperature is not 20°C, make a temperature correction (unless you already did). If the refractive index is still not close enough, repeat the measurement more carefully. If that doesn't help, check the calibration of the instrument using distilled water or another suitable liquid. If the instrument is calibrated correctly, your product is probably wet or otherwise impure. Purify it and try again.

OPERATION **36**

Optical Rotation

Principles and Applications

Light can be considered a wave phenomenon with vibrations occurring in an infinite number of planes perpendicular to the direction of propagation. When a beam of ordinary light passes through a *polarizer*, such as a Nicol prism, only the light-wave components that are parallel to the plane of the polarizer can pass through. The resulting beam of *plane-polarized light* has light waves whose vibrations are restricted to a single plane (see Figure F12).

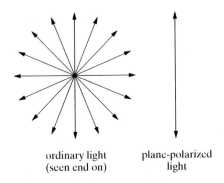

ordinary light
(seen end on)

plane-polarized
light

Figure F12 Schematic representations of ordinary and plane-polarized light

When a beam of plane-polarized light passes through an optically active substance, molecules of the substance interact with the light so as to rotate its plane of polarization. The angle by which a sample of an optically active substance rotates the plane of that beam is called its *observed rotation*, α. Measurement of such angles of rotation is known as *polarimetry* and the instrument that measures them is a *polarimeter*. The observed rotation of a sample depends on the length of the light path through the sample and the concentration of the sample, as well as its identity. The first two variables are not intrinsic properties of the sample itself, so the observed rotation is divided by these variables to yield its *specific rotation*, $[\alpha]$, as shown in this equation.

$$[\alpha] = \frac{\alpha}{lc} \qquad \textbf{(1)}$$

$[\alpha]$ = specific rotation

α = observed rotation

l = length of sample, in decimeters (dm)

c = concentration of solution, in g solute per mL solution

(for a neat liquid, substitute the density in g/mL)

The specific rotation of a pure substance is an intrinsic property of the substance and can be used to characterize it.

The *enantiomeric excess* (also called *optical purity*) of a chiral substance is calculated by dividing its observed specific rotation by the literature value for the appropriate pure enantiomer and multiplying by 100%.

$$\text{enantiomeric excess} = \frac{[\alpha]_{\text{observed}}}{[\alpha]_{\text{pure}}} \times 100\%$$

For example, the specific rotation of pure $(-)$-menthol is 50°, so the entamiomeric excess of a sample of $(-)$-menthol having a specific rotation of 30° is 60%

$$\text{enantiomeric excess} = \frac{30°}{50°} \times 100\% = 60\%$$

This means that the sample contains the equivalent of 60% $(-)$-menthol (which produces all of the optical rotation) and 40% racemic $(+/-)$-menthol, half of which (or 20% of the sample) is $(-)$-menthol and half $(+)$-menthol. So the total percentage of $(-)$-menthol in the sample is 60% + 20%, or 80%.

The composition of a mixture of two optically active substances can be calculated from its specific rotation using the equation:

$$[\alpha] = [\alpha]_A\, X_A + [\alpha]_B\, (1 - X_A) \tag{2}$$

In this equation, $[\alpha]$ is the specific rotation of the mixture, $[\alpha]_A$ is the specific rotation of component A, X_A is the mole fraction of component A, and $[\alpha]_B$ is the specific rotation of component B. For example, an equilibrium mixture of α-D-glucose ($[\alpha] = 112°$) and β-D-glucose ($[\alpha] = 18.7°$) has a specific rotation of 52.7°C. The mole fraction of α-D-glucose in the mixture can be calculated by substituting these values into Equation **2** and solving for X_A.

$$52.7° = (112°)X_A + (18.7°)(1 - X_A)$$
$$X_A = 0.364$$

Since both forms of glucose have the same molecular weight, the equilibrium mixture contains 36.4% α-D-glucose and 63.6% β-D-glucose by mass.

Experimental Considerations

The specific rotation of an optically active sample can be determined using a polarimeter such as the one illustrated in Figure F13. A polarimeter includes a light source, a polarizing prism called the *polarizer*, a sample cell

Figure F13 Cole-Parmer EW–81205 polarimeter

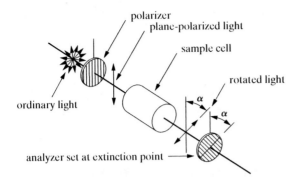

Figure F14 Schematic diagram of a polarimeter

to hold the sample, and a second polarizing prism called the *analyzer* (see Figure F14). The traditional light source is a sodium lamp, which produces monochromatic light with a wavelength of 589 nm, but modern polarimeters may use a halogen lamp with an orange filter to provide light of that wavelength. The polarizer is fixed in place and light that passes through it is polarized in the plane of its vertical axis—in other words, all light waves *except* those parallel to the polarizer's axis are blocked out. The analyzer can be rotated so that its axis is at an angle to the polarizer's axis. When the axis of the analyzer is perpendicular to that of the plane-polarized light, the light is essentially blocked out so that its intensity is at a minimum; this condition is called the *extinction point.* It is easier for the human eye to match two intensities than to recognize a point of minimum intensity, so in a precision polarimeter the analyzer contains two or more prisms set at a small angle to each other, and it is rotated until the prisms bracket the extinction point, transmitting dim light of equal intensity. The angle by which the analyzer must be rotated to reach this point equals the angle by which the sample rotated the beam of plane-polarized light—its observed rotation, α. Using Equation **1**, we can convert the observed rotation of the sample to its specific rotation given the sample's concentration and the length of the sample cell. The most commonly used sample cells have lengths of 1 dm (10 cm) and 2 dm (20 cm).

The optical rotation of a substance is usually measured in solution. Water and ethanol are common solvents for polar compounds, and dichloromethane can be used for less polar ones. Often a suitable solvent will be listed in the literature, along with the light source and temperature used for the measurement. The volume of solution required depends on the size of the polarimeter cell, but 10–25 mL is usually sufficient. A solution for polarimetry ordinarily contains about 1–10 g of solute per 100 mL of solution and should be prepared using an accurate balance and a volumetric flask. If the solution contains particles of dust or other solid impurities, it should be filtered [OP-15]. When possible, the concentration should be comparable to that reported in the literature. For example, the *CRC Handbook of Chemistry and Physics* reports the specific rotation of (+)-menthol as "+49.2 (al, $c = 5$)," where the concentration (c) is given in grams per 100 mL of an alcoholic (al) solution (c is defined differently here than in Equation **1**). Thus, a (+)-menthol solution suitable for polarimetry can be prepared by accurately weighing about 1.25 g of (+)-menthol, dissolving it in ethyl alcohol in a 25-mL volumetric flask, adding more alcohol up to the calibration mark, and mixing the solution thoroughly.

Because the solvent you use may alter the observed rotation, you should always run a *solvent blank* by measuring the optical rotation of the pure solvent, and subtract its rotation from that of your sample. Be sure to take account of + and − signs; for example, if your sample's optical rotation is +20.5° and the blank's optical rotation is −0.3°, the corrected rotation is +20.5° −(−0.3°) = +20.8°.

General Directions for Polarimetry

Equipment and Supplies

> polarimeter
> light source
> 1-dm or 2-dm sample cell
> 10-mL or 25-mL volumetric flask
> sample and solvent

 Standard Scale and Microscale

This procedure applies to a Zeiss-type polarimeter with a split-field image. Experimental details may vary somewhat for different instruments. If the light source is a sodium lamp, make sure that it has ample time to warm up (some require 30 minutes or more). Weigh the sample accurately and use a volumetric flask (or another appropriate volumetric container) to prepare a solution of the sample in a suitable solvent. For standard scale work, you can prepare about 25 mL of solution and use a 2-dm sample cell. For microscale work, prepare about 10 mL of solution and use a 1-dm or smaller sample cell.

Remove the screw cap and the glass end plate from one end of a clean sample cell (do not get fingerprints on the end plate) and rinse the cell with a small amount of the solution to be analyzed (for microscale work, rinse it with the solvent used to prepare the solution and let it dry). Stand the polarimeter cell vertically on the bench top and fill it with the solution, rocking the cell if necessary to shake loose any air bubbles. Add the last milliliter or so with a dropper so that the liquid surface is convex. Carefully slide on the glass end plate, trying not to leave any air bubbles trapped beneath it. If the cell has a bulge at one end, a small bubble can be tolerated; in that case, tilt the cell so that the bubble migrates to the bulge and stays out of the light path. Screw on the cap just tightly enough to provide a leakproof seal — overtightening it may strain the glass and cause erroneous readings. Place the sample cell in the polarimeter trough, close the cover, see that the light source is oriented to provide maximum illumination in the eyepiece, and focus the eyepiece if necessary. Set the analyzer scale to zero and (if necessary) rotate it a few degrees in either direction until a dark and a light field are clearly visible. You may see a vertical bar down the middle and a background field on both sides, as shown in Figure F15, or two fields divided

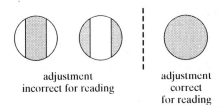

adjustment
incorrect for reading

adjustment
correct
for reading

Figure F15 Split-field image of polarimeter

down the middle. Focus the eyepiece so that the line(s) separating the fields are as sharp as possible. Rotate the analyzer scale away from zero in a clockwise direction, then (if necessary) in a counterclockwise direction, until you reach a point at which both fields are of nearly equal intensity. Then back off a degree or so toward zero and use the fine adjustment knob (if there is one) to rotate the scale *away* from zero until the entire visual field is as uniform as possible and the dividing lines between fields have all but disappeared. If you overshoot the final reading, move the scale back a few degrees so that you approach it again going away from zero. Read the rotation angle from the analyzer scale, using the Vernier scale (if there is one) to read fractions of a degree, and record the direction of rotation, + for clockwise or − for counterclockwise. Obtain another reading of the rotation angle, this time approaching the final reading going *toward* zero. For accurate work, you should take a half-dozen readings or more, reversing the direction each time to compensate for mechanical play in the instrument, and average them.

If you have rotated the scale about 90° and see a very bright visual field, you have gone too far; return to zero and try again.

Rinse the cell with the solvent used in preparing the solution. Fill the cell with that solvent (or use a solvent blank provided) and determine the solvent's rotation angle by the same procedure as before. Remove the solvent and let the cell drain dry, or clean it as directed by your instructor. Subtract the average observed rotation of the solvent blank (note + and − signs) from that of the sample to obtain the observed rotation of the sample, and calculate its specific rotation using Equation **1**. When reporting the specific rotation, specify the temperature, light source, solvent, and concentration.

Summary

1. Prepare solution of compound to be analyzed.
2. Rinse sample cell and fill with solution; place in polarimeter.
3. Adjust light source; focus eyepiece.
4. Rotate analyzer scale until optical field is uniform; read rotation angle.
5. Repeat step 4 several times, reversing direction each time, and average readings.
   IF solvent blank has been run, GO TO 7.
   IF not, GO TO 6.
6. Place solvent blank in polarimeter; GO TO 4.
7. Clean cell, drain dry, dispose of solution and solvent as directed.
8. Calculate observed rotation and specific rotation of sample.

When Things Go Wrong

If, when you look through the eyepiece of the polarimeter, you see a gap at the top of the visual field, there is an air bubble in the polarimeter tube. Tilt the tube until the bubble migrates to the bulge in the tube. If it has no bulge, add more solution as described in the General Directions until the tube is completely filled with liquid.

If, when you look through the eyepiece, the visual field is so dark that you can't compare the light intensities accurately, try the following remedies in order. Check to see that the sodium lamp (if you are using one) has warmed up enough to produce a bright light, and that is positioned properly for maximum illumination of the visual field. Check to see whether there is a movable ring (possibly marked +/−) at the far end of the polarimeter that, when rotated, adjusts the light intensity; if so, adjust it for a optimum intensity. Examine the sample in the polarimeter tube for turbidity (suspended solids), and if isn't water-clear, filter it by gravity filtration [OP-15].

If the specific rotation you calculate from your observed value is significantly different from the expected value, make sure that you subtracted the observed rotation of a solvent blank from that of your sample and that your calculation is correct. Also check to see whether the solvent you used and the temperature of your measurement are similar enough to those used to obtain the expected value to make a comparison valid. If you were analyzing a chiral compound provided by a chemical supply company, it may not be enantiomerically pure, which would lead to a low specific rotation; check the label on the bottle the chemical came in (it should list the value of $[\alpha]$) or ask your instructor to do so. If you still think your result may be inaccurate, reread the General Directions and repeat the measurement more carefully. If necessary, prepare a fresh solution, making sure to weigh the sample and measure the solvent volume accurately.

G. Instrumental Analysis

Measurements of physical properties can help you identify substances whose identities you already suspect, and they often give at least a rough indication of the purity of a substance. But they don't reveal much about the molecular makeup of a compound, and they are of little use when you are trying to identify an unknown compound or analyze a mixture of compounds. For these purposes you can use instruments that provide different kinds of information about compounds and mixtures. A gas chromatograph [OP-37] can separate the components of a mixture of liquids and often help you identify some or all of the components; a high-performance liquid chromatograph [OP-38] can do the same with solid mixtures. An infrared spectrometer [OP-39] can detect the functional groups in an organic compound and provide additional information about its molecular structure, facilitating the identification of unknown compounds and confirming the identities of reaction products. A nuclear magnetic resonance spectrometer [OP-40] may provide enough detailed structural information about an unknown compound to reveal its entire molecular structure. An ultraviolet-visible spectrometer [OP-41] provides information about unsaturated systems in molecules and can be used to measure the concentrations of many compounds in solution. Mass spectrometers [OP-42] make it possible to identify many thousands of organic compounds by reference to large databases of spectral information. They can also be used to determine molecular formulas and some molecular structures or structural features.

Gas Chromatography   OPERATION **37**

Gas chromatography (GC) affords a powerful method for the separation and analysis of volatile components of mixtures. Like all chromatographic methods, its operation is based on the distribution of the sample components between a mobile phase and a stationary phase. *Gas–liquid chromatography* is the most useful form of gas chromatography. The mobile phase for gas–liquid chromatography is an unreactive *carrier gas* such as helium, and the stationary phase consists of a high-boiling liquid on a solid support, contained within a heated column. The components of the sample must have reasonably high vapor pressures so that their molecules will spend enough time in the vapor phase to travel through the column with the carrier gas. Thus, gas chromatography can be used to separate gases, liquids that vaporize without decomposing when heated, and some volatile solids. (Solids must first be dissolved in a suitable solvent.) Less volatile liquids and solids can be separated by high-performance liquid chromatography [OP-38].

    Analytical gas chromatography can be used for both the qualititative and quantitative analysis of mixtures; that is, to identify the components of a mixture and find out how much of each component is present. An analytical gas chromatograph requires only a tiny amount of sample, usually a microliter or so. *Preparative gas chromatography* is used to separate the components of a mixture and recover the pure components, or to purify a major component of a mixture. Because even a large GC column will not accommodate sample sizes greater than about 0.5 mL, preparative GC cannot be used to separate large volumes of liquids. In the undergraduate organic chemistry lab, it is most useful for purifying the products of reactions conducted at the microscale level.

Principles and Applications

In gas–liquid chromatography, the components of a mixture are partitioned between the liquid stationary phase and the gaseous mobile phase, the carrier gas. The time it takes for a component to pass through the column, called its *retention time*, depends on its relative concentrations in the liquid and gas phases. While the molecules of a component are in the gas phase, they pass through the column at the speed of the carrier gas. While they are in the liquid phase, they remain stationary. The more time a component spends in the gas phase, the sooner it will get through the column, and the lower its retention time will be. Conversely, the more time a component spends in the liquid phase, the longer its retention time will be. The time a component spends in the gas phase depends on its volatility (and thus on its boiling point) and the temperature of the column, while the time it spends in the liquid phase depends on the strength of the attractive forces between its molecules and the molecules of the liquid phase.

In order for any two components of a mixture to be separated sharply, they must have significantly different retention times, and their bands (the regions of the column they occupy) must be narrow enough that they do not overlap appreciably as they exit the column. The degree of separation depends on a number of factors, including the length and efficiency of the column, the carrier gas flow rate, and the temperature at which the separation is carried out. Modern gas chromatographs provide the means to vary these factors precisely, and they offer an almost endless variety of applications, from analyzing automobile emissions for noxious gases to detecting PCBs in lake trout.

There are several important differences between gas–liquid chromatography and liquid–solid chromatographic methods such as column chromatography [OP-21] and thin-layer chromatography (TLC) [OP-22]. Unlike the mobile phase in a liquid–solid separation, the mobile phase in gas chromatography doesn't interact with the molecules of the sample; its only purpose is to carry them through the column. Therefore, the separation of the components of a mixture depends mainly on (1) how strongly they are attracted to the stationary phase and (2) how volatile they are. If the components have similar polarities, they will be attracted to the stationary phase to about the same extent, so they will tend to be separated in order of relative volatility, with the lower boiling components having the shorter retention times. If the components have different polarities, their retention times will depend on their relative polarities and the polarity of the stationary phase. A polar stationary phase will attract polar components more strongly than nonpolar ones, giving polar components longer retention times than nonpolar components. A nonpolar stationary phase will attract nonpolar components more strongly, giving them longer retention times.

A column chromatography or TLC separation is usually conducted at room temperature, and the temperature stays about the same throughout the separation. In contrast, a GC separation is usually carried out at an elevated temperature, and the temperature can be changed throughout the separation according to a preset program. Increasing the temperature of the column increases the fraction of each component in the gas phase, thus decreasing its retention time. The column temperature also affects the separation efficiency, because at excessively high temperatures components will tend to spend most of their time in the vapor phase, causing their bands to overlap.

Instrumentation

In addition to the column and carrier gas, a gas chromatograph must have some means of vaporizing the sample, controlling the temperature of the separation, detecting each constituent of the sample as it leaves the column, and recording data from which the composition of the sample can be determined. A typical analytical gas chromatograph includes the components diagrammed in Figure G1 and described here.

Injection Port. The injection port is the starting point for a sample's passage through the column. The sample, in a microliter syringe, is injected into the column by inserting the needle through a rubber or silicone septum and depressing the plunger. The sample size for a packed column is typically about 1–2 μL, but may vary from a few tenths of a microliter to 20 μL. With an open tubular column (described later), a sample splitter delivers about 1 nL (10^{-3} μL) of the injected sample to the column; the rest is discarded. The injector port is heated to a temperature sufficient to vaporize the sample, usually about 50°C above the boiling point of its least volatile component.

Packed Columns. A packed column is a long tube packed with a finely divided solid support whose particles are coated with the high-boiling liquid phase. The tube is usually made of stainless steel or glass. A typical column may be 1.5–3.0 m long with an inner diameter (i.d.) of 2–4 mm, and it is usually bent into a coil to fit inside a column oven, where its temperature

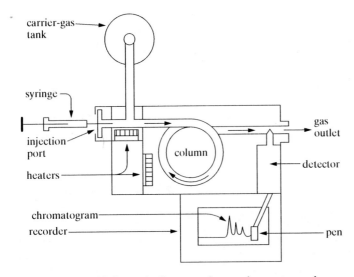

Figure G1 Schematic diagram of a gas chromatograph

is controlled. Although such columns can be packed by the user, most columns are purchased ready-made. The composition of the packing is described on a tag attached to each column. For example, a column packing described as "10% DEGS/Chromosorb W 80/100" contains 10% by mass of the liquid phase, diethylene glycol succinate, coated onto a Chromosorb W (crushed diatomaceous earth) support having a particle-size range from 80 to 100 mesh (0.17 to 0.15 mm). As for a fractional distillation column [OP-32], the efficiency of a chromatography column can be measured in terms of the number of theoretical plates it provides. Most packed columns have efficiencies ranging from 500 to 1000 theoretical plates per meter.

The mesh number is the number of meshes per inch in the finest sieve that will let the particles pass through.

Open Tubular Columns.   An *open tubular column* or *capillary column* consists of an open tube coated on the inside with the liquid phase. A very thin film of the liquid phase (ranging from 0.05 μm to 1.0 μm or more in thickness) is adsorbed on or chemically bonded to the inner wall of the column, which is a long capillary tube made of glass or fused silica. Fused silica open tubular (FSOT) columns are flexible enough to be bent into coils 15 cm or so in diameter. The columns range in size from "microbore" columns that may have an i.d. of 0.05–0.10 mm and a length of 20 meters to "megabore" columns having an i.d. of about 0.50 mm and a length of 100 meters or more. Packed columns are generally cheaper, less fragile, and easier to use than open tubular columns, and they accommodate much larger sample sizes, but open tubular columns are faster and less likely to react with the sample, and they provide unparalleled resolution of sample components—a microbore column may have an efficiency of more than 10,000 theoretical plates per meter. Since an open tubular column may be 20–50 times longer than a typical packed column, it is usually hundreds of times more efficient than a packed column containing the same liquid phase.

Column Oven.   The column is mounted inside a heated chamber, the *column oven*, that controls its temperature. Under *isothermal operation*, the oven temperature is kept constant throughout a separation; under *program operation*, it is varied according to a programmed sequence. As a general rule, the column temperature for isothermal operation should be approximately equal to or slightly above the average boiling point of the sample. If the sample has a broad boiling range, it is best separated by program operation. The more volatile components are eluted during the low-temperature end of the program, and less volatile components are eluted at its high-temperature end.

Carrier Gas.   A chemically unreactive gas is used to sweep the sample through the column. Helium is the most widely used carrier gas, but nitrogen and argon are used with some detectors. Carrier gases usually come in pressurized gas cylinders, and various devices are used to measure and control their flow rates.

Detector.   The detector is a device that detects the presence of each component as it leaves the column and sends an electronic signal to a recorder or other output device. A separate oven is used to heat the detector enclosure so that the sample molecules remain in the vapor phase as they pass the detector. The most widely used detectors for routine gas chromatography are *thermal conductivity* (*TC*) detectors and *flame ionization* (*FI*) detectors. A TC detector responds to changes in the thermal conductivity of the carrier gas stream as molecules of the sample pass through. With an FI detector, the component molecules are pyrolyzed (transformed by heating) in a hydrogen–oxygen flame, producing short-lived ions that are captured and used to generate an electrical current. Thermal conductivity detectors are comparatively simple and inexpensive, and they respond to a wide variety of organic and inorganic species. Flame ionization detectors are much more sensitive than TC detectors, so they are used with open tubular columns (for which TC detectors are unsuited) as well as packed columns. An FI detector is somewhat inconvenient to use because it requires hydrogen and air to produce the flame, and it cannot be used for preparative gas chromatography because it destroys the sample.

Data Display. A typical gas chromatograph intended for student use may have a mechanical *recorder* that records a peak on moving chart paper as each component band passes the detector. The resulting *gas chromatogram* consists of a series of peaks of different sizes, each produced by a different component of the mixture. The recorder may be provided with an *integrator* that measures the area under each component peak. Modern research-grade gas chromatographs are interfaced with computers and other hardware devices that display the peaks on a monitor screen, print or plot copies of the gas chromatogram, and provide digital readouts of peak areas, retention times, and various instrumental parameters.

Liquid Phases

The factor that most often determines the success or failure of a gas chromatographic separation is the choice of a liquid phase. In general, polar liquid phases are best for separating polar compounds, and nonpolar liquid phases for separating nonpolar compounds. However, there are hundreds of liquid phases available, and choosing the right one for a particular separation may not be an easy task. Some typical liquid phases and their maximum operating temperatures are listed in Table G1 in order of polarity (lower to higher). General-purpose liquid phases such as dimethylpolysiloxane (dimethylsilicone) can separate a variety of compounds successfully, usually in approximate order of their boiling points. Other liquid phases

Heating a column above its maximum operating temperature will cause the liquid phase to vaporize.

Table G1 Selected liquid phases for gas chromatography

Stationary phase	Maximum T, °C	X'	Y'	Z'
squalane	150	0	0	0
dimethylpolysiloxane (OV-1)	350	16	55	44
diphenyl/dimethylpolysiloxane (OV-17)	350	119	158	162
polyethylene glycol (Carbowax 20M)	250	322	536	368
diethylene glycol succinate (DEGS)	225	496	746	590
dicyanoallylpolysiloxane (OV-275)	275	629	872	763

Note: McReynold's numbers (X', Y', Z') apply to the commercial stationary phase in parentheses. Commercial designations for the same kind of stationary phase vary widely; for example, AT-1, DB-1, Rtx-1, SE-30, and DC-200 are all similar to OV-1.

have more specialized applications; for example, diethylene glycol succinate is a high-boiling ester that is used mainly to separate the esters of long-chain fatty acids, as in Experiment 54.

The selection of a liquid phase can be facilitated by the use of *McReynolds numbers*, which indicate the affinity of a liquid phase for different types of compounds; the higher the number, the greater is the affinity. X' is a McReynolds number measuring the relative affinity of a liquid phase for aromatic compounds and alkenes; Y' measures its affinity for alcohols, phenols, and carboxylic acids; and Z' measures its affinity for aldehydes, ketones, ethers, esters, and related compounds. For example, the polar liquid phase DEGS has a much higher Z' value (590) than does OV-1 (44), so it will retain an ester much longer. McReynolds numbers are particularly useful for selecting a liquid phase to separate compounds of different chemical classes, such as alcohols from esters. For example, a Carbowax 20M column should separate isopentyl acetate from isopentyl alcohol satisfactorily because its Y' and Z' values are very different; but an OV-17 column would not be a good choice for such a separation. Chromatography supply companies often provide detailed information about the kinds of separations their columns can accomplish.

Qualitative Analysis

The retention time (R) of a component is the time it spends on the column, from the time of injection to the time its concentration at the detector reaches a maximum. The retention time corresponds to the distance on the chromatogram, along a line parallel to the baseline, from the injection point to the top of the component's peak, as shown in Figure G2. This distance can be converted to units of time if the chart speed is known. If all instrumental parameters (temperature, column, flow rate, etc.) are kept constant, the retention time of a component will be characteristic of that compound and may be used to identify it. You can seldom (if ever) identify a compound with certainty from its retention time alone, but you can sometimes use retention times to confirm the identity of a compound whose identity you suspect. For example, if a known compound has the same retention times as an unknown compound on two or more different stationary phases, under the same operating conditions, the compounds are likely to be identical.

When you carry out a reaction and obtain a gas chromatogram of the product mixture, you may be able to guess which GC peak corresponds to which reactant or product, especially if the components have significantly different boiling points and the stationary phase is known to separate compounds in order of boiling point. However, it is always a good idea to back up your guess with proof. One way to identify the components of a product mixture is to "spike" the mixture with an authentic sample of a possible component and see which GC peak increases in relative area after spiking. For example, suppose that you carry out a Fischer esterification of acetic acid by ethanol and record a gas chromatogram of the product mixture.

$$\underset{\text{acetic acid}}{CH_3COOH} + \underset{\text{ethanol}}{CH_3CH_2OH} \rightleftharpoons \underset{\text{ethyl acetate}}{CH_3COOCH_2CH_3} + H_2O$$

Since this is an equilibrium reaction, the product mixture would be expected to contain some unreacted starting materials as well as the product. To

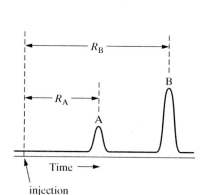

Figure G2 Retention times of two components, A and B

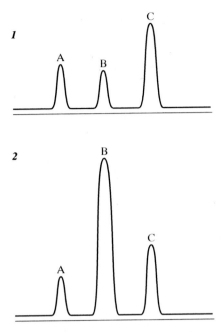

Figure G3 Use of gas chromatography for qualitative analysis: (1) chromatogram before adding ethanol; (2) chromatogram after adding ethanol

detect the presence of unreacted ethanol, you can add a small amount of pure ethanol to the sample and record a second gas chromatogram. The peak that increases in relative area is the ethanol peak, as shown in Figure G3. The process can be repeated using different pure compounds to identify the other components, if necessary.

Another way to identify the components of a mixture using GC is to obtain the spectra of the components after they pass through the column. As described in OP-42, a mass spectrometer coupled to a gas chromatograph can act as a "superdetector," identifying the components as they come off the column. If the GC column has a large enough capacity that individual components can be collected at the column outlet, they can often be identified from their IR or NMR spectra.

Quantitative Analysis

The use of gas chromatography for quantitative analysis is based on the fact that, over a wide range of concentrations, a detector's response to a given component is proportional to the amount of that component in the sample. Therefore, the area under a component's peak can be used to determine its mass percentage in the sample. If a peak on a gas chromatogram is symmetrical, its area can be calculated with fair accuracy by multiplying its height (h) in millimeters (measured from the baseline) by its width at a point exactly halfway between the top of the peak and the baseline ($w_{\frac{1}{2}}$), as shown in Figure G4.

$$\text{approximate peak area} = h \times w_{\frac{1}{2}}$$

Many gas chromatographs are equipped with an integrator that automatically calculates the peak areas and displays them digitally. Peak areas can

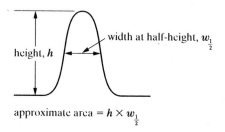

approximate area $= h \times w_{\frac{1}{2}}$

Figure G4 Measurement of approximate peak area

also be measured by making a photocopy of the chromatogram, accurately cutting out each peak with a sharp knife or razor blade (using a ruler to cut a straight line along the baseline), and weighing the peaks to the nearest milligram (or tenth of a milligram) on an accurate balance.

From the peak areas for a sample, you can sometimes estimate the percentage of each component in the mixture by dividing its area by the sum of the areas and multiplying by 100%. For example, suppose a gas chromatogram has three peaks, corresponding to three components, A, B, and C. If the area of peak A is 1227 mm^2, the area of peak B is 214 mm^2, and the area of peak C is 635 mm^2, the sum of the areas is 2076 mm^2, so the mass percentages of the components are approximately

$$\%A = \frac{1227}{2076} \times 100\% = 59.1\%$$

$$\%B = \frac{214}{2076} \times 100\% = 10.3\%$$

$$\%C = \frac{635}{2076} \times 100\% = \underline{30.6\%}$$

$$100.0\%$$

This kind of calculation is valid only if the detector responds similarly to each component, which may be the case if you are analyzing very similar compounds, as in a mixture of long-chain fatty acid esters. More often, however, a detector will respond differently to different compounds. Flame ionization detectors respond mainly to ions produced by certain reduced carbon atoms, such as those in methyl and methylene groups. For example, ethanol is only about 52% carbon by mass while heptane is 84% carbon, so the response of an FI detector to a gram of ethanol is only about three-fifths as great as its response to a gram of heptane. Thermal conductivity detectors also respond differently to different substances, but the variations are usually not as large as with FI detectors.

For accurate quantitative analysis of most mixtures, you should multiply each component's peak area by a *detector response factor* to obtain a corrected area that is proportional to its mass. Detector response factors can be obtained from the literature in some cases, or by direct measurement. For example, detector response factors for cyclohexane and toluene using a TC detector are reported to be 0.942 and 1.02, respectively (relative to benzene). Suppose that the gas chromatogram of a cyclohexane–toluene mixture is recorded using such a detector and has peak areas of 78.0 mm^2 for cyclohexane and 14.6 mm^2 for toluene. The corrected peak areas are then $(0.942 \times 78.0 \text{ mm}^2)$ = 73.5 mm^2 for cyclohexane and $(1.02 \times 14.6 \text{ mm}^2) = 14.9$ mm^2 for toluene. Dividing each corrected area by the sum of the corrected areas and multiplying by 100 gives mass percentages of 83.1% for cyclohexane and 16.9% for toluene.

If detector response factors for the components of a mixture are not known, you can determine them by the following method, which is described for two components, A and R. Component R is a reference compound, arbitrarily assigned a detector response factor of 1.00.

- Obtain a gas chromatogram of a mixture containing carefully weighed amounts of A and R.
- Measure the peak area for each component.
- Calculate the detector response factor for A using the following relationship:

$$\text{detector response factor for A} = \frac{\text{mass}_A}{\text{area}_A} \times \frac{\text{area}_R}{\text{mass}_R}$$

This method can be used to calculate detector response factors for any number of components simultaneously if their peaks are well resolved. Each component (including the reference compound) is carefully weighed, a gas chromatogram of the mixture is recorded, all of the peak areas are measured, and the detector response factors are calculated from the resulting masses and areas using the previous equation.

Preparative Gas Chromatography

Most preparative gas chromatographs are too expensive to be used routinely in undergraduate laboratories, but some inexpensive analytical gas chromatographs, such as the Gow-Mac 69-350, can be operated with preparative columns using an adapter that is connected to the exit port on the gas chromatograph. A special GC collection tube, provided in some microscale lab kits, is inserted into the adapter when the desired component's peak begins to appear on the chromatogram, and is removed just after all of the peak has been recorded. The condensed liquid is then transferred to a small ($\sim$100 μL) conical vial by centrifugation. Cotton or glass wool packing is used to support the conical vial and to center the top of the collection tube in the centrifuge tube. The liquid can then be used to obtain an IR spectrum [OP-39] or NMR spectrum [OP-40] of the compound.

If an adapter and specialized collection glassware are not available, a 2-mm outer diameter (o.d.) glass tube about 8 cm in length can be packed with glass wool and inserted into the GC exit port while the desired component's peak is being recorded. The component should condense on the surface of the glass wool, from which it can be washed off with a small amount of solvent (if the component is quite volatile, the tube should be chilled first). An IR or NMR spectrum of the compound can then be obtained in a solution of the solvent used, or the solvent can be evaporated.

General Directions for Recording a Gas Chromatogram

 Standard Scale and Microscale

Equipment and Supplies

> gas chromatograph and recorder
> 10-μL microsyringe
> screw-cap vial containing sample
> lab tissues
> syringe-washing solvent

Do not attempt to operate the instrument without prior instruction and proper supervision. The directions that follow are for a typical student-grade gas chromatograph connected to a mechanical recorder. For other kinds of instruments, follow your instructor's directions. Consult the instructor if the instrument doesn't seem to be working properly or if you have questions about its operation. It will be assumed that all instrumental parameters have been preset, that an appropriate column has been installed, that the carrier gas is flowing at the right rate, and that the column oven will be operated isothermally. If not, the instructor will show you what to do. Before you begin, be sure you have read the section in OP-5 about the use of syringes.

If the sample is a volatile solid, dissolve it in the minimum volume of a suitable low-boiling solvent; otherwise, use the neat liquid. Rinse a microsyringe with the sample a few times, and then partially fill it with the sample. A microsyringe is a very delicate instrument, so handle it carefully and avoid using excessive force that might bend the needle or plunger. If there are air bubbles inside the syringe, tap the barrel with the needle pointing up, or eject the sample and refill the syringe more slowly. Hold the syringe with the needle pointing up and expel excess liquid until the desired volume of the sample (usually 1–2 μL) is left inside. Wipe the needle dry with a tissue and pull the plunger back a centimeter or so to prevent prevaporization of the sample.

Take Care! The injection port is hot! Don't touch it.

Set the chart speed, if necessary, and switch on the recorder-chart drive (and activate the integrator, if there is one). Carefully insert the syringe needle into the injection port by holding the needle with its tip at the center of the septum and pushing the barrel slowly, but firmly, with the other hand until the needle is as far inside the port as it will go. Inject the sample by *gently* pushing the plunger all the way in just as the recorder pen crosses a chart line (the starting line); use as little force as possible to avoid bending the plunger. Withdraw the needle and promptly mark the starting line, from which all retention times will be measured. Let the recorder run until all of the anticipated component peaks have been recorded. Then turn off the chart drive and tear off the chart paper using a straightedge. Inspect the chromatogram carefully; if it is unsuitable because the significant peaks are too small, poorly resolved, or off-scale, repeat the analysis after taking measures to remedy the problem. If there is evidence of prevaporization (as indicated by an extraneous small peak preceding each major peak at a fixed interval), be sure that the plunger is pulled back before you inject the sample for the next run. Injecting the sample immediately after you insert the needle into the injector port will also prevent prevaporization, but this technique requires good timing.

Before you analyze a different sample, clean the syringe and rinse it thoroughly with that sample. When you are finished, rinse the syringe with an appropriate low-boiling solvent such as methanol or dichloromethane, remove the plunger, and set it on a clean surface to dry.

Summary

1. Rinse syringe, fill with designated volume of sample.
2. Start chart drive and integrator (if applicable).
3. Insert syringe needle into injector port, inject sample when pen crosses chart line.
4. Withdraw needle; mark starting line.
5. Stop chart drive after last peak is recorded.
6. Remove chromatogram, clean and dry syringe.

When Things Go Wrong

(*See your instructor before you make any adjustments to the instrument as suggested here, or let the instructor make the adjustments.*)

If you have injected your sample in the gas chromatograph but no peaks have appeared after 10 minutes or so, consider the following possibilities and, if one seems likely, refer to the indicated section or sections. You didn't fill the syringe properly or its needle is plugged and there is little if any sample in it. (See the General Directions.) You didn't insert the syringe needle far enough into the injection port, or the injection port is not hot enough. (See *Injection Port*, General Directions.) The column oven is not turned on or is not set to the right temperature. (See *Column Oven*.) The carrier gas is not turned on or its flow rate has not been adjusted correctly. (See *Carrier Gas*.) The filament current for a TC detector is not turned on, the flame of an FID detector is not lit, or the detector enclosure is not at the right temperature. (See *Detector*.) The recorder is not turned on or its chart is not advancing. (See *Data Display*, General Directions.) The liquid phase in the column is not appropriate for your sample. (See "Liquid Phases.")

If the peaks on your gas chromatogram are too small, you may not have injected enough sample. Try again, making sure that the liquid fills the syringe up to the desired mark, with no air bubbles (see OP-5). If that makes no difference the recorder gain may have been set too low; see your instructor.

If the peaks on your gas chromatogram are too large, flattening out near the top of the chart paper, you may have injected too much sample or the recorder gain may have been set too high.

If you recorded a gas chromatograph but some of its peaks are "doubled," with a small peak preceding a larger one at a fixed interval, you forgot to pull back the plunger of the syringe after you filled it, leaving sample in the needle that vaporized too soon. (See the General Directions.)

If you recorded a gas chromatograph but some of the peaks overlap significantly, consider the following possibilities and, if one seems likely, refer to the indicated section. You injected too much sample or injected the sample too slowly. (See the General Directions.) The column oven temperature is too high or the programmed temperature sequence is not appropriate. (See *Column Oven.*) The carrier gas flow rate is too high. (See *Carrier Gas.*) The liquid phase in the column is not appropriate for your sample. (See "Liquid Phases.")

Suppose you recorded a gas chromatogram but it has one or more extraneous peaks (peaks that you wouldn't expect to be there) that are not "doubled" as described previously. If the extraneous peaks are quite small, they may have came from traces of impurities in your product or from solvents and other chemicals you used to prepare the product. Check to see whether students who are analyzing the same product have chromatograms with similar peaks; if so, don't worry about them. If a small extraneous peak precedes all of your sample peaks, it may be an air peak, due to air that was in the syringe needle, or the peak of a volatile syringe-washing solvent (such as dichloromethane) that hadn't completely evaporated before you filled the syringe with your sample. If you find out (how?) that the peak is from a syringe-washing solvent, it's a good idea to run another sample using a clean, completely dry syringe. If one or more extraneous peaks are quite large, you may have injected your sample before a previous sample had completely passed out of the column. Wait a minute or so after the last peak of your chromatogram has appeared and then inject a fresh sample. If that doesn't make a difference and none of the previous possibilities seem likely, your product may need further purification.

OPERATION 38   High-Performance Liquid Chromatography

Principles and Applications

High-performance liquid chromatography (HPLC) can be regarded as a hybrid of column chromatography and gas chromatography, sharing some features of both methods. As in column chromatography, the mobile phase is a solvent or solvent mixture (the eluant) that carries the sample through a column packed with fine particles that interact with the components of the sample to different extents, causing them to separate. As in gas chromatography, the sample is usually injected onto the column and detected as it leaves the column, and its passage through the column is recorded as a series of peaks on a chromatogram. The stationary phase may be a solid adsorbent, as in column chromatography, but it is more often an organic phase that is bonded to tiny beads of silica gel. The silica beads are, in effect, coated with a very thin layer of a liquid organic phase. Each component of the sample is partitioned between a liquid mobile phase and the liquid stationary phase according to a ratio—the *partition coefficient*—that depends on its solubility in each liquid. The components of a mixture generally have different partition coefficients in a given liquid phase, so they pass down the column at different rates.

Instrumentation

HPLC was developed as a means of improving the efficiency of a column chromatographic separation by reducing the particle size. Most column chromatography packings contain particles with diameters in the 75–175 μm range, while most modern HPLC packings have particle sizes in the 3–10 μm range, increasing separation efficiency dramatically; but solvents will not easily flow through such small particles by gravity alone, so a powerful pump is needed to force them through the column at pressures up to 6000 psi (~400 atm).

The basic components of an HPLC system are diagrammed in Figure G5. The instrument ordinarily has several large solvent reservoirs, each of which can be filled with a different eluting solvent. *Isocratic elution* is elution

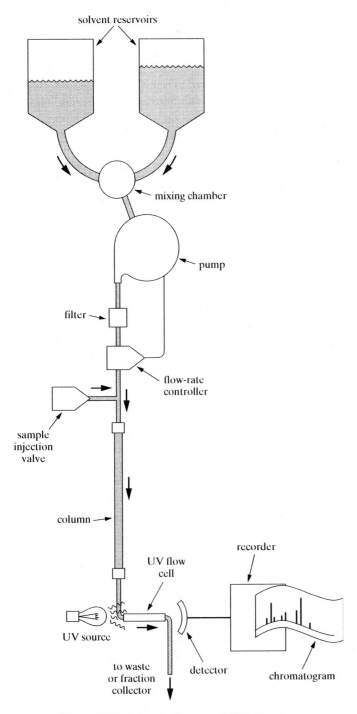

Figure G5 Schematic diagram of HPLC system

using a single solvent. *Gradient elution* utilizes two or more solvents of different polarity and varies the solvent ratio throughout a separation according to a programmed sequence. Gradient elution can reduce the time needed for a separation and increase the separation efficiency.

Preparative HPLC systems, which require special wide-bore columns, are equipped with fraction collectors to collect the eluant as it comes off the column; the fractions are evaporated to yield the pure components. Analytical HPLC systems, which are used to determine the compositions of mixtures, require much smaller samples and the components are not recovered. A typical analytical HPLC column is a straight stainless steel tube with a length of 10–25 cm and an inner diameter (i.d.) of 2.1–4.6 mm. A microbore analytical column may have an i.d. of 1 mm, while a large preparative column may have an i.d. of 50 mm or so. Samples are introduced onto the column by means of a syringe or sampling valve and are carried through it by the pressurized eluant mixture. Because the particles of the stationary phase are so small, a column can easily be plugged by particulate matter or adherent solutes introduced by either the eluant or the sample. To remove anything that might harm the column, the eluant is forced through one or more filters and the sample may be introduced through a short *guard column*.

As each component of a sample leaves the column, the detector responds to some property of the component, such as its ability to absorb ultraviolet (UV) radiation. The detector then sends an electronic signal to a recorder, which traces the component's peak on a chart. The resulting chromatogram is a graph of some property of the components, such as UV absorbance, plotted against the volume of the mobile phase. Because different solutes may have greatly different UV absorptivities at the wavelength used by an *ultraviolet absorbance detector*, for example, a detector response factor (see OP-37) must be determined for each component before its percentage in the mixture can be calculated. Components that do not absorb UV radiation will not be detected by a UV detector; in that case, the components can be converted to derivatives that do absorb UV radiation, or a different type of detector can be used. Any component whose refractive index is different from that of the eluant can be detected by a *refractive index (RI) detector*, but RI detectors are less sensitive than UV detectors.

High-performance liquid chromatography can be used to separate mixtures containing proteins, nucleic acids, steroids, antibiotics, pesticides, inorganic compounds, and many other substances whose volatilities are too low for gas chromatography. Because HPLC separations can be run at room temperature, there is little danger of decomposition reactions or other chemical changes that sometimes occur in the heated column of a gas chromatograph. These advantages have made HPLC the fastest growing separation technique in chemistry, but its use in undergraduate laboratories is limited because of the high cost of the instruments, columns, and high-purity solvents required (HPLC-grade water costs about $50 a gallon!).

Stationary Phases

Some HPLC columns contain a solid adsorbent such as silica gel or alumina. Such a stationary phase is much more polar than the eluant, so nonpolar components are eluted faster than polar ones. Most modern HPLC columns contain a *bonded liquid phase*—an organic phase that is chemically bonded to particles of silica gel. Silica gel contains silanol (—Si—OH) groups to which long hydrocarbon chains, such as octadecyl groups, can be attached by reactions such as the following:

$$-\text{Si}-\text{OH} \xrightarrow{\text{R}_2\text{SiCl}_2} -\text{Si}-\text{O}-\overset{\displaystyle \text{R}}{\underset{\displaystyle \text{R}}{\text{Si}}}-\text{Cl} \xrightarrow{\text{H}_2\text{O}} \xrightarrow{(\text{CH}_3)_3\text{SiCl}}$$

$$-\text{Si}-\text{O}-\overset{\displaystyle \text{R}}{\underset{\displaystyle \text{R}}{\text{Si}}}-\text{O}-\text{Si}(\text{CH}_3)_3 \qquad \text{R} = \text{CH}_3(\text{CH}_2)_{17}- \text{ (octadecyl)}$$

The bonded stationary phase in this example is *less* polar than the eluant, which may be a mixture of water with another solvent such as methanol, acetonitrile, or tetrahydrofuran. Therefore, polar components will spend more time in the eluant than in the stationary phase and will be eluted faster than nonpolar ones, reversing the usual order of elution. This mode of separation is called *reverse-phase chromatography.*

Other stationary phases operate by still different mechanisms. The stationary phase for *size-exclusion chromatography* is a porous solid that separates molecules based on their effective size and shape in solution. Small molecules can enter even the narrowest openings in the porous structure, larger molecules find fewer openings they can get into, and still larger molecules may be completely excluded from the solid phase. As a result, large, bulky molecules pass down the column faster than smaller molecules. A stationary phase for *ion-exchange chromatography* has ionizable functional groups that carry a negative or positive charge and therefore attract certain ionic solutes. In organic chemistry, such stationary phases are used mainly to separate ionizable organic compounds such as carboxylic acids, amines, and amino acids. Chiral stationary phases that can separate enantiomers and determine their optical purity are also available.

Hundreds of different stationary phases are used for HPLC separations. Some of the most popular reverse-phase packings contain silica gel bonded to methyl ($-\text{CH}_3$), phenyl ($-\text{C}_6\text{H}_5$), octyl [$-(\text{CH}_2)_7\text{CH}_3$], octadecyl [$-(\text{CH}_2)_{17}\text{CH}_3$], cyanopropyl [$-(\text{CH}_2)_3\text{CN}$], and aminopropyl [$-(\text{CH}_2)_3\text{NH}_2$] groups. Size-exclusion stationary phases contain silica, glass, and polymeric gels of varying porosity. Ion-exchange stationary phases include (1) styrene—divinylbenzene copolymers to which ionizable functional groups (such as $-\text{SO}_3\text{H}$ and $-\text{NR}_4{}^+\text{OH}^-$) are attached; (2) beads with a thin surface layer of ion-exchange material; and (3) bonded phases on silica particles.

General Directions for HPLC

 ## Standard Scale and Microscale

Do not attempt to operate the instrument without prior instruction and proper supervision. Because HPLC systems vary widely in construction and operation, only a very general outline of the procedure is provided here. It will be assumed that all instrumental parameters have been preset, that an appropriate reverse-phase column has been installed, and that the elution will be isocratic. If not, the instructor will provide additional directions.

Prepare an approximately 0.1% stock solution of the sample in the same solvent or solvent mixture as the one being used for the elution and dilute an aliquot of this solution further, if necessary. Be certain that you are using HPLC-grade solvents for preparing the solution and eluting it through the column. The solvents should be purged with helium prior to elution to remove dissolved gases. Filter the solution through a 1.0-μm membrane filter to remove any particulate matter. Degas it, if necessary, as directed by your instructor. Inject the sample through the injection port, or use a sampling valve to introduce it into the system. (Using a guard column to protect the analytical column from particles is advisable.) Wait until no more peaks appear on the chromatogram, and then inspect the chromatogram to see if the components are well resolved. If they are not, repeat the determination using a higher percentage of the more polar solvent (usually water) in the solvent mixture.

OPERATION 39

Infrared Spectrometry

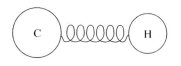

Figure G6 "Ball-and-spring" model of a chemical bond

Principles and Applications

The atoms of a molecule behave as if they were connected by flexible springs rather than by rigid bonds resembling the connectors of a ball-and-stick molecular model (see Figure G6). A molecule's component parts can oscillate in different *vibrational modes*, which are vividly described by such terms as stretching, rocking, scissoring, twisting, and wagging. When infrared (IR) radiation is passed through a sample of a pure compound, its molecules can absorb radiation of the energy and frequency needed to bring about transitions between vibrational ground states and vibrational excited states. For example, if a molecule contains a C—H bond that vibrates 90 trillion times a second in its vibrational ground state, the molecule must absorb IR radiation having exactly that frequency (9.0×10^{13} Hz) to jump to an excited state in which the C—H bond vibrates twice as fast.

The frequency of IR radiation is usually given in *wave numbers* ($\bar{\nu}$). The wave number in cm^{-1} of a vibration is the number of peak-to-peak waves per centimeter. The relationship between wave number and frequency (ν) in hertz (s^{-1}) is given by

$$\nu = c \cdot \bar{\nu}$$

where c is the speed of light ($\sim 3.0 \times 10^{10}$ cm s^{-1}). For example, the wave number of 9.0×10^{13} Hz radiation is

$$\bar{\nu} = \frac{\nu}{c} = \frac{9.0 \times 10^{13}\ \text{s}^{-1}}{3.0 \times 10^{10}\ \text{cm s}^{-1}} = 3000\ \text{cm}^{-1}$$

The wave number of an IR band is the inverse of its wavelength, which is generally measured in micrometers (μm); the two can be interconverted using the following relationship:

$$\text{wave number (in cm}^{-1}) = \frac{10^4\ \mu\text{m/cm}}{\text{wavelength(in } \mu\text{m})}$$

For example, the wave number of a 5.85 μm C=O bond vibration is $(10,000/5.85)$ cm^{-1} = 1710 cm^{-1}.

The *infrared (IR) spectrum* of a compound plots the amount of IR radiation the compound absorbs over a broad range of wavelengths, usually about

$2.5–17$ μm $(4000–600$ cm$^{-1})$. An IR spectrum is obtained by placing a sample of the compound in an *infrared spectrometer*, which measures the sample's response to IR radiation of different wavelengths. On a typical IR spectrum, wavelengths increase and wave numbers decrease going from left to right; so wave-number ranges are usually reported with the higher value first. When a sample absorbs IR radiation of a given wavelength, the recorder pen on a dispersive IR spectrometer (described below) moves downward by a distance that depends on the amount of IR radiation absorbed. Therefore, a typical IR spectrum consists of a series of inverted peaks, with each peak corresponding to a different kind of bond vibration. Because vibrational transitions are usually accompanied by rotational transitions in the frequency region scanned, the inverted peaks appear as comparatively broad "valleys" called *IR bands*, rather than sharp peaks like those seen in NMR spectra [OP-40].

Infrared spectrometry is most often used to detect the presence of specific functional groups and other structural features from band positions and intensities, and to show whether an unknown compound is identical to a known compound whose IR spectrum is reproduced in literature sources or computerized databases. The *fingerprint region* of an IR spectrum $(1250–670$ cm$^{-1})$ is best for establishing that two substances are identical, since the bands found in this region are often characteristic of the molecule as a whole and not of isolated bonds. IR spectrometry is particularly useful in synthetic organic chemistry, because comparing the spectra of the reactant and product should clearly show whether or not a reaction took place, and will often reveal the identity of the product.

IR spectrometry may also be used to assess the purity of a compound, monitor the rate of a reaction, measure the concentration of a solution, and study hydrogen bonding and other phenomena.

A spectrometer is an instrument that measures and records the components of a spectrum in order of wavelength, mass, or some other property.

Instrumentation

In a conventional *dispersive infrared spectrometer*, IR radiation is broken down into its component wavelengths by a diffraction grating or prism and beamed through the sample, one wavelength interval at a time. As the spectrum is scanned from lower to higher wavelength (higher to lower wave number), the radiation that passes through the sample at each wavelength is detected and its intensity is recorded on a chart. The resulting IR spectrum is a graph of *percent transmittance* (the percentage of radiation that passes through the sample) on the *y*-axis versus wavelength and wave number on the *x*-axis.

In a *Fourier-transform infrared (FTIR) spectrometer*, the sample is irradiated briefly with an intense beam that contains the entire spectrum of IR radiation in the instrument's range. The multiple-wavelength radiation that exits the sample is then converted by a microprocessor to an IR spectrum that closely resembles a scanned spectrum. To understand how an FTIR spectrometer works, refer to the diagram in Figure G7. The heart of the instrument is a *Michelson interferometer*, which causes two beams of IR radiation to interfere with one another. The radiation generated by a heated filament or another IR source passes through a *beam splitter* that sends half of the radiation to a fixed mirror and half to a moving mirror. The beam that reflects off the fixed mirror always travels the same distance from the source to the sample; the distance the other beam travels is varied continuously by the moving mirror. The beams from both mirrors are recombined before the IR radiation passes

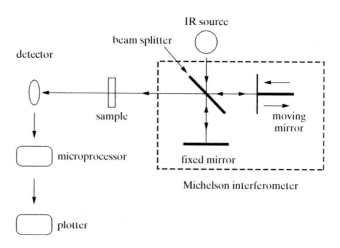

Figure G7 Schematic diagram of an FTIR spectrometer

through the sample. Suppose for a moment that the IR beam coming from the source is of a single wavelength, λ. When the moving and fixed mirrors are the same distance from the beam splitter, both beams will travel the same distance before they recombine. As a result, the peaks and troughs of their waveforms will be aligned and will interfere constructively with one another, sending a beam of high intensity to the sample. When the moving and fixed mirrors are not the same distance from the beam splitter, the beams will travel different distances and their waveforms will usually not be aligned peak to peak when they recombine. For example, if the peaks of one beam exactly align with the troughs of the other, the beams will interfere destructively with one another, sending no radiation to the sample. In other alignments, the intensity of the combined beam will be somewhere between these two extremes.

As a result of constructive and destructive interference, the intensity of a light beam of wavelength λ will vary, with a sinusoidal wave pattern, at the frequency of the moving mirror. This pattern is called an *interferogram*. Because light beams of a different wavelength will align differently when they recombine, each different wavelength of IR radiation will generate a different interferogram. Interferograms are additive, so the interferogram that reaches the sample will be the sum of all the interferograms of all of the wavelengths. If IR radiation of a particular wavelength is absorbed as it passes through the sample, the intensity of that wavelength's interferogram — and thus its contribution to the overall interferogram — will decrease, changing the waveform of the overall interferogram that exits the sample. As a result, the interferogram that exits the sample will differ from the one that entered it in a way that depends on the amount by which the intensity of each of its component interferograms was reduced by the sample. In other words, this interferogram contains all of the intensity information for all of the different wavelengths that passed through the sample — there is no need to break down the IR radiation into its component wavelengths and measure the intensity of the transmitted radiation at each individual wavelength, as for a dispersive IR spectrometer.

When the interferogram impinges on a detector, it generates an electronic signal that is sent to a microprocessor. The microprocessor uses a mathematical technique called *Fourier-transform analysis* to "decode" the interferogram and recover the intensity data for each wavelength. These data are then plotted as an FTIR spectrum.

Unlike a continuous-wave IR spectrometer, which takes 5 minutes or more to record a spectrum, an FTIR spectrometer can acquire a complete spectrum in a few seconds, making it possible to accumulate many spectra of the same sample in a minute or less. The data from these spectra are averaged to yield a composite spectrum that is much cleaner and better resolved than a spectrum obtained by a dispersive instrument, because it lacks the electronic background noise that can distort IR bands in a scanned spectrum. This averaging capability makes it possible to obtain good spectra using very small samples (1 mg or less). In addition, a helium–neon laser is used as a standard against which the frequencies of the radiation are measured, so the wave numbers displayed by an FTIR spectrum are much more accurate than those on a scanned spectrum.

Experimental Considerations

An IR spectrometer is a precision instrument that may cost more than a Jeep Grand Cherokee, but does not respond as well to rough handling. Treat it with respect and follow instructions carefully to prevent damage and avoid the need for costly repairs. Never unplug or switch off the instrument unless directed otherwise—some IR spectrometers must be kept on continually to prevent damage to their optical parts. Keep water and aqueous solutions away from an IR spectrometer, since some of its components may be water sensitive. Never move the chart holder, drum, or pen carriage on a dispersive instrument while the instrument is in operation or before it has been properly reset at the end of a run—this can damage mechanical components. Do not leave objects lying on the bed of a flat-bed recorder, as they might jam the chart drive.

Background and Reference Scans. Before running the IR spectrum of a sample in an FTIR spectrometer the instructor or user must perform a *background scan*, which is automatically subtracted from the spectrum of the sample to remove IR bands (such as those from atmospheric CO_2) that don't belong to the sample. The same background scan can be used for many samples, but a new background should be recorded occasionally to compensate for changes in the lab's atmosphere. Alternatively, a *reference scan* can be run, saved in the instrument's memory, and subtracted from the sample's spectrum before it is printed or plotted.

When a neat liquid sample is being analyzed in a sample cell, the background scan is run either with no cell in the sample compartment or (sometimes) with a cell containing the salt plates to be used with the sample. The background scan then contains signals due to carbon dioxide and water vapor from the atmosphere and certain optical components of the instrument. When a liquid sample is being analyzed using a disposable IR card, the blank card to which the liquid will be applied is used for the background or reference scan. When a solution is to be analyzed, the background scan is run with a sample cell containing the pure solvent. When a KBr disk spectrum is being run in a Mini-Press (described in the sample preparation section), the empty press can be placed in the sample compartment for the background scan (this may not be necessary for routine spectra).

See the following section on sample preparation for information about sampling devices.

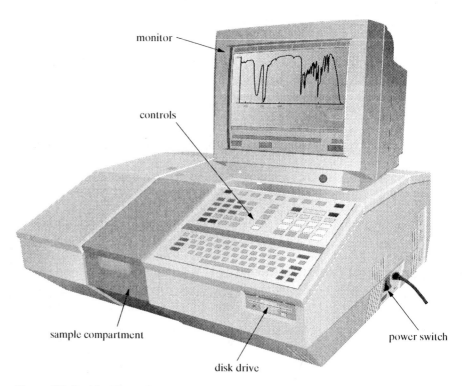

monitor

controls

sample compartment

disk drive

power switch

Figure G8 Perkin–Elmer Spectrum RXI FTIR spectrometer

FTIR Spectrometers.   To obtain an IR spectrum using an FTIR spectrometer, (Figure G8) the user places a cell or other sampling device containing the sample into the sample compartment and selects the desired
number of scans (usually 4–16). Unless a large number of scans is selected,
the spectrum should appear on the monitor in less than a minute. A copy of
the spectrum can then be obtained by using a plotter or a computer printer.
Most FTIR spectrometers can print the wave number of each band directly
on the spectrum; with some instruments, you may need to determine the
wave numbers by using a cursor and then write them on the spectrum.
FTIR spectrometers are often provided with a library of IR spectra that can
be searched for comparison with the spectrum of a reaction product or an
unknown compound.

Dispersive IR Spectrometers.   A typical dispersive IR spectrometer
with a flat-bed recorder is shown in Figure G9. The chart paper on which
spectra are recorded, which has wave number and wavelength values printed
on it, is clipped to a moveable paper carriage. With other instruments, the
chart paper may be wrapped around a moveable drum. The user may need
to align the chart paper by matching a wave number on the chart (for instance,
4000 cm^{-1}) with an alignment mark on the recorder bed or drum. If it isn't
aligned properly, the IR bands will appear at the wrong locations on the
chart paper, and their wave numbers will be incorrect. The user then moves
the paper carriage or drum to its initial position (with the pen at the far left
of the chart paper), moves the pen to a point near the top of the chart paper
with an attenuator control to set the baseline, and scans the spectrum.

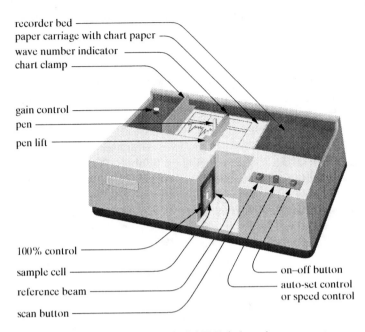

recorder bed
paper carriage with chart paper
wave number indicator
chart clamp

gain control
pen
pen lift

100% control
sample cell
reference beam
scan button

on–off button
auto-set control
or speed control

Figure G9 Perkin–Elmer Model 710b infrared spectrometer

Sample Preparation

Most IR spectra are obtained by using sample cells into which the sample is introduced with a syringe or by some other means. Certain sampling accessories, such as attenuated total reflectance devices (ATRs), make it possible to record the IR spectrum of a liquid or solid without using a sample cell. Since sample cells and other sampling accessories vary widely in construction and application, most sampling techniques should be demonstrated by the instructor.

Care of Infrared Windows. In most sample cells, the sample is placed between two *windows* made by compressing a metal halide (sodium chloride, silver chloride, etc.) to form a transparent round or rectangular crystal. These windows are very fragile and most are water-soluble. A window should be touched only on the edges with clean, dry hands or gloves and handled with great care to avoid damage. Sodium chloride and potassium chloride windows must not be exposed to moisture; even breathing on such a window can cause some etching because of moisture in your breath. Therefore, all samples and solvents that will come into contact with NaCl and KCl windows must be dry, and the windows must be stored in a dry atmosphere when not in use. After use, NaCl and KCl windows must be washed with a dry, volatile solvent such as dichloromethane, sodium chloride-saturated absolute ethanol, or a 1:1 mixture of absolute ethanol and low-boiling petroleum ether. They are then allowed to air dry or carefully blotted dry with nonabrasive tissues, and stored in a desiccator. Silver chloride windows can be washed with ethanol or acetone. These windows are unaffected by water, but they must be stored in the dark because AgCl eventually turns black when exposed to light.

Thin Films.

Thin films cannot be used for very volatile liquids, because they may evaporate before the spectrum is complete; with such liquids, use a spacer as described later for volatile neat liquids. To prepare a thin film of a neat (undiluted) liquid using a *demountable cell* such as the one illustrated in Figure G10, layer a gasket and a window on the back plate and place 1–2 drops of the liquid on the window. (If the cell has round windows, use an O-ring rather than a gasket.) Then position the upper window by touching an edge to the corresponding edge of the lower window and carefully lowering it into place. Press the windows together so that the liquid fills the space between them, taking care to exclude air bubbles. Place another gasket (or O-ring) on the top window, add the front plate, and tighten the cinch nuts just enough to hold the cell components together securely. Be careful not to overtighten the nuts, as this may make the liquid film too thin or even break a window. After you have recorded a spectrum, disassemble the cell and wash, dry, and store the windows as described previously.

The demountable silver chloride mini-cell illustrated in Figure G11 is especially convenient for microscale work, because less sample is required; a small drop from a Pasteur pipet is usually sufficient. After the sample is spread between the flat sides of the recessed AgCl windows, the cell is assembled by placing the windows, protected by two O-rings, inside the plastic window holder and carefully screwing on the plastic retainer. When an infrared spectrum is recorded, the cell is supported in the special cell holder shown.

Volatile Neat Liquids.

If a liquid is very volatile, its spectrum can be run in a demountable, sealed, or sealed–demountable cell, using a spacer approximately 0.015–0.030 mm thick. Fill a demountable cell by placing the spacer on the lower window, adding sufficient liquid to fill the cavity in the spacer, positioning the upper window, and reassembling the cell as described

Take Care! Hold the windows by their edges.

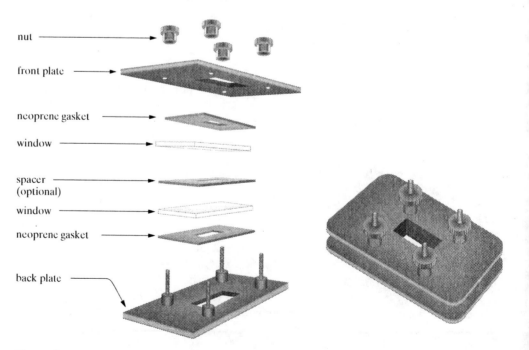

Figure G10 Demountable cell for rectangular windows

nut

front plate

neoprene gasket

window

spacer (optional)

window

neoprene gasket

back plate

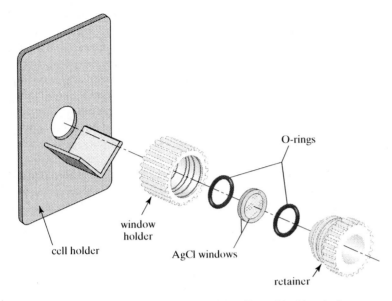

Figure G11 Demountable mini-cell for silver chloride windows

in the previous section. After you have recorded a spectrum, disassemble the cell and wash, dry, and store the windows as described previously.

A *sealed cell* or *sealed–demountable cell* (Figure G12) is filled by injecting the sample into one of its filling ports with a Luer-Lok syringe body. If you are using a Sealed–demountable cell that has not been assembled, assemble it using Figure G12 as a guide, starting with the back plate. Remove both plugs from the filling ports, draw about 0.5 mL of liquid into the syringe,

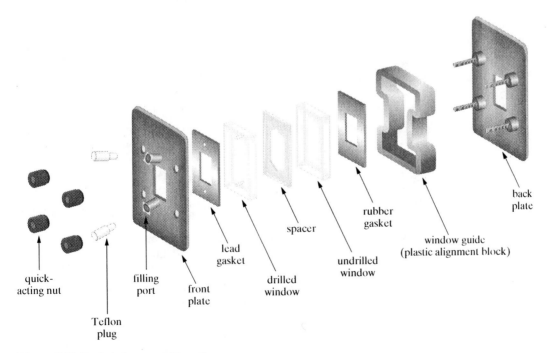

Figure G12 Sealed–demountable cell

Take Care! The syringes are fragile and break easily.

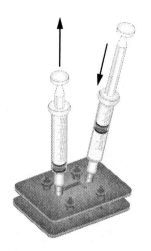

Figure G13 Push–pull method for flushing a sealed or Sealed–demountable cell

and carefully insert the syringe tip into one of the ports with a slight twist. Holding the cell upright with the syringe port at the bottom, depress the syringe's plunger until the space between the windows is filled and a small amount of liquid appears at the upper port. If there is much resistance to filling, try the push–pull method described next. If there are air bubbles between the windows, they can sometimes be removed by tapping gently on the metal frame of the cell. Put the cell on a flat surface, remove the syringe, and insert a plug in the upper port with a slight twist. Remove any excess solvent in the lower port with a piece of tissue paper or cotton and close that port with another plug.

After the spectrum has been run, clean the cell by removing most of the liquid with a syringe and flushing the cell several times with a dry, volatile solvent. To flush a cell using the push–pull method illustrated in Figure G13, lay it on a flat surface and insert a syringe filled with solvent into one port and an empty syringe into the other port. Twist the syringes slightly as you insert them so that they will not pull out easily. Slowly push on one plunger while pulling on the other, as shown in the figure, to draw washing liquid through the cell from one syringe to the other. Do this several times; then remove the excess liquid with an empty syringe and dry the cell by passing clean, dry air or nitrogen between the windows. An ear syringe or a special cell-drying syringe can be used to force air (gently!) through the cell, or it can be dried by attaching a trap and aspirator to one port and a drying tube filled with desiccant [OP-12a] to the other. Store the cell in a desiccator [OP-26b].

Solutions. Solutions of solids or liquids in a suitable solvent can be analyzed in a cell that has a spacer 0.1 mm or more in thickness. An inexpensive solution cell, constructed as described in *J. Chem. Educ.* **1991**, *68*, A124, is particularly useful for microscale work, as it requires only a small volume of solution. The solvent should be dry, relatively nonpolar, unreactive with the solute, and transparent to IR radiation in the regions of interest. It should ordinarily dissolve enough of the solute to yield a 5–10% solution; more dilute solutions can be used with FTIR spectrometers. Only a few solvents, such as carbon tetrachloride, chloroform, and carbon disulfide, meet these criteria with a wide range of solutes. These solvents (particularly carbon tetrachloride) are all hazardous to human health, and carbon disulfide is extremely flammable, so they are to be used only with appropriate safety equipment (hood, gloves, safety goggles, etc.) and under close supervision by an instructor. It is best to use spectral-grade solvents so that impurities will not give rise to extraneous peaks.

When the sample is run, the IR spectrum of the solvent is, in effect, subtracted from the solution spectrum to give the spectrum of the solute. Even then, strong solvent peaks will obscure certain portions of the spectrum. For example, the spectrum of a solute run in chloroform will yield no useful information in the $1250–1200$ cm^{-1} and $800–650$ cm^{-1} regions because the chloroform absorbs nearly all of the IR radiation there. Sometimes spectra are run separately in two solvents that absorb at different wave numbers, such as chloroform and carbon disulfide, to obtain the equivalent of a complete spectrum.

With an FTIR spectrometer, an IR cell is first filled with the pure solvent and the solvent's spectrum is run as a background or reference scan. When the spectrum of the sample is run in the same cell, the background or reference spectrum is subtracted to yield the spectrum of the solute. The background spectrum is stored in the instrument until another background is recorded, so a number of samples can be run in the same cell using the same background. With a dispersive IR spectrometer, one cell is filled with the solution and another identical cell is filled with the solvent, using spacers of the same thickness. The spectrum is then run with the solvent cell in the reference beam.

To prepare a sample cell for a solution spectrum, make up a 5–10% solution of the substance to be analyzed. Usually, 0.1–0.5 mL of solution will be required. A 0.1-mm spacer can be used unless the solution is quite dilute. If the cell is ported, fill it using a Luer-Lok syringe body as described under "Volatile Neat Liquids." When the spectrum has been recorded, clean the cell by removing the excess solution and flushing it (as described under *Volatile Neat Liquids*) with the pure solvent used in preparing the solution. If necessary, rinse the cell with a more volatile solvent before drying it, and store it in a desiccator.

A silver chloride mini-cell can be used for solutions by filling the recessed side of one AgCl window with the solution and covering it with the flat side of the other window, so as to exclude air bubbles.

Mulls.   A solid sample can be prepared as a *mull* in Nujol (a kind of mineral oil) or another mulling oil. The sample spectrum should be compared with a spectrum of the mulling oil so that peaks due to the oil can be identified and disregarded during interpretation. When Nujol is used, the aliphatic $C-H$ stretching and bending regions (3000–2850, 1470, 1380 cm^{-1}) cannot be interpreted, but most functional groups and other structural features can be identified. If it is necessary to examine the entire spectrum, another sample can be prepared using a complementary mulling oil such as Fluorlube. Nujol is essentially transparent at wave numbers lower than 1300 cm^{-1}, while Fluorlube is transparent above 1300 cm^{-1}. Some scattering of IR radiation by particles of the solid will reduce transmittance at the high wave-number end of the spectrum, so the baseline of a dispersive IR spectrometer should not be set at that end, but wherever the transmittance is highest.

Preparing a usable mull takes some practice, so follow the directions carefully and be ready to prepare additional samples if the first one doesn't work out. Grind about 10–20 mg of the solid in an agate or mullite mortar until it coats the inner surface of the mortar and has a glossy appearance (at least 5 minutes of grinding is recommended). The particles must be ground to an average diameter of about 1 μm to avoid excessive radiation loss by scattering. Add a drop or two of mulling oil and grind the mixture for a minute or so after all of the solid that coated the inside of the mortar is incorporated into a paste. The paste should have about the consistency of petroleum jelly. Transfer most of the mull to the lower window of a demountable cell using a rubber policeman. Spread the mull into a uniform thin film by sliding or rotating the top window, taking care to exclude air bubbles. The mull, as viewed through the cell windows, should be translucent or transparent, not opaque or grainy. Assemble the cell, run the spectrum, and clean the windows as for a neat liquid, using low-boiling petroleum ether or another suitable solvent. Then dry and store the windows as described previously.

Potassium Bromide Disks.   A potassium bromide (KBr) disk is prepared by mixing a solid with *dry* spectral-grade potassium bromide and using a die to press the mixture into a more or less transparent wafer. Potassium bromide absorbs moisture from the atmosphere, so it must be kept in a

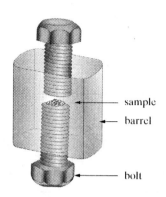

sample

barrel

bolt

Figure G14 Potassium bromide Mini-Press

If another kind of press is to be used, follow the manufacturer's or your instructor's directions for preparing the disk.

See J. Chem. Educ. **1977**, *54*, 287 *for additional suggestions regarding the preparation of KBr disks.*

tightly capped container and stored in an oven or desiccator. It is advisable to dry it before use by heating it in a 110°C oven for several hours. Even then it will usually pick up enough moisture during sample preparation to give rise to O—H bands around 3450 cm^{-1} and 1640 cm^{-1}.

It takes time and effort to prepare a usable KBr disk by manual grinding because it is difficult to get the particles fine enough; even then, there may be some spectral degradation due to scattering of IR radiation in the sample. (A ball mill, as described in the next paragraph, works better.) Grind 0.5–2.0 mg of the solid very finely in a dry agate or mullite mortar as if you were preparing a mull; then add about 100 mg of dry potassium bromide, mix it thoroughly with the sample, and grind it very finely. Use about half of this mixture to prepare a disk with a Mini-Press, as described below. If a pneumatic press will be used, a larger sample may be necessary.

If a small vibrating ball mill is available, such as the Wig-L-Bug used by some dentists, mix about 1 mg of the solid with 200 mg of dry KBr (you can scale up the quantities if necessary), put the mixture in the ball mill, and run the mill for 5 minutes. Use about 1/4 of this mixture to prepare a disk with a Mini-Press.

To use a Mini-Press (Figure G14), screw in its bottom bolt five full turns and introduce the indicated amount of sample mixture into the barrel. Keeping the open end of the barrel pointed up, tap it gently against the bench top to level the mixture, brush down any material on the threads with a soft brush, and screw in the top bolt. Alternately tap the bottom bolt on the bench top and screw in the top bolt with your fingers several times to level the sample further. When the bolt is finger tight, clamp the bottom bolt in a vise and gradually tighten the top bolt as far as you can with a heavy wrench (or to 20 ft-lbs with a torque wrench). Leave the die under pressure for a minute or two. Remove both bolts, leaving the KBr disk in the center of the barrel, and check to see that it is translucent or nearly transparent and homogeneous; if not, prepare a new disk. A disk may be cloudy or inhomogeneous because the sample is wet, not thoroughly ground, or not completely mixed; the disk is too thick; the sample size is too large; or the top bolt was not tightened enough.

Place the block containing the disk on a holder (provided with the Mini-Press) in the spectrometer sample beam. As for a mull, scattering will reduce transmittance at the left end of the spectrum, so for a dispersive instrument set the baseline wherever the transmittance is highest. After you have run the spectrum, punch out the KBr disk using the eraser end of a pencil. Wash the barrel and bolts with water, rinse them with acetone or methanol, and store the clean, dry press in a desiccator.

Melts. IR spectra of low-melting solids can sometimes be obtained by one of the following procedures.

A. Spread a thin, uniform layer of the finely powdered solid on a silver chloride window, cover it with another silver chloride window, and heat the assembly *slowly* on a hot plate. As soon as the solid melts to produce a uniform film between the windows, remove the windows, press them together with forceps until the sample solidifies, install them in a demountable cell, and run the spectrum as for a thin film. If the spectrum is distorted because of excessive light scattering, try method **B**.

B. Heat a pair of sodium chloride windows in an oven to a temperature at least 20°C above the melting point of the solid, then set one window on the back plate of a demountable cell (don't forget to use a gasket or O-ring). Without delay, place about 0.1 g of the solid on that window; position the other window so that the solid, as it melts, fills the space between the windows; and assemble the cell. If possible, run the spectrum while the sample is still liquid. You may still be able to get a good spectrum if the liquid crystallizes as a glassy film or very small crystals, but large crystals produce excessive light scattering.

Disposable IR Cards. If your laboratory has disposable IR cards available, you can obtain a spectrum by using the blank card for a background or reference scan and then applying the sample to the same card and scanning its spectrum. Apply the sample, as a neat liquid or a solution of a liquid or solid in a volatile solvent such as dichloromethane, to the matrix (a thin polymer film) in the middle of the card and allow the solvent (if any) to evaporate. After running the spectrum, subtract the saved reference spectrum, if one was recorded. When necessary to obtain a satisfactory spectrum, another spectrum can be recorded using the same card, after excess sample has been removed or more sample has been added. Excess sample may result from using a liquid neat rather than in solution; the excess can be removed by rinsing the matrix with a small amount of volatile solvent and letting the solvent evaporate. The card can be reused if it is rinsed with enough solvent to remove all traces of the previous sample (you may have to run another spectrum to be sure of that).

One disadvantage of this method is that the IR bands of the card's matrix, which is made of polyethylene (PE) or polytetrafluoroethylene (PTFE), may alter or obscure some bands from your product, particularly those in the C—H region if a PE matrix is used. Also, liquids that boil below 130°C may evaporate too fast to yield a satisfactory spectrum.

Attenuated Total Reflectance (ATR) Sampling. Some FTIR instruments are equipped with ATR accessories that greatly simplify the process of recording an IR spectrum. The sample is placed on a crystal of zinc selenide, germanium, diamond, or some other material having a high refractive index. An infrared beam is directed through the crystal at the angle at which total reflection occurs from both of its faces so that the beam, in effect, bounces off the sample numerous times to yield what is called an evanescent wave. At each point of contact, the wave penetrates a few micrometers into the sample. At frequencies where the sample absorbs energy, the IR beam is attenuated or altered. It is then directed to the instrument's detector, where the spectrum is generated.

A typical ATR accessory suitable for liquid samples has a recessed crystal surface in a horizontal plate. After a background scan is performed, the crystal is covered with a thin layer of the liquid and the spectrum is recorded. Solids are best analyzed on an ATR accessory having a diamond surface; the solid is forced onto the surface with a pressure tip that holds it in close contact with the crystal. In either case the crystal surface must be cleaned thoroughly with a solvent-soaked tissue or cotton swab after a spectrum is run.

General Directions for Recording an Infrared Spectrum

 Standard Scale and Microscale

Equipment and Supplies

(Starred items are needed only for the sample types indicated in parentheses.)

IR spectrometer

sample

Pasteur pipet

IR sample cell(s) and windows (or other sampling device)

*Luer-Lok syringe, spacer (volatile neat liquids)

*spectral-grade solvent, Luer-Lok syringe, spacer (solutions)

*mortar & pestle, mulling oil, glass rod with rubber policeman (mulls)

*dry potassium bromide, mortar & pestle, Mini-Press, vise, wrench (KBr disks)

*hot plate and forceps or oven (melts)

*disposable IR card (IR cards)

solvent for cleaning windows, etc.

tissues

desiccator for storing windows

Do not attempt to operate the instrument without prior instruction and proper supervision. The construction and operation of commercial IR spectrometers vary widely, so the following is meant only as a general guide to assist you in recording an IR spectrum. Specific operating techniques must be learned from the instructor, the operating manual, or both. It will be assumed that the necessary operational parameters have been set beforehand; if not, your instructor will show you what to do. If you are using an ATR accessory, your instructor should show you how to apply your sample. If you are using an FTIR spectrometer, follow Procedure **A**; for a dispersive IR spectrometer, follow Procedure **B**.

A. FTIR Spectrometer. Your instructor will tell you which keys to use for such processes as scanning a spectrum, using a cursor, and printing or plotting a spectrum. If your compound is in solution, run a background or reference spectrum with a cell containing the pure solvent in the sample compartment and use the same cell for the solution. If you are using a disposable IR card, scan its spectrum before you apply the sample. Otherwise run the background or reference spectrum (if one is necessary) with the sample compartment empty just before you record your spectrum. Prepare a sample cell, IR card, or KBr disk by one of the methods described previously and place the cell, card, or KBr disk holder in the sample cell holder. Select the desired number of scans (four is usually sufficient for a routine spectrum) and start scanning the spectrum. Wait until a spectrum appears on the monitor. If the instrument does not automatically subtract a background or reference spectrum, do so as directed by your instructor. If the spectrum doesn't look right—for example, if the low or high wave-number end is missing or you are seeing only a small part of the total spectrum—display the "normal" spectrum using the appropriate key(s) (some instruments use *Rerange* and *Rescale* keys for this purpose). The strongest bands should extend nearly to the bottom of your spectrum; if they do not, use a vertical-scale expansion key to improve the appearance of the spectrum. Enter additional data to your spectrum, such as wave numbers, as directed by your instructor. See that the printer or plotter is turned on and properly adjusted, and then press the appropriate key to print or plot the spectrum. If the instrument you are using doesn't not record wave numbers directly on the spectrum, use the cursor arrow keys to move the cursor to the significant bands and write the displayed wave numbers below the corresponding IR bands on your spectrum. Reset the instrument to display a normal spectrum, if necessary. Then remove the sample cell and close the sample compartment. Clean, dry, and store the cell windows (if you are using them) as described previously.

B. Dispersive IR Spectrometer.   Prepare a sample cell or KBr disk containing the sample by one of the methods described previously and place the cell or KBr disk assembly in the sample cell holder. If you are running a solution spectrum, place an identical cell containing the solvent in the reference compartment; otherwise, leave it empty. If necessary, place chart paper on the paper carriage (or wrap it around a drum); then align the paper properly and move the paper carriage or drum to the starting position. Set the 100% transmittance control so that the pen is at 85–90% *T*, lower the pen onto the chart paper, and scan the spectrum at "normal" or "fast" speed. Examine the spectrum to see that the absorption bands show satisfactory intensity and resolution. Ideally, the strongest absorption band should have a maximum transmittance of 5–10%. If your first spectrum is not acceptable, try varying the following parameters, depending on the sample preparation method:

> **Take Care!** Make sure the instrument has been reset correctly before attempting to move the drum or carriage.

- *Neat liquid*—Vary cell path length or film thickness.
- *Solution*—Vary concentration or cell path length.
- *KBr disk*—Vary amount of sample or thickness of disk.
- *Mull*—Vary amount of sample or film thickness.

Then run another spectrum, using a slower speed if desired. Remove the sample and spectrum and reset the instrument, if necessary. Clean, dry, and store the sample cell or cell windows as described previously.

Summary

1. Put sample in sample cell or prepare IR card or KBr disk.
2. Put sample cell, card, or KBr disk holder in instrument's sample holder.
   IF you are using an FTIR spectrometer, GO TO 6.
3. Align chart paper, move drum or carriage to starting position.
   IF you are not analyzing a solution, GO TO 5.
4. Put solvent cell in reference beam.
5. Adjust 100% transmittance control, lower pen to paper.
6. Scan spectrum; subtract any background or reference spectrum if necessary.
   IF you are using a dispersive IR spectrometer, GO TO 8.
7. Print or plot spectrum; record wave numbers if necessary.
8. Disassemble cell; clean, dry, and store cell windows.

Note: Additional steps are required for running solution spectra.

When Things Go Wrong

If you scan an IR spectrum and see significant absorption near 2350 cm^{-1} (due to CO_2 from the atmosphere), a background scan has probably not been run recently. Perform a background scan as described in the General Directions.

If you scan a spectrum and see nothing but a flat line near the 100% transmittance level, you (or someone else) probably did a background scan with the sample in the cell compartment. Perform a background scan with nothing (or the solvent, if you are analyzing a solution) in the cell compartment, as described in the General Directions.

If you scan an FTIR spectrum and see only a partial spectrum or a display that doesn't look like a normal spectrum, use the scale- and range-setting keys on the spectrometer to display a normal spectrum.

If the baseline of your mull, KBr disk, or melt spectrum begins (on the high frequency end) well below the 100% transmittance level and slopes upward before flattening out near the 2500 cm^{-1} region, and especially if your peaks "tail off," (are distorted along their trailing edges), your sample (for a mull or KBr disk) was not ground finely enough or the melt formed unsuitably large crystals. Prepare a new mull or KBr disk, or try method B for a melt. If you still can't get a good spectrum, you may have to use a different sampling method.

If your IR spectrum shows significant absorption bands near 3500 cm^{-1} and 1650 cm^{-1}, especially if your compound does not contain OH or NH bonds, your sample is probably wet. If you are running a liquid film, solution spectrum, or mull, clean the IR windows immediately and thoroughly so that they won't be etched by the water, Then dry your product and run another spectrum. If you are using a KBr disk the presence of a little water may be unavoidable, but preparing a new disk using oven-dried KBr and avoiding extended exposure to the atmosphere should help.

If some bands on your IR spectrum "bottom out" near 0% transmittance (they will usually be somewhat flattened out at the bottom), your sample film is too thick or (for a solution) the solution is too concentrated. On an FTIR spectrometer you may be able to get a decent spectrum by using the "Autex" key or an equivalent scale expansion control. Otherwise you will have to prepare the sample cell again, using less sample or a less concentrated solution.

Suppose the peaks on your IR spectrum are broad and indistinct (lacking fine structure). If your sample is a neat liquid, it has probably been evaporating or flowing out of the infrared beam area. Look at the infrared windows to see if there are areas where there is no liquid between them. If so, prepare another sample cell, using an additional drop of liquid, and try to record the spectrum more rapidly. If you prepared a mull or KBr disk with a solid sample, you probably didn't grind the solid finely enough. The spectrum of a mull may also have a sloping baseline in this case. See *Mulls* or *Potassium Bromide Disks* and prepare another mull or disk.

If you run a spectrum with a dispersive IR spectrometer and all of the wave numbers seem to vary by the same amount from the expected values, the chart paper wasn't aligned properly or the spectrometer wasn't calibrated properly. Check to see that the appropriate wave number on the chart paper is directly opposite the alignment mark on the instrument, and if not, align it and run another spectrum. If the chart paper was aligned properly, ask your instructor to have the spectrometer recalibrated (or to help you do it) and then run another spectrum.

Interpretation of Infrared Spectra

Because most IR bands are associated with specific chemical bonds, it is usually possible to deduce the functional class of an organic compound from its IR spectrum. The *stretching* vibrations of chemical bonds resemble the vibrations of springs in that stronger bonds have higher vibrational energies and frequencies than weaker ones. Thus, triple bonds generally absorb at higher wave numbers than double bonds, and double bonds absorb at higher wave numbers than single bonds. However, because of the comparatively low mass of a hydrogen atom, single bonds to hydrogen ($C-H, O-H, N-H$, etc.) have even higher vibrational frequencies than double and triple bonds. Stretching bands involving single bonds to hydrogen occur at the high frequency (left) end of an IR spectrum, in the region between 3700 and 2700 cm^{-1} (2.7–3.7 μm). Triple bonds usually absorb between 2700 and 1850 cm^{-1} (3.7–5.4 μm), and double bonds and aromatic bonds absorb between 1950 and 1450 cm^{-1} (5.1–6.9 μm). Most IR bands between 1500 and 600 cm^{-1} (6.7–16.7 μm) are produced by *bending* vibrations or single-bond stretching vibrations. It takes less energy to bend a bond than to stretch it, so bending vibrations tend to have comparatively low frequencies. An absorption band in the 1500–600 cm^{-1} region may be associated with more than one bond; for example, the so-called acyl–oxygen stretching band of an ester arises from the vibration of $C-C-O$ units rather than isolated $C-O$ bonds.

What to Look for in an IR Spectrum

Consider the IR spectrum of 2-methyl-1-propanol (isobutyl alcohol) in Figure G15. At first glance, it may seem indecipherable—just a series of dips and rises in a graph. But each "dip" (IR band) arises from a stretching or bending vibration of one or more bonds in the 2-methyl-1-propanol molecule, and some of the bands can tell you a great deal about the molecules that gave rise to them. First look at the WAVENUMBERS scale at the

The IR spectra in this book are reproduced from the Aldrich Library of FT-IR Spectra, Edition II, *with permission.*

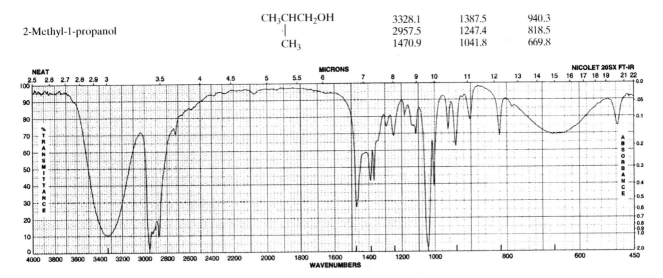

2-Methyl-1-propanol	CH$_3$CHCH$_2$OH	3328.1	1387.5	940.3
		2957.5	1247.4	818.5
	CH$_3$	1470.9	1041.8	669.8

Figure G15 Infrared spectrum of 2-methyl-1-propanol

bottom of the spectrum; from this scale you can read off the wave number, in cm^{-1}, of each band. For example, the first large band in the spectrum is between 3600 and 3000 cm^{-1}, and its minimum is at approximately 3330 cm^{-1}. Bands designated by tick marks (short lines) at the bottom of a spectrum have their exact wave numbers listed, so you can find a more accurate wave number for this band, 3328.1 cm^{-1}, in the list of numbers above the right-hand end of the spectrum.

Another numerical scale at the top of the spectrum indicates the wavelength in micrometers (microns). The numbers on the left side of the spectrum are transmittance values. For example, the minimum in the 3328 cm^{-1} band has a transmittance of ~10%, meaning that only 10% of the 3328 cm^{-1} IR radiation passed through the sample. The other 90% was absorbed during a change in the bond vibrational frequency of some bond in the compound—but which bond? Since the band is at the left end of the spectrum, the bond must have a high vibrational energy, and we have already noted that bands in the 3700–2700 cm^{-1} region of the spectrum arise from vibrations of bonds to hydrogen atoms. There are only two bonds of this type in the molecule, C—H bonds and O—H bonds. Now note that there are two bands in the 3700–2700 cm^{-1} region—the strong, broad, symmetrical one on the left, and a rather ragged band with several minima (centered around 2900 cm^{-1}) on the right. The ragged band is actually composed of several overlapping bands, arising from vibrations of several different bonds of the same general type. Since there is only one O—H bond in the molecule while there are nine C—H bonds, it is reasonable to assume that the overlapping bands on the right arise from C—H stretching vibrations and that the band on the left is the O—H band. Note that the size of a band has little to do with the number of bonds that give rise to it; the single O—H bond has a much broader band than the nine C—H bonds.

So the bond responsible for a given IR band can often be identified from its *location* on the spectrum (as indicated by its wave number), its *intensity* (relative strength), and its *shape*. Most O—H bands, like the one in the previous spectrum, are very broad and strong (a *strong* band is one whose minimum is near the bottom of the spectrum). Most C—H bands are relatively strong and give rise to a ragged array of overlapping bands.

Another very strong band appears at 1042 cm^{-1} in the Figure G15 spectrum. This band, which is in the wave-number region for single-bond stretching vibrations (except those involving hydrogen), arises from stretching vibrations of the C—O single bond. The presence of both a C—O and an O—H band in the IR spectrum of a compound is good evidence that the compound is an alcohol (or possibly a phenol), since all alcohols contain a C—O—H grouping in their molecules.

Although we have now located bands corresponding to every kind of bond in the 2-methyl-1-propanol molecule (except the C—C single bonds, which don't give prominent bands), there are still a number of bands left. This is because the same kind of bond can undergo different kinds of vibrations. For example, most of the bands just to the left of the C—O band arise from *scissoring*, *wagging*, and *twisting* vibrations of CH_2 and CH_3 groups (see Figure G16), and the very broad, weak band centered at 670 cm^{-1} arises from an O—H bending vibration.

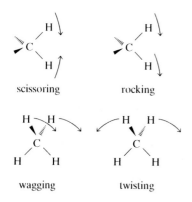

Figure G16 Some carbon–hydrogen vibrations

Spectral Regions

You can see that some IR bands, particularly the stretching bands we have discussed, are more easily recognized than others and are more useful in revealing the presence of functional groups. Many of the other bands can be ignored for the time being, although they may provide useful information to a chemist skilled in spectral interpretation. The key to efficient IR spectral interpretation is *knowing where to look* for the more useful bands. Examining the following regions of the IR spectrum will help you locate the most useful IR bands quickly:

Region 1: $3600–3200$ cm^{-1} ($2.8–3.1$ μm). Bands in this region can arise from O—H and N—H stretching vibrations of alcohols, phenols, amines, and amides. O—H bands are generally very strong and broad; N—H bands are somewhat weaker, and in the case of primary amines and amides, they have two peaks.

Region 2: $3100–2500$ cm^{-1} ($3.2–4.0$ μm). This region contains most of the C—H stretching vibrations. A strong band in the $3000–2850$-cm^{-1} region, arising from C—H bonds to sp^3 carbon atoms, is present for most organic compounds. The sp^2 C—H bonds associated with aromatic hydrocarbons and alkenes absorb at higher frequencies ($3100–3000$ cm^{-1}), and the C—H bonds of aldehyde (CHO) groups absorb at lower frequencies. The O—H bond of a carboxylic acid gives rise to a very broad absorption band in this region.

Region 3: $1750–1630$ cm^{-1} ($5.7–6.1$ μm). This region contains most of the carbonyl (C=O) stretching bands of aldehydes, ketones, carboxylic acids, amides, and esters. The carbonyl band is usually strong and quite unmistakable. Unsaturated compounds may have a C=C stretching band in the $1670–1640$-cm^{-1} region, but this band is nearly always weaker and narrower than a carbonyl band.

Region 4: $1350–1000$ cm^{-1} ($7.4–10.0$ μm). This region is usually cluttered with many C—H bending bands and other bands, but it is often possible to identify the C—O stretching bands of alcohols, phenols, carboxylic acids, and esters, and some C—N stretching bands of amines and amides.

These four spectral regions are shaded on the IR spectrum in Figure G17. In this spectrum, the absence of any band in Region 1 (or a broad band in Region 2) eliminates from consideration all compounds containing O—H and N—H bonds, including alcohols, phenols, and primary or secondary amines and amides. In Region 2, the appearance of a weak "shoulder" on the C—H band at 3050 cm^{-1} indicates an sp^2 C—H bond associated with either an aromatic ring or a carbon–carbon double bond. Region 3 has a strong C=O band at 1719 cm^{-1}, and Region 4 shows a strong C—O band at 1276 cm^{-1} as well as a weaker one at 1109 cm^{-1}. The absence of an O—H or N—H band and the presence of the C=O and two C—O bands suggest that the compound responsible for this spectrum is an ester. The compound is, in fact, the aromatic ester ethyl benzoate, whose structure and bond wave numbers are shown in the margin.

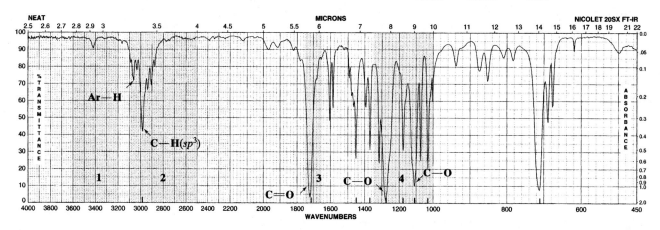

2981.9	1367.2	1108.5
1718.5	1275.8	1028.5
1451.4	1175.2	710.3

Figure G17 Classification of a compound from its IR spectrum

1719
ethyl benzoate
1276   1109

Table G2 summarizes the locations and characteristics of important absorption bands from the four spectral regions and tells you where to look for other bands that may help you confirm the presence of a particular functional group.

The best way to become proficient at identifying IR bands is to study spectra that contain those bands, such as the spectra in your lecture textbook, in the "Characteristic Infrared Bands" section that follows, and in collections of spectra described in Category F of the Bibliography. When you have learned to recognize the most important IR bands, you can take some shortcuts that will help you identify functional groups quickly—or at least eliminate the functional groups that aren't there. The following flow chart should help you do that. Just start at the top and work your way down, following the yes–no arrow that answers each question.

Possible family   C=O present?   *Possible family*
                              yes ↙   ↘ no

carboxylic   ←yes— O—H present?   O—H present? —yes→   alcohol
acid   ↓ no   no ↓   or phenol

amide (1°, 2°) ←yes— N—H present?   N—H present? —yes→   amine (1°, 2°)
                            ↓ no   no ↓

ester   ←yes— C—O present?   C=C present? —yes→   alkene
                            ↓ no   no ↓

aldehyde   ←yes— C—H at ~2700 cm⁻¹?   Ar—H present? —yes→   aromatic
                            ↓ no   no ↓   hydrocarbon

ketone   ←yes— None of the above?   None of the above? —yes→ alkane
or 3° amide   or 3° amine

Flow chart for detecting functional classes from IR bands

This flow chart is intended as a rapid screening device and is not infallible; some bands (such as the C=C stretching band) are hard to identify with certainty, and the locations of other bands may vary widely. Moreover, some

Table G2 Important bands in Regions 1–4 of infrared spectra

Region	Frequency range (cm^{-1})	Bond type	Family	Comments
1	3500–3200	N—H	amine, amide	weak–medium. 1°C: 2 bands; 2°C: 1 band; 3°C: no bands; see also Region 3.
	3600–3200	O—H	alcohol, phenol	broad, strong; see Region 4.
2	3300–2500	O—H	carboxylic acid	very broad, strong, centered around 3000; see Regions 3 and 4.
	3100–3000	C—H	aromatic hydrocarbon, alkene	may be shoulder on stronger sp^3 C—H band.
	2850–2700	C—H	aldehyde	weak to medium, usually two sharp bands; see Region 3.
3	1740–1685	C=O	aldehyde	strong; see Region 2.
	1750–1660	C=O	ketone	strong.
	1725–1665	C=O	carboxylic acid	strong; see Regions 2 and 4.
	1775–1715	C=O	ester	strong; see Region 4.
	1695–1615	C=O	amide	strong; see Region 1.
4	1350–1210	C—O	carboxylic acid	medium–strong; see Regions 1 and 3.
	1300–1180	C—O	phenol	strong; see Region 1.
	1200–1000	C—O	alcohol	strong; see Region 1. Wave numbers in order 3° > 2° > 1°.
	1310–1160	C—O	ester	strong; see Region 3. Accompanied by weaker C—O band as for alcohol.

Note: Tentative classifications must be confirmed by referring to the following descriptions of specific families.

compounds may contain more than one functional group; thus hydroxyacetone (CH_3COCH_2OH) has both an O—H and a C=O band in its spectrum, but it is not a carboxylic acid, as you could tell from the location of its O—H band. When you arrive at a tentative conclusion about the nature of the compound responsible for an IR spectrum, you should refer to Table G2 to see whether other bands in the spectrum are consistent with your initial choice. Then study the spectral characteristics of the appropriate class of compounds to confirm (or disprove) your tentative classification. The IR correlation chart on the back endpaper of this book may also help you identify some infrared spectral bands.

For example, suppose that the flow chart suggests that your compound may be an alcohol or phenol. You can first check Table G2 to see if any other bands characteristic of alcohols and phenols appear in its spectrum, such as a C—O band. If so, you should read the "Characteristic Infrared Bands" section about alcohols and phenols to find out whether your compound is an alcohol or a phenol. If you find that your compound is an alcohol, you should then study its spectrum for clues to its structure. The frequency of its C—O band may tell you whether it is primary, secondary, or tertiary. By consulting the sections on aromatic hydrocarbons and alkenes (which also apply to other compounds that contain aromatic rings and C=C bonds), you can find out whether your alcohol is aromatic or contains a carbon–carbon double bond. Of course, if you find that your compound is *not* an alcohol or phenol, you should continue down the chart or start back from the beginning.

Characteristic Infrared Bands

This section contains information about the most useful IR bands of the most commonly encountered kinds of organic compounds—alkanes, alkenes, aromatic hydrocarbons, alcohols, phenols, aldehydes, ketones, carboxylic acids, esters, amines, amides, and organic halides. Note that compounds other than hydrocarbons may contain bands characteristic of alkanes, alkenes, or aromatic hydrocarbons, so you can check the sections for these hydrocarbons when you are interpreting the spectra of other kinds of compounds. For each family of organic compounds, a summary of the main spectral features that characterize the family is followed by a description of individual bond vibrations and a representative IR spectrum. The wave-number ranges given are for solids (in Nujol mulls or KBr discs) or neat liquids; values for solutions may differ somewhat. Although the wave-number ranges apply to most of the organic compounds in each class, compounds with certain structural features (such as highly strained rings) may have bands outside of the ranges indicated.

On the spectra, absorption bands are designated either as stretching (ν) or bending (δ) bands. Only those bands that are most useful for identifying functional groups or structural features are labeled. Note that the exact wave numbers of significant bands (designated by tick marks along the lower edge of the spectra) are listed with each spectrum.

Alkanes.   Alkanes are identified primarily by the absence of any IR bands characteristic of functional groups. Their spectra are quite simple, containing only the C—H stretching and bending vibrations characteristic of sp^3 hybridized carbon atoms. Since nearly all other organic compounds contain such C—H bonds, their spectra will also contain some or all of the bands described for alkanes. (See Figure G18.)

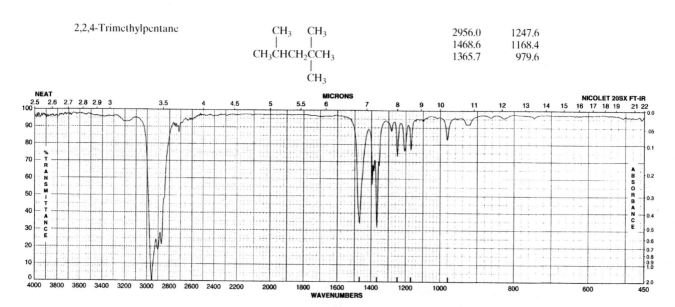

Figure G18 IR spectrum of an alkane, 2,2,4-trimethylpentane

C—H *stretch:* 3000–2800 cm^{-1} (multiple overlapping bands, strong to weak). CH$_3$ bands are near 2960 cm^{-1} and 2870 cm^{-1}. CH$_2$ bands are near 2925 cm^{-1} and 2850 cm^{-1}. Nearly always to the *right* of 3000 cm^{-1}.

C—H *bend:* 1465–720 cm^{-1} (moderate to weak). CH$_3$ bands are near 1450 cm^{-1} and 1375 cm^{-1}. CH$_2$ bands occur near 1465 cm^{-1}, between 1350 cm^{-1} and 1150 cm^{-1} (several weak bands), and sometimes around 720 cm^{-1}. The 720 cm^{-1} band is characteristic of unbranched alkanes having seven or more carbon atoms.

Alkenes. Most alkenes contain the same kinds of bands as alkanes, plus additional bands associated with carbon–carbon double bonds and vinylic (=C—H) carbon–hydrogen bonds. The presence of one or two strong bands in the 1000–650 cm^{-1} region and a sharp band near 1650 cm^{-1} suggests an alkene functional group, especially if the compound is not aromatic. (See Figure G19.)

=C—H *stretch:* 3125–3030 cm^{-1} (moderate to weak). May appear as a shoulder on a stronger sp^3 C—H band, but nearly always to the *left* of 3000 cm^{-1}.

C=C *stretch:* 1675–1600 cm^{-1} (moderate to weak; narrow). May be absent for symmetrical alkenes. Conjugation moves band to lower wavelengths.

=C—H *out-of-plane bend:* 1000–650 cm^{-1} (usually strong). Position depends on type of substitution: RCH=CH$_2$ has bands at 995–985 and 915–905 cm^{-1}; *cis*-RCH=CHR a band at 730–665 cm^{-1}; *trans*-RCH=CHR a band at 980–960 cm^{-1}; and R$_2$C=CH$_2$ a band at 895–885 cm^{-1} (R = alkyl or aryl substituent).

Aromatic Hydrocarbons. Most compounds containing benzene rings are characterized by (1) aromatic C—H (Ar—H) stretching bands near 3070 cm^{-1}, (2) a distinctive pattern of weak bands in the 2000–1650-cm^{-1} region, (3) two sets of bands near 1600 cm^{-1} and 1515–1400 cm^{-1}, and (4) one or more strong absorption bands in the 900–675-cm^{-1} region. The

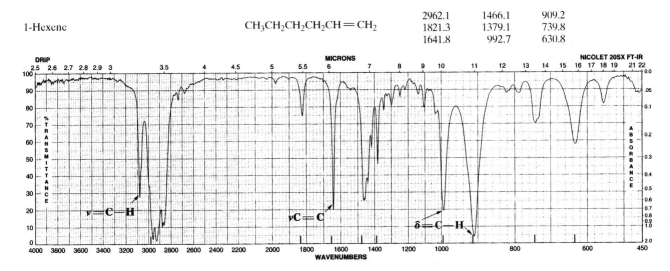

Figure G19 IR spectrum of an alkene, 1-hexene

Isopropylbenzene

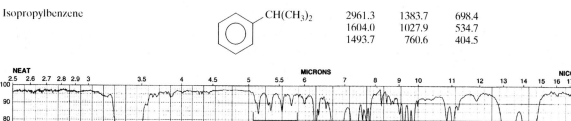

	2961.3	1383.7	698.4
	1604.0	1027.9	534.7
	1493.7	760.6	404.5

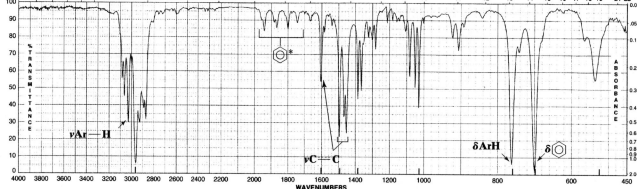

Figure G20 IR spectrum of an aromatic hydrocarbon, isopropylbenzene. The bands marked ⬡* are aromatic overtone-combination bands

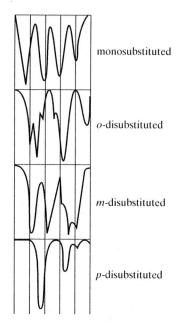

monosubstituted

o-disubstituted

m-disubstituted

p-disubstituted

Figure G21 Typical absorption patterns of substituted aromatic compounds in the 2000–1650-cm^{-1} region

presence of such bands and the absence of absorption bands characteristic of functional groups suggest an aromatic hydrocarbon. (See Figure G20.)

Ar—H *stretch:* 3100–3000 cm^{-1} (moderate to weak). May appear as a shoulder on a stronger sp^3 C—H band, but nearly always to the *left* of 3000 cm^{-1}.

Overtone-combination vibrations: 2000–1650 cm^{-1} (multiple bands, weak). The band pattern is related to the kind of ring substitution as shown in Figure G21.

C≕C *stretch:* 1615–1585 cm^{-1} and 1515–1400 cm^{-1} (variable).

Ar—H *out-of-plane bend:* 910–730 cm^{-1} (strong). The band frequency varies with the number of adjacent ring hydrogens:

two adjacent hydrogens: 855–800 cm^{-1}

three adjacent hydrogens: 800–765 cm^{-1}

four or five adjacent hydrogens: 770–730 cm^{-1}

Monosubstituted, *meta*-disubstituted, and some trisubstituted benzenes show an additional ring-bending band around 715–680 cm^{-1}. For example, a *meta*-disubstituted benzene has three adjacent ring hydrogens, so it should have bands in the 800–765-cm^{-1} and 715–680-cm^{-1} regions.

Alcohols. The presence of a strong, broad band centered around 3300 cm^{-1} and a strong C—O band in the 1200–1000-cm^{-1} region is good evidence for an alcohol. (See Figure G22.) A C—O band above 1200 cm^{-1} may suggest a phenol, particularly when it is accompanied by bands indicating an aromatic ring structure.

O—H *stretch:* 3600–3200 cm^{-1} (strong, broad). Usually centered near 3300 cm^{-1}.

2-Methyl-1-propanol

$$CH_3CHCH_2OH$$
$$\overset{|}{CH_3}$$

3328.1	1387.5	940.3
2957.5	1247.4	818.5
1470.9	1041.8	669.8

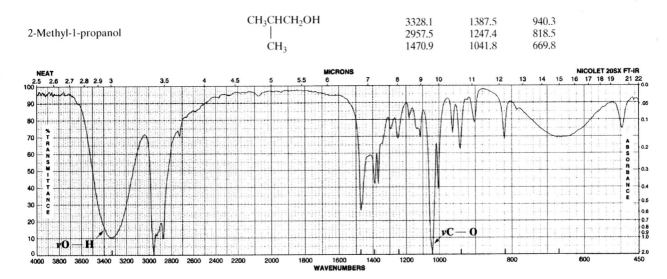

Figure G22 IR spectrum of a primary alcohol, 2-methyl-1-propanol

C—O *stretch:* 1200–1000 cm^{-1} (strong to moderate). Most saturated aliphatic alcohols absorb near 1050 cm^{-1} if they are primary, near 1110 cm^{-1} if they are secondary, and near 1175 cm^{-1} if they are tertiary. Alicyclic alcohols and alcohols with aromatic rings or vinyl groups on the carbon that is bonded to OH absorb at wave numbers about 25–50 cm^{-1} lower than these.

Phenols. Phenols are characterized by a strong, broad band centered around 3300 cm^{-1} and a strong band near 1230 cm^{-1}, accompanied by bands indicating an aromatic structure. (See "Aromatic Hydrocarbons" and Figure G23.)

Phenol

OH

3372.6	1224.4	751.8
1595.3	1168.0	689.8
1498.9	809.8	506.0

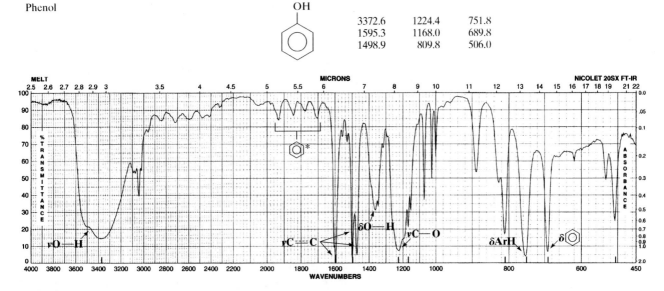

Figure G23 IR spectrum of phenol. *Note:* The bands marked ⬡* are aromatic overtone-combination bands

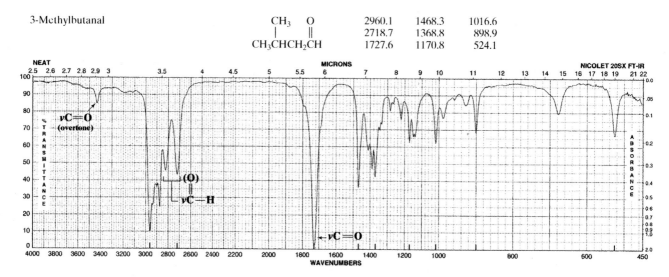

3-Methylbutanal

CH$_3$ O			
CH$_3$CHCH$_2$CH	2960.1	1468.3	1016.6
	2718.7	1368.8	898.9
	1727.6	1170.8	524.1

Figure G24 IR spectrum of an aldehyde, 3-methylbutanal

O—H *stretch:* 3600–3200 cm^{-1} (strong, broad).

O—H *bend:* 1390–1315 cm^{-1} (moderate).

C—O *stretch:* 1300–1180 cm^{-1} (strong). Usually close to 1230 cm^{-1}. This band may be split, with several distinct peaks.

Aldehydes. The presence of a sharp, medium-intensity band near 2720 cm^{-1} and a strong carbonyl band near 1700 cm^{-1} is good evidence for an aldehyde. (See Figure G24.)

$$\overset{\text{(O)}}{\underset{}{\overset{\|}{\text{C}}}}\!\!-\!\!\text{H}$$ *stretch:* 2850–2700 cm^{-1} (moderate to weak). From the carbonyl C—H bond. Most aldehydes have two bands near 2850 and 2720 cm^{-1}, with the low-frequency band well separated from other aliphatic C—H bands.

C=O *stretch:* 1740–1685 cm^{-1} (strong). Most unconjugated aldehydes absorb near 1725 cm^{-1}; conjugation of the carbonyl group with an aromatic ring or another unsaturated system shifts the band to the 1700–1685-cm^{-1} region. A weak overtone of this band may appear near 3400 cm^{-1}.

Ketones. The presence of a strong carbonyl band around 1700 cm^{-1} is good evidence for a ketone if other bands described in Table G2 (O—H, N—H, C—O, and aldehyde C—H) are absent. One or more bands in the 1300–1100-cm^{-1} region arise from C—C—C vibrations involving the carbonyl carbon. Such bands are generally weaker and narrower than C—O bands, for which they might otherwise be mistaken. (See Figure G25.)

C=O *stretch:* 1750–1660 cm^{-1} (strong). Most unconjugated aliphatic ketones absorb around 1715 cm^{-1}, and conjugated ketones absorb near 1670 cm^{-1}. A weak C=O overtone band is usually evident near 3400 cm^{-1}.

2-Pentanone

$$O$$
$$\parallel$$
$$CH_3CCH_2CH_2CH_3$$

2963.9	1366.0	1170.7
1717.4	1295.5	727.0
1422.9	1235.5	591.9

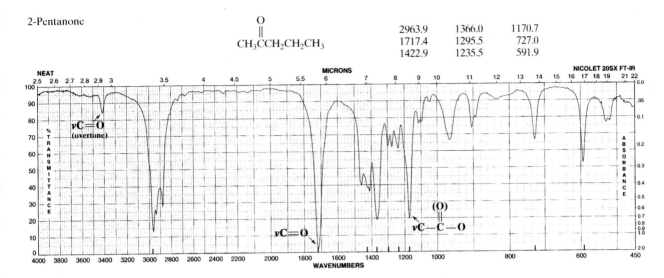

Figure G25 IR spectrum of a ketone, 2-pentanone

$$(O)$$
$$\parallel$$
$$C—C—C \quad stretch–bend:$$ 1300–1100 cm^{-1} (moderate). Often multiple bands. Unconjugated ketones absorb around 1230–1100 cm^{-1}; conjugated ketones absorb around 1300–1230 cm^{-1}.

Carboxylic Acids. The presence of a very broad band centered near 3000 cm^{-1} and a carbonyl band around 1700 cm^{-1} is good evidence for a carboxylic acid. (See Figure G26.)

Hexanoic acid

$$O$$
$$\parallel$$
$$CH_3(CH_2)_3CH_2COH$$

3191.2	1710.7	1293.4
2959.4	1467.5	1213.2
2669.9	1413.8	939.2

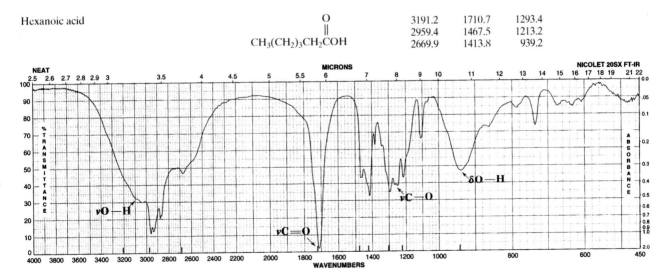

Figure G26 IR spectrum of a carboxylic acid, hexanoic acid

O—H *stretch:* 3300–2500 cm^{-1} (strong, very broad). C—H stretching bands are usually superimposed on this band.

C=O *stretch:* 1725–1665 cm^{-1} (strong). Unconjugated acids absorb around 1725–1700 cm^{-1}; conjugated acids absorb around 1700–1665 cm^{-1}.

C—O *stretch:* 1350–1210 cm^{-1} (strong). Long-chain acids may have a number of sharp peaks in this region.

O—H *bend:* 950–870 cm^{-1} (moderate, broad).

Esters. The presence of a strong carbonyl band around 1740 cm^{-1} and an unusually strong C—O band in the 1310–1160-cm^{-1} region is good evidence for an ester, especially if there is no O—H band. (See Figure G27.)

C=O *stretch:* 1775–1715 cm^{-1} (strong). Near 1770 cm^{-1} for phenyl esters (RCOOAr) and vinyl esters, 1740 cm^{-1} for most unconjugated esters, and 1730–1695 cm^{-1} for formates and conjugated esters.

C—O *stretch (acyl–oxygen):* 1310–1160 cm^{-1} (strong, broad). Occurs near 1310–1250 cm^{-1} for conjugated esters, 1240 cm^{-1} for unconjugated acetates, and 1210–1165 cm^{-1} for most other unconjugated esters. Both the "acyl–oxygen" and "alkyl–oxygen" bands arise from coupled vibrations involving C—C—O groupings.

C—O *stretch (alkyl–oxygen):* 1200–1000 cm^{-1} (moderate). Occurs in the same region as alcohol C—O bands and varies in the same way with changes in the alkyl group's structure. Esters of phenols absorb at higher wave numbers.

Amines. Primary amines are characterized by a medium-intensity, two-pronged band near 3350 cm^{-1} and two medium–strong bands near 1615 and 800 cm^{-1}, the latter one being very broad. Secondary amines have a single weak band near 3300 cm^{-1} and a broad band near 715 cm^{-1}.

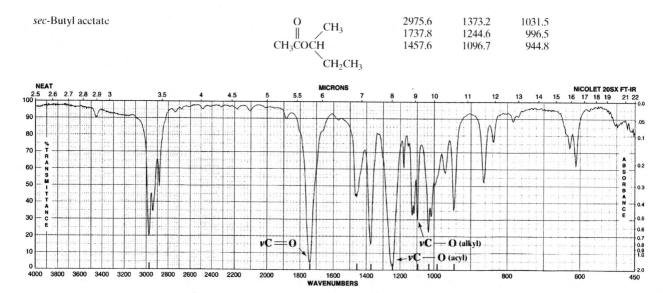

sec-Butyl acetate

2975.6	1373.2	1031.5
1737.8	1244.6	996.5
1457.6	1096.7	944.8

Figure G27 IR spectrum of an ester, *sec*-butyl acetate

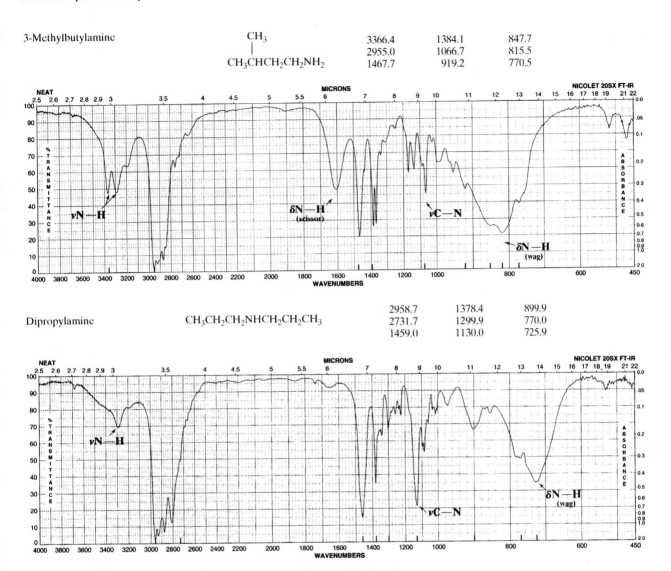

Figure G28 IR spectra of a primary amine, 3-methylbutylamine, and a secondary amine, dipropylamine

Tertiary amines can sometimes be distinguished by the presence of a C—N band. (See Figure G28.)

N—H *stretch:* 3500–3200 cm^{-1} (moderate to weak, broad). Primary aliphatic amines give rise to a two-pronged band centered near 3350 cm^{-1}, secondary aliphatic amines have one weak band near 3300 cm^{-1}, and tertiary amines have none. Primary and secondary aromatic amines absorb near 3400 and 3450 cm^{-1}, respectively.

N—H *bend (scissoring):* 1650–1500 cm^{-1} (strong to moderate). Usually near 1615 cm^{-1} for primary amines. Seldom observed for secondary aliphatic amines; secondary aromatic amines absorb near 1515 cm^{-1}.

N—H *bend (wagging):* 910–660 cm^{-1} (strong to moderate, broad). Often strong and very broad; around 910–770 cm^{-1} for primary amines and near 715 cm^{-1} for secondary amines.

2-Methylpropanamide

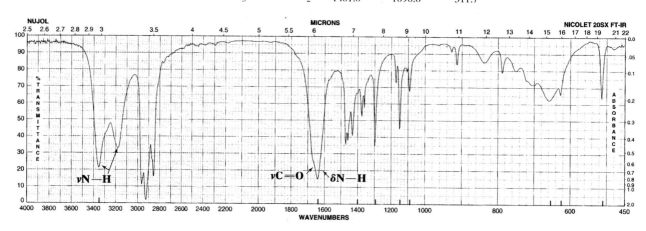

Figure G29 IR spectrum of an amide, 2-methylpropanamide

C—N *stretch:* 1340–1020 cm^{-1} (strong to moderate). Around 1340–1250 cm^{-1} for aromatic amines, and 1250–1020 cm^{-1} for aliphatic amines. As with an alcohol C—O band, the frequency of an aliphatic C—N band varies with changes in the structure of the attached alkyl group.

Amides. The presence of a carbonyl band near 1640 cm^{-1} and two bands (or peaks) in the 3500–3000 cm^{-1} region is good evidence for an amide. (See Figure G29.)

N—H *stretch:* 3450–3300 cm^{-1} and 3225–3180 cm^{-1} (one or two bands, strong to moderate). Primary amides have two bands (or two prongs on a broad band) near 3400 and 3200 cm^{-1}. Secondary amides have a single N—H stretching band near 3340 cm^{-1}, with an N—H bending overtone near 3080 cm^{-1}. Tertiary amides have no N—H stretching bands.

C=O *stretch:* 1695–1615 cm^{-1} (strong). Usually centered near 1640 cm^{-1}.

N—H *bend:* 1655–1615 cm^{-1} (primary) or 1570–1515 cm^{-1} (secondary) (strong to moderate). This band usually overlaps the carbonyl band on the spectra of primary amides obtained using KBr disks or mulls; it appears at lower frequencies on the spectra obtained in solution. The band is near 1540 cm^{-1} for most secondary amides, and an overtone can sometimes be seen at about 3080 cm^{-1}.

Organic Halides. Alkyl chlorides and bromides show fairly strong absorption between 800 and 500 cm^{-1}. Additional chemical evidence is usually needed to characterize organic halides. (See Figure G30.)

C(X)—H *bend:* 1300–1150 cm^{-1} (moderate). Observed only for halides with terminal halogen atoms (—CH$_2$X).

C—Cl *stretch:* 850–550 cm^{-1} (strong to moderate). Two bands near 725 and 645 cm^{-1} when the chlorine is terminal, and below 625 cm^{-1}

1-Chloropentane $CH_3(CH_2)_4Cl$

2959.3	1282.1	789.3
1467.0	1037.2	730.8
1380.3	925.7	653.7

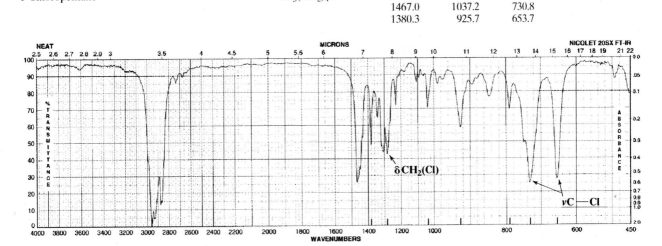

Figure G30 IR spectrum of an alkyl chloride, 1-chloropentane

otherwise—unless several chlorine atoms are on the same or adjacent carbons. Ar—Cl bonds absorb around 1175–1000 cm^{-1}.

C—Br *stretch:* 760–500 cm^{-1} (strong to moderate). Near 645 cm^{-1} when the bromine is terminal. Ar—Br bonds absorb around 1175–1000 cm^{-1}.

Nuclear Magnetic Resonance Spectrometry   OPERATION **40**

The theoretical principles underlying nuclear magnetic resonance (NMR) spectrometry can be found in most textbooks of organic chemistry and in appropriate sources listed in Category F of the Bibliography in this book. Here, we will review only those principles that are needed to gain a working knowledge of NMR spectrometry.

Nuclear magnetic resonance spectrometry is based on the magnetic properties of certain nuclei that possess a quality known as *spin*. The nucleus of a ^1H atom, which is a single proton, has spin. The nuclei of ^{13}C atoms also have spin, but the nuclei of ^{12}C atoms, which are nearly 100 times more abundant than ^{13}C atoms, do not. An atom with spin behaves like a tiny bar magnet. When placed in a strong magnetic field, it tends to become aligned with the field. For convenience, we will refer to a nucleus that is aligned with the external magnetic field as being in an **up** spin state and a nucleus that is aligned against the field as being in a **down** spin state. If a sample containing magnetic nuclei is placed in a magnetic field and exposed to radio frequency (RF) radiation of just the right frequency, some of its **up** nuclei will flip over, into the **down** spin state. This transition is illustrated in Figure G31. Since a nucleus in the **down** state is less stable (contains more energy) than a nucleus in the **up** state, the spin transition results in an absorption of energy by the nucleus. Such a transition is possible only if the energy of an RF photon, $h\nu$ is exactly equal to the energy of the transition,

If you are not familiar with the principles and terminology of NMR spectrometry, read the section "Interpretation of ^1H NMR Spectra" or consult your lecture text.

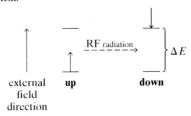

Figure G31 Spin transition of a magnetic nucleus

ΔE, so that $\nu = \Delta E/h$. When this is the case, the *resonance condition*—the condition under which nuclei of a given kind can undergo spin transitions—is fulfilled. The transition energy, ΔE, is directly proportional to the strength of the external magnetic field, H_o, so ν is also proportional to H_o. This means that the resonance condition for a nucleus can be attained either by adjusting the frequency of the RF radiation or by adjusting the strength of the external field.

a. ^{1}H NMR Spectrometry

Instrumentation

There are two fundamentally different ways of obtaining an NMR spectrum. With a *continuous-wave (CW) NMR spectrometer*, the sample is irradiated continuously with RF waves as the magnetic field or RF frequency is varied, and the electromagnetic signals generated by nuclei as they change spin states are converted to peaks on a moving chart. With a *Fourier-transform NMR (FT-NMR) spectrometer*, the sample is irradiated with intense pulses of full-spectrum RF radiation that displace the nuclei from their equilibrium distribution. Their response to the displacement is monitored, generating data that is converted by a microprocessor to an NMR spectrum.

Continuous-Wave NMR. In a continuous-wave NMR spectrometer, a glass tube containing the sample is placed between the poles of a large magnet and irradiated with RF radiation from a transmitter coil as the magnetic field is "swept" (varied continuously) over a preset range. In an instrument of the type diagrammed in Figure G32, the magnetic field is swept from low to high field (*downfield* to *upfield*) by varying the strength of an electric current passing through the sweep coils. When the resonance condition for a particular kind of nucleus in the sample is met, nuclei of that kind flip from the **up** state to the **down** state. As they do so, they generate a small fluctuating magnetic field that can be detected by a receiver coil encircling the sample tube. The receiver coil sends an electronic signal to an

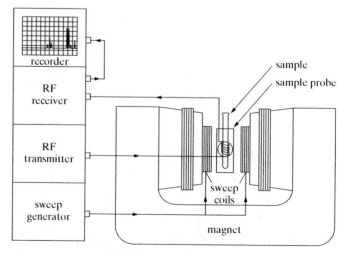

Figure G32 Schematic diagram of a continuous-wave NMR spectrometer

RF receiver, which amplifies and modifies the signal so that it can be displayed on a recorder as part of an NMR spectrum.

An NMR spectrum is a record of all the signals generated by all of the different kinds of nuclei in the sample that absorb RF radiation over the range swept by the instrument. If the sweep range is one in which the resonance conditions for ^1H nuclei (protons) are met, the spectrum should display different signals for protons that are in different molecular environments. For example, protons on the benzene ring in *para*-xylene are in a different molecular environment than protons on the methyl groups, so the NMR spectrum of *p*-xylene will display two signals, one for each kind of proton, at different positions on the spectrum, as shown in Figure G33. The position of a signal relative to the position of a reference signal, usually that of tetramethylsilane (TMS), is called its *chemical shift*. A chemical shift, represented by the Greek letter δ, is ordinarily measured in parts per million (ppm). Since the TMS signal ($\delta = 0$) is on the right (upfield) side of a spectrum, chemical shifts increase from right to left.

A typical CW-NMR spectrometer suitable for use by undergraduate students may operate at a frequency of 60 MHz and a magnetic field strength of approximately 1.4 tesla (14,000 gauss). When a ^1H NMR spectrum is recorded using a 60-MHz spectrometer, the magnetic field is swept over a range of about 1.4×10^{-5} tesla (0.14 gauss), which is only 10 millionths of the external field strength, or 10 parts per million (ppm). This sweep range can be extended to 15 ppm or so to detect protons whose resonance conditions occur outside this range.

Fourier-Transform NMR. A Fourier-transform NMR (FT-NMR) spectrometer is capable of producing spectra with better resolution and a much higher signal-to-noise ratio than any CW instrument. In an FT-NMR instrument, the sample (in an appropriate sample tube) is placed between the poles of a powerful electromagnet and irradiated with a short ($\sim 10\ \mu s$) pulse of RF radiation that covers the entire frequency range of interest. The pulse is so intense that it raises all of the absorbing nuclei into the high-energy **down** state. As the high-energy nuclei return to their equilibrium

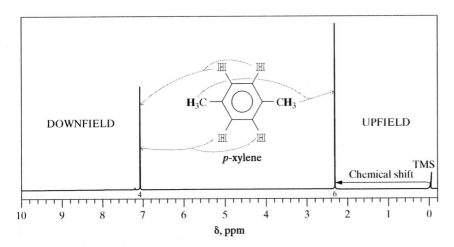

Figure G33 ^1H NMR spectrum of *p*-xylene

state, they generate a *free induction decay (FID)* signal that contains information about the nuclei whose resonance conditions were met by any of the RF frequencies in the pulse. The FID signal, which is equivalent in information content to a complete NMR spectrum, is detected and sent to a microprocessor that accumulates and averages the FID signals from a series of pulses. The microprocessor then "decodes" the averaged FID signal by Fourier-transform analysis and converts it to a conventional NMR spectrum. The FID signal generated by one pulse takes less than a second to acquire, so an FT-NMR spectrometer can accumulate and average the equivalent of several hundred NMR spectra in the 2–5 minutes it takes a CW instrument to record a single spectrum. The resulting averaged spectrum has very little electronic noise and is thus much "cleaner" than a conventional CW spectrum.

Adding the data generated by successive pulses improves the quality of the NMR spectrum because a signal increases in intensity with each addition, while electronic noise, being random, tends to cancel out.

A typical research-grade FT-NMR spectrometer uses an electromagnet whose components are cooled with liquid helium. At the temperature of liquid helium (4 K, $-269°C$), the coils of wire that generate the magnetic field are electrical superconductors, making it possible to attain very high field strengths of 14 tesla or more. Increasing the field strength of an NMR spectrometer causes the NMR signals to spread out, reducing overlap between adjacent signals. This and the high signal-to-noise ratio make complex ^1H NMR spectra generated on an FT instrument easier to interpret than those obtained with a CW instrument.

Chemical-Shift Reagents

The amount of structural information that can be obtained from a CW-NMR spectrum is often limited by the presence of overlapping signals. Increasing the magnetic-field strength reduces overlapping by increasing the chemical shifts (in Hz) of all the signals by the same amount. Using a *chemical-shift reagent* also changes the chemical shifts of NMR signals, but it affects different signals differently; some are shifted more than others, and some may not be shifted at all. Nevertheless, an appropriate chemical-shift reagent can often be used to separate the signals of interest and facilitate spectral interpretation.

Chemical-shift reagents are organometallic complexes of certain paramagnetic rare earth metals. These complexes can coordinate with the oxygen and nitrogen atoms of alcohols, amines, carbonyl compounds, and other Lewis bases. The local magnetic field produced by a paramagnetic metal atom shifts the signals of nearby protons to an extent that varies with distance; the closer a nucleus is to the metal atom, the more its chemical shift will change. Different chemical-shift reagents have different effects on a spectrum; thus tris(dipivaloylmethanato)europium(III) [Eu(dpm)$_3$] causes signals to shift to the left (downfield) on an NMR spectrum, while the corresponding complex of praseodymium [Pr(dpm)$_3$] causes signals to shift to the right (upfield).

Sample Preparation

Most substances analyzed by NMR are first dissolved in a suitable solvent. Liquids that are no more viscous than water can sometimes be analyzed neat, but neat liquids may give broadened peaks and other spectral distortions due to intermolecular interactions. FT-NMR spectrometers yield good proton NMR spectra with solution concentrations as low as 0.1% (w/v). CW-NMR instruments usually require concentrations on the order of 5–20% (w/v), although satisfactory spectra may be obtained with lower concentrations when the amount of sample is limited. The liquid or solution is placed in a special thin-walled *NMR tube*, which is closed with a tight-fitting cap to prevent evaporation. A typical NMR tube has an o.d. of 5 mm, a length of 17.5 cm, and is both fragile and expensive. The NMR tube should be straight and uniform; a tube that wobbles when it is rolled down a slightly inclined glass plate will give large spinning sidebands, as discussed under the heading *Spinning Rate*. For a routine ^{1}H NMR analysis on a CW instrument, you can prepare the sample as described here.

- Dissolve 50–100 mg of your compound (10–50 mg for microscale work) in 0.5–0.8 mL of a suitable solvent.
- Filter the solution directly into the NMR tube through a Pasteur pipet containing a small plug of tightly packed glass wool (Figure G34).
- Add 5–15 μL of a reference standard, ordinarily TMS (tetramethylsilane).
- Cap the NMR tube carefully.
- Invert the tube several times to mix the components thoroughly.

Some commercial deuterated solvents contain added TMS, in which case the third step is omitted. The NMR tube should be filled to a depth of at least 2.5 cm, but should be no more than three-fourths full. TMS boils near room temperature, so it should be kept in a refrigerator and added with a *cold* syringe or fine-tipped dropper. For very high-resolution spectra, the sample should be *degassed* by bubbling a fine stream of pure nitrogen through it for one minute; degassing is not necessary for routine spectra.

NMR Solvents

A solvent suitable for ^{1}H NMR analysis should have no protons that produce intense signals of their own, since they might obscure signals from the sample. Therefore, hydrogen-containing solvents such as chloroform and acetone are used in their completely deuterated forms. Deuterium (^{2}H) undergoes resonance at about $6\frac{1}{2}$ times the field strength required for ^{1}H, so an isotopically pure deuterated solvent doesn't interfere with a proton NMR spectrum. Most deuterated solvents, however, contain a significant amount of the protic form, giving rise to one or more small signals. For example, the NMR spectrum of a deuterochloroform (chloroform-*d*, $CDCl_3$) solution has a small $CHCl_3$ signal at 7.27 ppm, but this signal usually doesn't interfere with the solute's signals. A solvent for FT-NMR analysis *must* contain deuterium, because the instrument locks onto the resonance signal of deuterium to help the user adjust the controls for maximum spectral resolution.

A good NMR solvent should also have a low viscosity, a high solvent strength, and no appreciable interactions with the solute. Deuterochloroform is the most widely used NMR solvent because its polarity is low enough to prevent significant solute–solvent interactions, and most organic compounds are sufficiently soluble in it for NMR analysis. When a more polar solvent is required, dimethyl-*d*$_6$ sulfoxide can be used, often in mixtures with deuterochloroform. It is convenient to add 1–3% TMS to the bulk solvent so that it doesn't have to be added during sample preparation; some commercial solvents already contain TMS.

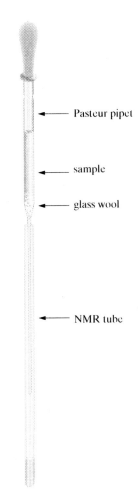

Figure G34 Filtering an NMR solution

Table G3 Properties of some NMR solvents

Solvent	^1H δ, ppm	Solvent strength	Freedom from interactions	Viscosity
carbon disulfide	none	good	good	low
cyclohexane-d_{12}	1.4	poor	good	medium
acetonitrile-d_3	2.0	good	fair	low
acetone-d_6	2.1	good	poor	low
dimethyl-d_6 sulfoxide	2.5	very good	poor	high
1, 4-dioxane-d_8	3.5	good	fair	medium
deuterium oxide	~5.2(v)	good	poor	medium
chloroform-d	7.3	very good	fair	low
pyridine-d_5	7.0–8.7	good	poor	medium
trifluoroacetic acid-d	~12.5(v)	good	poor	medium

Note: δ is the chemical shift of the protic form of the solvent; v = variable; solvent strength refers to the ability to dissolve a broad spectrum of organic compounds.

Table G3 compares the properties of some deuterated NMR solvents and gives the approximate chemical shifts (δ) of their ^1H NMR signals.

Instrumental Parameters

The following information applies mainly to CW-NMR spectrometers. FT-NMR spectrometers may be operated quite differently.

Spinning Rate. To average out the effect of magnetic-field variations in the plane perpendicular to the axis of the NMR tube, an NMR sample is rotated at a rate of 30–60 revolutions per second while its NMR spectrum is being recorded. It is important to use an appropriate spinning rate. Excessively high rates create a vortex that may extend into the region of the receiver coil—this is most likely when there is not enough solution in the sample tube. Spinning rates that are too low can cause *spinning sidebands* or signal distortion (see Figure G35). Spinning sidebands are small peaks that are symmetrically spaced on either side of a main peak at a distance equal to the spinning rate; thus an NMR tube spun at 30 cycles per second can give rise to sidebands 30 Hz from each main peak. Spinning sidebands can also be caused by field inhomogeneity and wobbling NMR tubes or sample spinners. To find out whether small signals are spinning sidebands or impurity peaks, change the spinning rate and scan over the peaks again to see if their positions change.

Field Homogeneity. Recording a good NMR spectrum requires that the magnetic field be homogeneous (uniform) at the sample. The most important homogeneity control, usually called the Y control, is adjusted to produce a uniform field along the axis of the sample tube. For routine work on a previously tuned CW-NMR spectrometer, the Y control can be set by placing a blank sheet of paper over the chart paper and repeatedly scanning a strong peak in the spectrum of the sample (or of a standard acetaldehyde solution), each time making small adjustments in the Y control until the peak is as tall and narrow as possible and shows a good "ringing" (beat) pattern. Figure G36 shows an excellent ringing pattern for the quartet

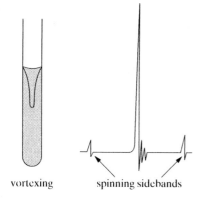

vortexing   spinning sidebands

Figure G35 Effects of spinning rate

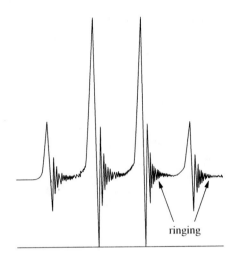

Figure G36 Ringing pattern for the quartet of acetaldehyde

(4-peak signal) of acetaldehyde, characterized by the high amplitude, long duration, and exponential decay of the "wiggles" following the main peaks.

An FT-NMR spectrum does not show a ringing pattern; magnetic-field homogeneity is adjusted by maximizing the intensity of a deuterium lock signal. Ordinarily the instructor or a lab technician performs such adjustments.

Signal Amplitude. The amplitude (height) of the signals on an NMR spectrum is adjusted with two controls. The *spectrum amplitude* control changes the amplitude of both the signals and the baseline noise. The *RF power* control increases signal height without increasing baseline noise up to the point at which *saturation* begins. At that point, the number of nuclei in both spin states is so nearly equal that increasing the intensity of the RF radiation no longer increases the number of transitions; instead it can cause distortion and reduce the signal size. Unless high sensitivity is required, the RF power level is usually set to a value at which there is little likelihood of saturation (usually about midrange), and the spectrum amplitude control is then adjusted so that the strongest peak in the spectrum extends nearly to the top of the chart paper.

Sweep. There are four sweep controls on a typical CW-NMR spectrometer; these control the reference point of the spectrum, the sweep rate (sweep width divided by sweep time), and the portion of the spectrum to be scanned. The *sweep zero* control is used to set the signal for the reference compound to the proper value (zero for TMS). The *sweep width* control sets the total chemical-shift range to be scanned, usually 600–1000 Hz when the entire spectrum is being scanned on a 60-MHz instrument. A 600-Hz (10-ppm) range can be used if the sample is known to contain no protons that absorb downfield of 10 δ. A CW-NMR spectrum is often scanned at a rate of 1 Hz per second, so the *sweep time* can be set numerically equal to the sweep width (for example, 600 s for a 600-Hz sweep width). The *sweep offset* control is used when only a specific portion of the spectrum is to be scanned; it sets the upfield limit of the scan. For example, if a scan between 350 and 500 Hz is desired, the sweep offset should be 350 Hz and the sweep range 150 Hz.

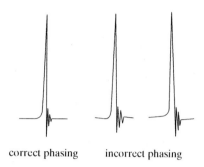

correct phasing   incorrect phasing

Figure G37 Effects of phasing on baseline

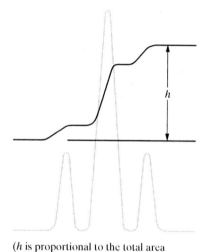

h

(*h* is proportional to the total area under the triplet)

Figure G38 Measuring signal areas

Phasing. The *phasing* control should be adjusted to obtain a straight baseline before and after a signal (see Figure G37). Correct phasing is much more important when an NMR spectrum is being integrated than when it is being recorded.

Integrating a Spectrum

When an NMR spectrum recorded by a CW instrument is *integrated*, the recorder pen traces a horizontal line until it reaches a signal; then it rises a distance that is proportional to the signal's area as it crosses the signal. Since the area of a signal on a proton NMR spectrum is proportional to the number of protons responsible for the signal, integrating the spectrum makes it possible to determine how many protons give rise to each signal.

Here is a summary of the steps in the integration of a typical ^{1}H NMR spectrum. If you will be expected to integrate your NMR spectrum, your instructor will provide more detailed directions.

1. The RF power is optimized to provide an acceptable signal-to-noise ratio.
2. The instrument is switched to the integral mode.
3. While the spectrum is scanned rapidly, the integral amplitude control is adjusted until the integrator trace spans the vertical axis of the chart.
4. With the sweep offset and sweep width controls set to scan a region free from NMR signals, the balance control is adjusted during a slow scan of that region to give a horizontal line.
5. While scanning a signal, the phasing control is adjusted to make the integrator traces before and after the signal as nearly horizontal as possible.
6. The integral over the entire spectrum is recorded (preferably once in each direction) using a sweep time about one-fifth to one-tenth that for the normal spectrum. The pen should be returned to the baseline after each scan.
7. The relative peak areas are determined by measuring the vertical distances between the integrator traces before and after each signal, and the results for successive scans are averaged (see Figure G38).

General Directions for Operating a CW-NMR Spectrometer

Standard Scale and Microscale

Equipment and Supplies

NMR spectrometer
sample
deuterated NMR solvent
tetramethylsilane (TMS)
NMR tube
Pasteur pipet
glass wool
tissues
washing solvent

Do not attempt to operate the instrument without prior instruction and proper supervision. Do not make any adjustments other than the ones specified, except at the instructor's request and under his or her supervision. Some of the adjustments described here may be made in advance by the instructor or a lab technician. The following procedure applies to a typical 60-MHz CW-NMR spectrometer; operating procedures for other instruments may vary considerably.

Prepare a solution of the sample in a suitable NMR solvent as described under "Sample Preparation," and add 1–3% TMS, if necessary. Fill the NMR tube to a depth of about 3 cm with this solution and cap the tube. Wipe the outside of the NMR tube carefully with a tissue paper or lint-free cloth, insert it in the sample spinner using a depth gauge to adjust its position, wipe it again, and carefully place the assembly into the sample probe between the magnet pole faces. Adjust the air flow to spin the sample at 40–50 Hz; it may need to be readjusted later to minimize spinning sidebands. Align the chart paper on the recorder and cover it with a sheet of scrap paper. Set the sweep controls to scan the desired range at a suitable rate. Typical settings for a 60-MHz instrument are: sweep offset, 0; sweep width, 600 Hz; and sweep time, 600 s. Set the RF power to about midrange and the filter response time to 1 s or less. Set the spectrum amplitude control to about midrange, and scan the spectrum to find the tallest peak. Readjust the spectrum amplitude to keep that peak on scale near the top of the chart during a scan. If necessary, optimize peak shape and ringing and adjust the phasing as directed by your instructor. Set the TMS peak to 0.0 δ with the sweep zero control; you may have to sweep through the TMS signal several times, adjusting the control each time, until it is lined up with the zero on the chart paper. Set the recorder baseline, if necessary, to a convenient location near the bottom of the spectrum. Remove the scrap paper and record the spectrum. If you are to integrate your spectrum, cover it with scrap paper while you make the adjustments described under "Integrating a Spectrum"; then remove the paper and record the integral on the original spectrum.

Take Care! Handle NMR tubes with great care; they are fragile and may break.

When you have finished scanning and integrating your spectrum, remove the spectrum and record the control settings and other relevant information on it. Then remove the sample tube as demonstrated by your instructor; follow directions carefully or the tube may break. Dispose of the solution as directed by your instructor. Clean the sample tube immediately and thoroughly using a Pasteur pipet and an appropriate solvent. The protic form of the solvent in which the sample was dissolved is generally used for cleaning. For example, if the solvent was $CDCl_3$, rinse the tube with $CHCl_3$—*not* the much more expensive deuterated solvent. Invert the tube in a suitable rack and let it drain dry. Before being reused, an NMR tube should be dried in an oven to remove all traces of the wash solvent.

Summary

1. Prepare solution, add TMS if necessary.
2. Transfer solution to NMR tube, cap tube.
3. Wipe tube, insert in sample spinner.
4. Insert spinner in probe, adjust spinning rate.
5. Align chart paper, cover with scrap paper.
6. Set sweep, RF power, and filter response time controls.
7. Scan spectrum and adjust spectrum amplitude control.

8. Zero TMS signal.
9. Set baseline, remove scrap paper.
10. Scan spectrum.
11. Integrate spectrum.
12. Remove and clean sample tube, dispose of solution.

When Things Go Wrong

If the NMR spectrum of a sample run in $CDCl_3$ has a small extraneous peak at $\delta\ 7.3$, don't worry about it; the peak arises from the small amount of $CHCl_3$ in the solvent. If you are using some other deuterated solvent, check Table G3 to see where the signal of the protic form of that solvent occurs and disregard any small peaks at that location. (See "NMR Solvents.")

If each signal in your NMR spectrum is flanked by two considerably smaller peaks equidistant from the signal, the extraneous signals are probably spinning sidebands. Increase the sample spinning rate and scan over the same region. If the peaks change position but don't disappear, try to identify the cause and take corrective action. (See *Spinning Rate.*)

If the peaks in your NMR spectrum are distorted or unusually broad, the sample may be spinning too fast (forming a vortex), too slowly (causing field inhomogeneity), or not at all. Check the spinning rate and adjust it if necessary; you might also need to increase the sample size to prevent vortexing problems or decrease it to reduce inhomogeneity. (See *Spinning Rate.*) If the peaks are still too broad, check to see if the NMR tube is straight and uniform; if it isn't, transfer the sample to a different NMR tube. If the sample might contain ferromagnetic impurities, filter it as you transfer it to the other NMR tube. (See "Sample Preparation.") If that doesn't help, the spectrum amplitude control may be set too high or the magnetic field may be inhomogeneous; see your instructor for help. (See *Signal Amplitude, Field Homogeneity.*)

If some of the peaks in your spectrum extend to the top of the chart paper and flatten out there, your sample solution may be too concentrated. Dilute it by half and try again. (See "Sample Preparation.") If the sample concentration is appropriate, the signal amplitude control may be set too high. See your instructor about having it readjusted. (See *Signal Amplitude.*)

If the chemical shifts of all your peaks are off by the same amount, the reference peak was not positioned correctly. Use the sweep zero control to reposition the reference peak to zero for TMS, or to the appropriate value for a different reference compound. (See *Sweep.*) If there is no reference peak, add TMS and then adjust the sweep zero control. (See "Sample Preparation.")

If you are running the NMR spectrum of a carboxylic acid or enolic compound and can't find a signal for the OH proton, change the sweep range to include the region above 10 ppm. (See *Sweep.*)

If the baseline of your NMR spectrum is not straight and horizontal, adjust the phasing control until it is. (See *Phasing.*)

Interpretation of ^1H NMR Spectra

A proton NMR spectrum provides numerical data in the form of chemical shifts, signal areas, signal multiplicities, and coupling constants. Working out the structure of a molecule from these numbers is a fascinating mental

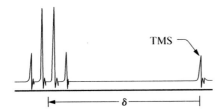

Figure G39 Chemical shift of a proton NMR signal

exercise comparable to the work of a cryptographer who reconstructs meaningful messages from coded symbols.

The *chemical shift* (δ) is the distance, measured in hertz or parts per million, from the center of a signal to some reference signal, usually that of tetramethylsilane (TMS). The TMS signal occurs farther upfield (to the right) than nearly all other proton signals, so the chemical shift of a signal is usually measured as its distance downfield (to the left) from that of TMS, as shown in Figure G39. Note that a signal may have more than one peak (4 peaks in this example).

The *signal area*, which is the sum of the areas under all of the peaks in a signal, is proportional to the number of protons giving rise to the signal. Signal areas can be determined by using an electronic integrator that traces a line across each proton signal after it is recorded. The area of the signal is proportional to the vertical rise of the integrator pen as it crosses the signal; that is, to the height of the "steps" drawn by the integrator pen, as shown in Figure G38. Integrated signal areas can be converted to proton numbers using the following relationship:

$$\text{number of protons responsible for signal} =$$
$$\text{total number of protons} \times \frac{\text{area under signal}}{\text{area under all signals}}$$

For example, suppose that a compound with the molecular formula $C_{10}H_{14}$ has four signals with relative areas of 42, 7, 14, and 35. The sum of the areas is 98, so the number of protons responsible for the first signal is

$$14 \times \frac{42}{98} = 6$$

By similar calculations, it can be shown that 1, 2, and 5 protons, respectively, are responsible for the other three signals. If the molecular formula of a compound is not known, relative proton numbers can be obtained by reducing the signal areas to the lowest ratio of integers.

The signal generated by a given set of protons may be split into several peaks as a result of *coupling* interactions with nearby proton sets (refer to your lecture textbook or see your instructor for an explanation of coupling). The *multiplicity* of a signal is simply the number of separate peaks it contains; its *coupling constant* is the distance between two adjacent peaks in the signal, measured in hertz (Hz). Figure G40 shows the signals of two sets of protons that are interacting with one another; the protons of set *a* have split the signal of the protons of set *b* into four peaks (a quartet), and the *b* protons have split the signal of the *a* protons into three peaks (a triplet).

Stop and Think: What is the most likely structure for this compound if it contains a benzene ring and one alkyl side chain?

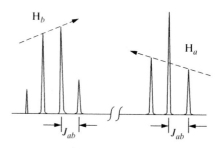

Figure G40 Signals of nearest neighbor protons

The coupling constant, which is equal for the two signals, is represented by J_{ab}. In the simplest case, the number of protons responsible for splitting the signal of a neighboring set of protons can be determined by subtracting 1 from the number of peaks in that signal. Thus the three peaks in the a signal are produced by two neighboring b protons, and the four peaks in the b signal by three neighboring a protons. An interacting triplet–quartet grouping of this kind is good evidence for an ethyl (CH_3CH_2-) group.

Ideal triplets and quartets should be symmetrical, having relative peak area ratios of $1:2:1$ and $1:3:3:1$, respectively. As shown in Figure G40, however, the signals in an actual spectrum are often somewhat distorted, giving paired peaks of unequal height. Note that the two signals in the figure are not perfectly symmetrical but appear to "lean" toward one another, with the peaks on the side facing the other signal being higher than predicted. This and the fact that their coupling constants are equal provide additional evidence that the protons responsible for the two signals are, in fact, coupling with one another and not with some other proton sets in the molecule.

The following general procedure should help you derive structural information from a proton NMR spectrum:

1. Measure the integrated area of each signal and use it to determine the number of protons responsible for the signal. Each set of equivalent protons (protons in the same molecular environment) gives rise to a signal, and the relative signal areas can tell you how many protons are in each set. For example, 3,3-dimethyl-2-butanone has nine hydrogen atoms on the three equivalent methyl groups to the left of the carbonyl group and three on the other methyl group, so its ^1H NMR spectrum has two signals with an area ratio of 3:1.

2. Determine the chemical shift of each signal on the delta scale by measuring the distance, in parts per million, from the center of the signal to the TMS reference peak. The chemical shift of a signal may indicate what kind of protons is responsible for the signal or may suggest their relative locations in the molecule. For example, alkyl hydrogen atoms that are remote from electron-withdrawing substituents should have a chemical shift of approximately 0.9 ppm if they are primary, 1.3 ppm if they are secondary, and 1.5 ppm if they are tertiary. Electron-withdrawing groups containing oxygen, nitrogen, or halogens tend to move ^1H NMR signals downfield, thereby increasing their chemical shifts. Benzene rings give rise to large downfield shifts, making it quite easy to recognize aromatic compounds from their ^1H NMR spectra. The correlation chart in Figure G41 summarizes chemical-shift data for

3,3-dimethyl-2-butanone

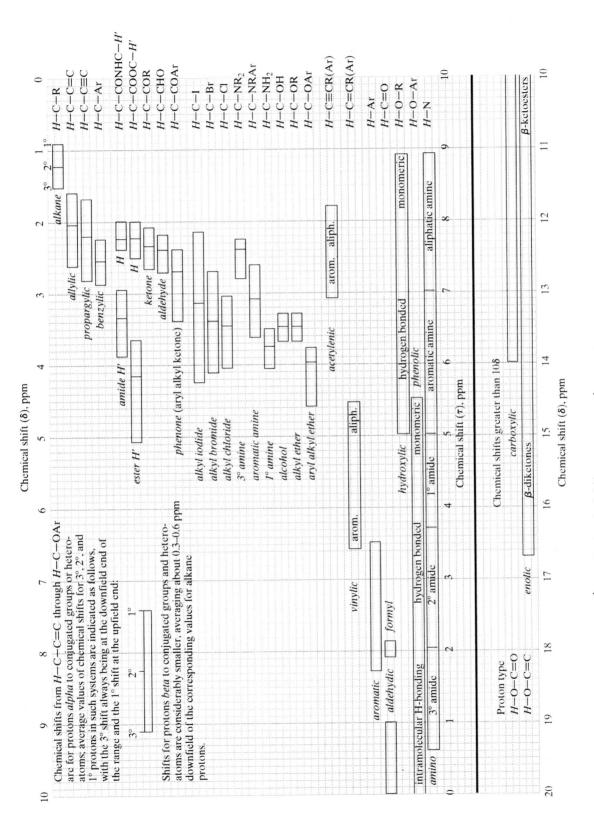

Figure G41 Correlation chart relating ¹H NMR chemical shifts to proton environments

a number of proton types. Chemical-shift ranges for compounds from the common families of organic compounds are given in Table G4.

3. Determine the multiplicity of each signal by counting the number of distinguishable peaks in the signal. (Very small peaks may be obscured by baseline noise.) If the signal of a proton set is reasonably symmetrical and contains evenly spaced peaks, it may be possible to determine how many nearby protons are coupled with the protons in that set by subtracting 1 from the multiplicity of the signal. Irregular signals and signals that have been split by several dissimilar proton sets must be analyzed by more advanced methods.

4. Measure the coupling constant of each signal that contains more than one peak and try to determine from the resulting values and the way each signal "leans" what other signals might be coupled with it.

All of this information can be used to build up the structure of a molecule piece by piece. For example, consider the ^{1}H NMR spectrum in Figure G42 of a ketone having the molecular formula $C_7H_{14}O$. The spectrum has only two signals, a and b, which have a relative area ratio of 6:1. Since the compound

Table G4 Approximate ^{1}H NMR chemical-shift ranges for different types of protons

Family	Proton type	Chemical shift Range (δ, ppm)
alcohol	**H**—O—R	1–5.5
	H—C—OH	3.4–4
phenol	**H**—O—Ar	4–12
	H—Ar	6–8.5
aldehyde	**H**—C=O	9–10
	H—C—CHO	2.2–2.7
ketone	**H**—C—COR	2–2.5
carboxylic acid	**H**—O—C=O	10.5–12
	H—C—COOH	2–2.6
ester	**H**—C—COOR	2–2.5
	H—C—OC=O	3.5–5
amine	**H**—N—R (aliphatic)	1–3
	***H**—N—R (aromatic)	3–5
	H—C—N	2.2–4
amide	***H**—N—C=O	5–9.5
	H—C—CO—N	2–2.4
	H—C—NC=O	3–4
halide	**H**—C—Br	2.5–4
	H—C—Cl	3–4
aromatic hydrocarbon	**H**—Ar	6–8.5
	H—C—Ar	2.2–3

Note: Signals of proton types marked with an asterisk are often very broad.

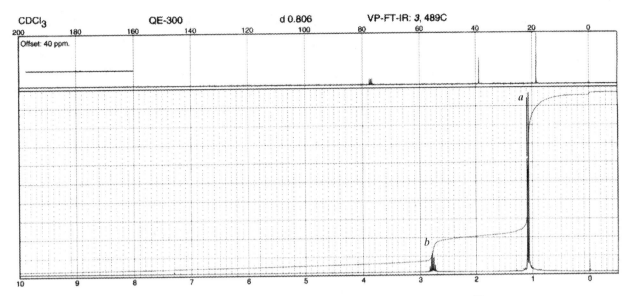

Figure G42 ^{13}C and ^1H NMR spectra of compound with molecular formula $C_7H_{14}O$. (The ^{13}C spectrum is in the narrow strip at the top.)

contains 14 protons, $\frac{6}{7}$ of them, or 12, must be responsible for signal *a*, and $\frac{1}{7}$ of them, or 2, for signal *b*. Signal *a* has only two peaks, indicating that the *a* protons have only one neighboring proton. The seven peaks in signal *b* (visible under a magnifying glass) indicate that the *b* protons have six neighbors, and the higher chemical shift of their signal suggests that they are close to the electron-withdrawing carbonyl group. The only alkyl group in which a single proton has six equivalent protons for neighbors is the isopropyl group, $(CH_3)_2CH—$. Two such groups provide the required total of 12 *a* and 2 *b* protons, and attaching them both to a carbonyl group gives the complete structure of the ketone, which is 2,4-dimethyl-3-pentanone. Note that the structure in the margin accounts for all of the features of the ^1H NMR spectrum: the 6:1 area ratio for the *a* and *b* protons; the higher chemical shift of the signal for the *b* protons resulting from their proximity to the carbonyl group; the splitting of the *a* signal into two peaks by each neighboring *b* proton; the splitting of the *b* signal into seven peaks by each group of six neighboring *a* protons; and the identical coupling constants of the two signals.

For further information about the interpretation of ^1H NMR spectra and information on advanced topics such as second-order effects, two-dimensional NMR, and computer simulation of NMR spectra, refer to your lecture textbook or to appropriate sources in Category F of the Bibliography.

b. ^{13}C NMR Spectrometry

Except where noted here, most of the principles and experimental methods described previously for ^1H NMR spectrometry also apply to ^{13}C NMR spectrometry. Because carbon-13 nuclei are much less abundant than hydrogen nuclei, signals from the ^{13}C nuclei in a typical sample are about 6000 times weaker than those from its ^1H nuclei. A CW-NMR spectrometer can hardly distinguish such weak signals from electronic baseline noise, so carbon-13 NMR spectra are ordinarily recorded using Fourier-transform NMR instruments. Since such instruments are highly complex and are seldom available for use by undergraduate students, no attempt will be made here to describe their operation.

A typical ^{13}C NMR spectrum is usually simpler and easier to interpret than the ^1H NMR spectrum of the same compound. Carbon–carbon splitting is unimportant because carbon-13 nuclei cannot couple with nonmagnetic carbon-12 nuclei and because there is only a slight chance that two carbon-13 atoms will be next to one another in the same molecule. Hydrogen nuclei couple strongly with carbon-13 nuclei, however—not only with the nuclei of the carbon atoms that they are bonded to but with those of more distant carbons as well. Since such coupling results in very complex spectra, it is usually prevented by various *decoupling* techniques. For *broad-band proton decoupling*, the sample is subjected to continuous broad-spectrum RF radiation covering the resonance frequencies of its protons, causing all of the protons to flip over (change spin states) so rapidly that their coupling effects on the adjacent carbon atoms average out to zero. In a broad-band decoupled ^{13}C NMR spectrum, each ^{13}C signal is a single peak rather than a multiplet. For an example, see the broadband decoupled ^{13}C NMR spectrum in Figure G42, which is displayed in a narrow strip above the ^1H NMR spectrum. (The three closely spaced peaks near 78 ppm arise from the solvent, chloroform-*d*.)

Another decoupling technique, *off-resonance decoupling*, yields *proton-coupled* spectra, in which only those hydrogens that are attached directly to a carbon atom split the signal of that carbon atom. Thus, the three protons of a methyl group will split the signal of the methyl carbon into a quartet, but will not split the signal of any other carbon in the molecule. Proton-coupled spectra are less common and harder to interpret than broad-band decoupled spectra, so we will not consider them further.

The area of a ^{13}C signal, unlike that of a ^1H signal, is not directly proportional to the number of carbon atoms responsible for the signal, so ^{13}C NMR spectra are not integrated. Consequently, three of the four parameters that can be obtained from ^1H NMR spectra are not present in a broadband decoupled ^{13}C NMR spectra. Since there is no integration, there are no signal areas, and since all of the signals are singlets, there are no multiplicities or coupling constants to interpret. That leaves only the chemical shifts which, as it happens, are extraordinarily useful.

The ^{13}C NMR spectrum of a compound gives us direct information about the compound's carbon "backbone"; information that is often not available from its ^1H NMR spectrum. A broad-band decoupled ^{13}C NMR spectrum of a compound contains one sharp peak for each kind of carbon atom in its molecules. Since many compounds have few, if any, magnetically equivalent carbon atoms, most of the peaks in a typical ^{13}C NMR spectrum are one-carbon peaks, each arising from a different carbon atom. Thus the absence of a linear relationship between signal area and the number of carbon atoms is not a serious handicap.

Table G5 ^{13}C NMR chemical-shift ranges for different types of carbon atoms

Type of carbon atom	Chemical-shift range (δ, ppm)
1° alkyl, RCH_3	0–40
2° alkyl, R_2CH_2	10–50
3° alkyl, R_3CH	15–50
alkene, $C{=}C$	100–160
alkyne, $C{\equiv}C$	60–90
aryl, 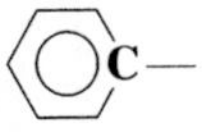 ⬡$C-$	100–170
alkyl halide, $C-X$ (X$=$Cl Br)	5–75
alcohol or ether, $C-O$	40–90
amine, $C-N$	10–70
aldehyde or ketone, $C{=}O$	180–220
carboxylic acid or ester, $O{=}C-O$	160–185
amide, $O{=}C-N$	150–180

Carbon-13 chemical shifts cover a much broader range than proton chemical shifts; about 250 ppm compared to 15 ppm or so for protons (see Table G5). As a result, the peaks on a ^{13}C NMR spectrum are usually well separated. The chemical shift of a carbon-13 atom is very sensitive to changes in its hybridization and molecular environment. Carbon atoms that are sp^2 hybridized have much higher chemical shifts (100–160 δ) than sp^3 carbons (0–60 δ), and the chemical shifts of sp carbons are somewhere in between (65–105 δ). As in ^{1}H NMR spectrometry, electron-withdrawing groups cause downfield chemical shifts at nearby carbon atoms, but they can cause upfield shifts at more distant carbon atoms. For example, a chlorine atom increases the chemical shift of an α-carbon atom by about 30 ppm and of a β-carbon atom by about 10 ppm, but it *decreases* the chemical shift of a γ-carbon atom by about 5 ppm.

$$\begin{array}{ccc} \gamma & \beta & \alpha \\ C- & C- & C-Cl \\ -5 & +10 & +30 \end{array}$$

Other electron-withdrawing groups have similar effects. Electron-donating groups have the opposite effect, decreasing the chemical shifts of α and β carbon atoms and increasing the chemical shifts of γ carbon atoms.

Table G5 shows chemical-shift ranges for some kinds of ^{13}C atoms. Such tables can be used to assign the peaks in a ^{13}C NMR spectrum to specific carbon atoms. For example, the ^{13}C NMR spectrum of methyl methacrylate has five peaks with the chemical shifts shown here.

J. Chem. Educ. **1987**, *64*, 915 *describes a method for estimating* ^{13}C *chemical shifts.*

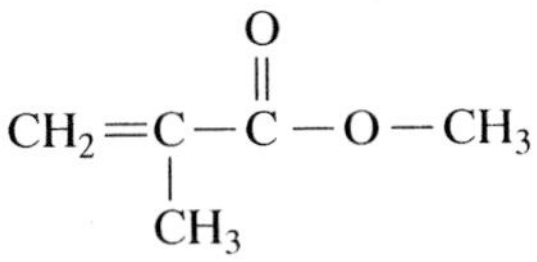

$$CH_2{=}C{-}\overset{\overset{\displaystyle O}{\|}}{C}{-}O{-}CH_3$$
$$\underset{CH_3}{|}$$

methyl methacrylate

1	18 ppm	4	137 ppm
2	52 ppm	5	167 ppm
3	125 ppm		

From Table G5, we find the following information for carbon atoms like those in methyl methacrylate:

ester carbonyl carbon: 160–185 ppm

alkene carbon: 100–160 ppm

carbon bonded to oxygen (C—O): 40–90 ppm

primary alkyl carbon: 0–40 ppm

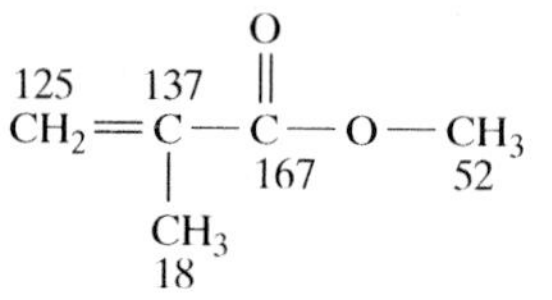

Note that the C—O chemical-shift range from the table is for alcohols and ethers, but a carbon atom on the alcohol portion of an ester is in a similar environment. From this information it is easy to match the peak at 167 ppm with the carbonyl carbon, the peaks at 125 ppm and 137 ppm with the alkene carbons, the peak at 52 ppm with the OCH$_3$ methyl group, and the peak at 18 ppm with the remaining methyl group. Since the alkene carbon *beta* to the two oxygen atoms would be expected to have a higher chemical shift than the alkene carbon *gamma* to them, we can assign chemical-shift values to the carbon atoms as shown in the margin.

Such assignments can often be confirmed by a technique called *distortionless enhanced polarization transfer (DEPT)*, which can pinpoint the carbon atom that is responsible for a particular signal.

OPERATION 41

Ultraviolet–Visible Spectrometry

Principles and Applications

Ultraviolet–visible (UV–VIS) spectrometers, which detect the absorption of radiation in the visible (~400–800 nm) and near ultraviolet (~200–400 nm) regions of the electromagnetic spectrum, are useful for both qualitative and quantitative analysis of organic compounds. Radiation in these regions may induce molecules to undergo transitions from an electronic ground state to one or more excited states. The energy required for an electronic transition is much greater than that needed to induce a vibrational or nuclear magnetic transition, so the wavelength of the radiation used is much shorter: 200–800 nm compared to about 2.5–50 μm for IR spectrometry and several meters for NMR. Most transitions involving electrons in single bonds and isolated double bonds require wavelengths shorter than 200 nm, but conventional UV–VIS spectrometers are not designed to scan this region because oxygen from the air absorbs UV radiation below 200 nm. Therefore, the electronic transitions of organic compounds that give rise to UV–VIS spectral bands usually involve pi electrons in aromatic and conjugated aliphatic systems, and certain nonbonded electrons. An example of such a transition is illustrated in Figure G43, in which a pi electron

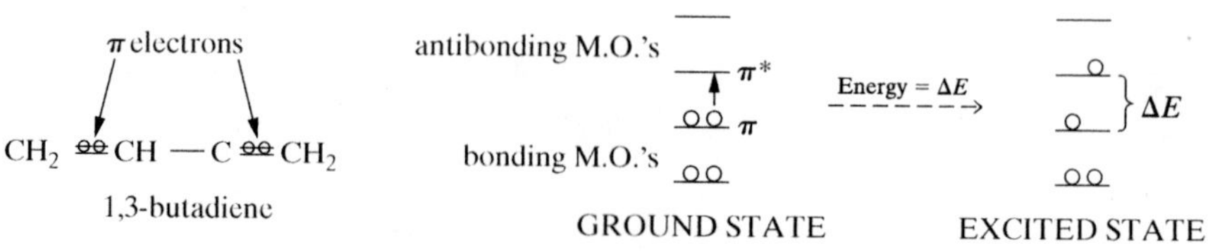

1 nanometer(nm) = 10^{-9}m. The older unit "millimicrons" (mμ) is sometimes used for nm.

Figure G43 Electronic energy transition in 1,3-butadiene

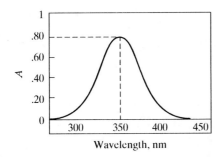

Figure G44 An ultraviolet absorption band

in the ground-state electron configuration of 1,3-butadiene jumps from its bonding molecular orbital (M.O.) to an unoccupied antibonding molecular orbital.

A UV–VIS spectrum is often quite featureless compared to an IR or NMR spectrum, and may consist of only one or two broad *absorption bands*. The broad band structure is caused by rotational and vibrational transitions that accompany each electronic transition; each different combination of rotational and vibrational transitions has a different energy, so collectively they span a broad range of wavelengths. The height of an absorption band above the baseline of a UV–VIS spectrum is measured in units of *absorbance, A*. The position of an absorption band is given by its wavelength of maximum absorbance, λ_{max}, which is measured from the top of the band. For example, the absorption band illustrated in Figure G44 has an absorbance of 0.80 and a λ_{max} of 350 nm. Absorbance is related to *transmittance (T)*, the fraction of incident radiation transmitted through a sample, by the following equation:

$$A = \log(1/T) = -\log T \tag{1}$$

Thus, the transmittance of a UV–VIS band is equal to 10^{-A}. For a band with $A = 0.80$, $T = 10^{-0.80} = 0.16$, meaning that about 16% of the light entering the sample passes through unchanged and the remaining 84% is absorbed by the sample.

a. UV–VIS Spectra

Sample Preparation

Routine UV–VIS spectra are nearly always obtained in solution. The solvent must be transparent (or nearly so) in the regions to be scanned. Water, 95% ethanol, methanol, dioxane, acetonitrile, and cyclohexane are suitable down to about 210–220 nm; many other solvents can be used at higher wavelengths. The preferred solvent is 95% ethanol, in part because it doesn't require additional purification; most other solvents must be purified or purchased as spectral-grade solvents. The solvent must not, of course, react with the solute. For example, alcohols should not be used as solvents for aldehydes.

If possible, the solution to be analyzed should produce a maximum absorbance of about 1 when the solute's strongest absorption band is scanned. Using Beer's law (see Equation **2** in OP-41b) we in can show that the molar concentration of such a solution, if analyzed in a 1-cm sample cell, should be less than or equal to $1/\varepsilon_{max}$, where we define ε_{max} as the maximum molar absorptivity of the solute over the wavelength range to be scanned. For example, if the solute's strongest band has a molar absorptivity of 10,000 at its λ_{max} value, the solution concentration should be about 1×10^{-4} M. To make up such dilute solutions accurately, it may be necessary to prepare a stock solution that is too concentrated by several powers of ten, and then measure an aliquot of this solution and dilute it. For example, to prepare a $\sim 1.0 \times 10^{-4}$ M solution of cinnamic acid (M.W. = 148), you could measure 0.15 g (~ 1.0 mmol) of the solid into a 100-mL volumetric flask and fill it to the mark with solvent; then transfer a 1-mL aliquot of this 0.010 M solution into another 100-mL volumetric flask and fill it to the mark with solvent. For microscale work, you could use 15 mg of the cinnamic acid and a 10-mL volumetric flask to make the stock solution and dilute a 0.1-mL aliquot of this solution to 10 mL in another volumetric flask. For qualitative work, when knowing the exact concentration of the solution is not necessary, you can omit the dilution step and use a 10-μL syringe to measure a specified or calculated amount of solute into about 25 mL of the solvent. If the molar absorptivity of the solute is not known, it may be necessary to find the optimum concentration by trial and error, starting with a more concentrated solution and diluting it as needed to bring all of the absorption bands on scale.

Sample Cells

The most commonly used *spectrophotometer cell* is a transparent rectangular container with a square cross section, having a path length of 1.00 cm and a capacity of about 3 mL. Cells with capacities of about 1 mL and 0.5 mL are available for microscale work. Two opposite sides of a typical cell are non-transparent (usually frosted) and the other two are transparent. Silica or quartz cells are used for the UV region; optical glass or plastic cells are suitable in the visible region. Cells must be scrupulously cleaned; they should never be touched on their transparent sides, since even a fingerprint can yield a spectrum.

Recording a Spectrum

The construction of UV–VIS spectrometers varies widely. Both single- and double-beam instruments are available, with and without recording capability. For a double-beam recording instrument, two identical spectrophotometer cells are filled about two-thirds full with (1) a solution of the compound being analyzed (the sample) and (2) the solvent used to prepare the solution. For a single-beam instrument, the same cell is used for both the sample and the pure solvent. A spectrophotometer cell filled with the sample or solvent is held by its nontransparent sides and inserted into the appropriate cell holder inside the instrument's sample compartment, oriented so that the light beam will pass through its transparent faces.

Before a spectrum is recorded, the user selects the wavelength region to be scanned and may also select a radiation source appropriate for that region. A tungsten lamp can be used between 300 and 800 nm and a hydrogen lamp between 190 and 350 nm. The absorbance range of some instruments can be preset; a typical range is from zero to one or two absorbance units. Modern computerized instruments allow the operator to apply a baseline correction, so that the instrument automatically subtracts any absorption due to the solvent as the spectrum is scanned. Such instruments have a monitor to display the spectrum and a printer or plotter to record it. Most older instruments record UV–VIS spectra on a roll of chart paper that feeds onto a flat recorder bed as the spectrum is scanned.

Once a spectrum has been recorded, the data it contains can be presented as a tabulation of λ_{max} values giving either the absorbance, molar absorptivity (ε), or log ε at each wavelength specified. For example, the UV spectrum of cinnamic acid is reported in one reference book as "λ^{al} 210 (4.24), 215 (4.28), 221 (4.18), 268 (4.31)." The numbers in parentheses are log ε values for peaks having the λ_{max} values (in nanometers) given, and "al" indicates that the spectrum was run in ethyl alcohol.

General Directions for Operating a Recording UV–VIS Spectrometer

 Standard Scale and Microscale

Equipment and Supplies

UV–VIS spectrometer
sample
solvent(s)
sample cell(s)
lens paper

Do not attempt to operate the instrument without prior instruction and proper supervision. UV–VIS spectrometers vary widely in construction and operation, so the following is intended only as a general guide and may not be applicable to the instrument you will be using. Specific operating instructions should be learned from in-class demonstrations or the operator's manual.

Prepare a solution of the sample in an appropriate solvent as described under "Sample Preparation." Unless otherwise instructed, clean a sample cell (or two for a double-beam instrument) by wiping its surfaces with a lens paper moistened with spectral-grade methanol or another appropriate solvent, and then let the solvent evaporate. This should leave the cell surfaces free of contaminants that may have accumulated since the cell was last used. Be sure that the instrument, recorder or printer, and source lamps are on and have had sufficient time to warm up. If necessary, select the appropriate radiation source for the desired wavelength range and set the absorbance range to 0–1 or another appropriate value. Set the starting and ending wavelengths. If the instrument scans from high to low wavelength, the starting wavelength will be the highest wavelength of the range to be scanned. If there is no provision for setting the ending wavelength, you will have to end the scan manually. Some instruments require manual adjustment of zero and 100% transmittance values (or infinite and zero absorbance values) before a spectrum or baseline is run; if so, make the adjustments as directed by your instructor. If the instrument provides for a baseline correction, fill the cell with the solvent, cap it, and place it in the sample compartment with a transparent side facing the light source. Double-beam instruments require two such cells, one in the sample beam and the other in the reference beam. Then close the compartment door and record the baseline as directed by your instructor.

For a single-beam instrument, remove the solvent from the sample cell, rinse it with a small amount of the solution to be analyzed, fill it with that solution, cap it, and place it in the cell holder in the sample compartment. For a double-beam instrument, place one capped cell, filled with the sample, in the sample-cell holder, and place another cell, filled with the solvent, in the reference-cell holder. Close the sample compartment door. If the instrument uses chart paper, position it so that the scan starts on an ordinate (vertical) line, label this line with the starting wavelength, lower the pen to the paper, and begin to scan the spectrum. For a computerized instrument, start the scan as directed by your instructor. If any absorption band goes off-scale so that its top is "chopped off," change the absorbance range or dilute the sample. If both ultraviolet and visible regions are to be scanned, change the radiation source if necessary (many instruments do this automatically) and scan the spectrum in the other region. If the scan doesn't stop automatically when the end of the wavelength range has been reached, stop it manually. If the instrument has a chart recorder, raise the pen from the chart and tear off the chart paper; then write down the wavelength and absorbance ranges along its x- and y-axes and record the wavelength interval between chart units. If the instrument has a printer or plotter, initiate printing or plotting of the spectrum. Measure and write down the λ_{max} and absorbance values of all significant bands.

Dispose of the used solvent and solution as directed by your instructor. Rinse the sample cell several times with an appropriate solvent. If necessary, clean it further using a liquid detergent or a special cleaning solution. Never use a *dry* lens paper, an abrasive cleanser, or any scrubbing implement (such as a pipe cleaner with a wire core) that might scratch the cell. Drain both cells of excess solvent and dry them as directed by your instructor.

Summary

1. Clean sample cell(s).
2. Select source, if necessary.
3. Set absorbance and wavelength ranges.
4. Record baseline, if necessary.

5. Scan spectrum of sample.
6. Record spectral parameters.
7. Dispose of used solvent and solution, clean sample cell(s).

b. Colorimetry

Inexpensive nonrecording single-beam UV–VIS spectrometers, often called *colorimeters*, are frequently used for routine quantitative analysis of compounds in solution. The Spectronic 20 illustrated in Figure G45 is a widely used colorimeter, and most other colorimeters are operated similarly. The solution to be analyzed is prepared as described for a recording spectrometer, and is then placed in a *cuvette* that is inserted into the sample compartment. The wavelength control is set to a wavelength at which the sample absorbs strongly and, after some preliminary adjustments, the absorbance (or percent transmittance) of the solution is read from the scale.

A cuvette looks like a small test tube— but it should never be used as one!

The concentration of a solution can be determined from its absorbance value using either Beer's law (Equation **2**) or a calibration curve of absorbance versus molar concentration.

Beer's law

$$c = \frac{A}{\varepsilon \cdot b}$$

c = molar concentration **(2)**

A = absorbance

b = cell path length, in cm

ε = molar absorptivity, in L mol^{-1} cm^{-1}

Equation **2** can be used to determine concentrations only when solutions of the solute obey Beer's law over the appropriate concentration range and when the absorptivity of the solute is known.

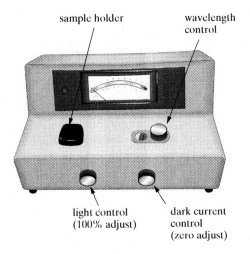

sample holder

wavelength control

light control (100% adjust)

dark current control (zero adjust)

Figure G45 Spectronic 20 spectrophotometer

General Directions for Operating a Colorimeter

 Standard Scale and Microscale

Equipment and Supplies

> colorimeter
> solution to be analyzed
> solvent
> cuvette(s)

Do not attempt to operate the instrument without prior instruction and proper supervision. These directions are for operation of the B & L Spectronic 20 or a similar instrument and may not be applicable to all such instruments.

Obtain or prepare the solution to be analyzed. Make sure that the instrument has been switched on and that adequate time has been allowed for warm-up. Set the wavelength to the desired value. Rotate the zero adjust control until the digital readout or the pointer of an analog dial indicates 0% transmittance. To read an analog scale, position your eyes so that the pointer is directly over its reflection in the mirror; this prevents parallax errors. Insert a clean cuvette containing the pure solvent (the one used to prepare the solution being analyzed) into the sample holder, making sure that the alignment mark on the cuvette is opposite the mark on the cell holder. Adjust the 100% control until the transmittance reading is 100%. Rinse the cuvette with a little of the solution to be analyzed, and then fill it with that solution. Replace it in the sample holder, being careful to position it in exactly the same alignment as before. If you are using an instrument with an analog dial, read the percent transmittance as accurately as possible; otherwise you can read the absorbance directly. Note that it is more accurate to read %*T* from a dial and convert it to absorbance than to read the absorbance directly from the nonlinear absorbance scale.

Different cuvettes are sometimes used for the solvent and sample, but errors will result if the cuvettes are not well matched; for precise work, it is best to use the same cuvette for all measurements.

If another solution containing the same solvent and solute is to be analyzed, empty the cuvette and rinse it with that solution before you make the next measurement. After the last measurement, rinse the cuvette with pure solvent, clean it, and let it air dry. Dispose of the used solvent and solution(s) as directed by your instructor. Convert the percent transmittance values to absorbance values, if necessary, using the equation $A = \log(100/\%T)$.

Summary

1. Prepare or obtain solution.
2. Clean cuvette.
3. Set wavelength and zero adjust control.
4. Fill cuvette with solvent, set 100% control.
5. Fill cuvette with solution, record %*T*.
6. Clean cuvette, dispose of used solvent and solution(s).

Mass Spectrometry

The other kinds of spectrometry described in this Instrumental Analysis section use some kind of electromagnetic radiation—radio frequency, infrared, ultraviolet, or visible—to gently probe the molecules of a sample and induce them to reveal their secrets. No molecules are damaged; once the spectrum is recorded, they return to their former states. By contrast, mass spectrometry uses a brute-force approach to determine molecular structures. Molecules that enter a mass spectrometer are pummeled by high-energy electrons and shattered into fragments, which are pushed and pulled along a curved path until they smash into an ion collector at journey's end. The fragments cannot be put back together to form the original molecule, so mass spectrometry is a destructive method of analysis.

Mass spectrometry is not really a spectroscopic method in the usual sense, in that no electromagnetic radiation is absorbed. But a mass spectrum does resemble a conventional spectrum in that it consists of a series of peaks of different amplitude plotted along a numerical scale. Although mass spectra are not easily interpreted and mass spectrometers are costly and complex instruments, mass spectrometry has become an increasingly valuable analytical tool for scientists and technologists in a variety of fields.

Principles and Applications

When a compound is bombarded with a beam of high-energy electrons in a mass spectrometer, each of its molecules (M) can lose an electron and form a *molecular ion*, $M^{\cdot+}$

$$M \longrightarrow M^{\cdot+} + e^-$$

Because a molecular ion contains both an unpaired electron and a positive charge, it is called a *radical cation*. If the energy of the electron beam is high enough, many of the molecular ions will have enough excess vibrational and electronic energy to break apart into fragments. Each pair of fragments consists of another positive ion (A^+), called a *daughter ion*, and a neutral molecule (X).

$$M^{\cdot+} \longrightarrow A^+ + X$$

The unpaired electron from the molecular ion may end up on either A^+ or X, depending on the kind of fragmentation. Each daughter ion may in turn break down, losing a neutral fragment to form yet another daughter ion, and so on.

$$A^+ \longrightarrow B^+ + Y$$

For example, the fragmentation of an ammonia molecule takes place as shown, forming ions (cations and radical cations) having approximate masses of 17, 16, 15, and 14 atomic mass units.

$$H-\overset{\circ\circ}{N}-H \xrightarrow{-e^-} H-\overset{\circ+}{N}-H \xrightarrow{-H^\circ} H-\overset{+}{\underset{\circ}{N}}\overset{\circ}{} \xrightarrow{-H^\circ} H-\overset{+}{\underset{\circ}{N}}\overset{\circ}{} \xrightarrow{-H^\circ} \overset{+}{\underset{\circ}{N}}\overset{\circ}{}$$

ammonia   molecular ion   daughter ions

The positive ions formed during these transformations are accelerated into an evacuated chamber in which they are separated according to their mass-to-charge ratios (m/e), usually by means of strong electric and magnetic fields (see Figure G46). As a beam of ions with a given m/e value impinges on an *ion collector*, it gives rise to an electrical current that is amplified and displayed on a monitor as a peak whose amplitude is proportional to the number of ions striking the detector. A *mass spectrum* is a record, usually printed out as a table of data or a computer-generated bar graph, of the relative abundances of all the ions arranged in order of their m/e values. Because most daughter ions have a charge of $+1$, the m/e value associated with a peak is nearly always equal to the mass of the ion that gave rise to that peak—or in rare cases, to one-half of its mass. A large molecule may be fragmented into several hundred different ions with different m/e values and relative abundances, so mass spectrometers are provided with microprocessors that record, store, and process the data.

Mass spectrometry is an extremely valuable tool for structural analysis; it can be used to identify or characterize a host of organic (and inorganic) compounds, including biologically active substances with very complex molecular structures. The mass spectrum of a compound usually gives the mass of its molecular ion, which is essentially equal to its molecular weight. It also provides the masses of smaller pieces of its molecules, which can often be identified with the help of published tables of molecular fragments. A high-resolution mass spectrum provides data that can be used to determine a compound's molecular formula. Such information often makes it possible to piece together the compound's molecular structure, or at least to learn more about its structural features.

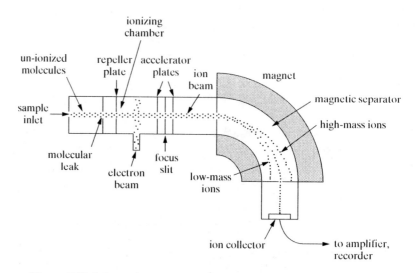

Figure G46 Schematic diagram of a single-focusing mass spectrometer

Different compounds yield distinctively different patterns of ion fragments, so an unknown compound can sometimes be identified by a comparison of its mass spectrum with mass spectral data from the scientific literature. Unfortunately, peak intensities on mass spectra are very sensitive to instrumental parameters, such as the energy of the electron beam. Therefore, there may be significant differences in the mass spectra recorded on different instruments, or even by different operators using the same instrument. Nevertheless, it is often possible to make a tentative identification from a literature comparison and then to confirm it by recording the mass spectra of the unknown and the most likely known compounds under identical operating conditions. This process can be facilitated by using a computer to compare the spectrum of the unknown with the mass spectra in a memory bank, which may contain tens of thousands of such spectra.

Instrumentation

Mass spectrometers come in a wide variety of sizes and configurations, ranging from high-resolution mass spectrometers that may take up most of an instrument room to compact "tabletop" mass spectrometers that can be used as detectors for gas and liquid chromatographs. Although research-grade mass spectrometers are very expensive, tabletop mass spectrometers are within the equipment budgets of some undergraduate chemistry programs.

In a conventional *single-focusing* mass spectrometer (diagrammed in Figure G46) the sample is introduced into a sample inlet system that is heated to keep some or all of its molecules in the vapor state. These molecules find their way into the *ionizing chamber* through an aperture called a *molecular leak*, which may be a tiny hole in a piece of gold foil. The ionizing chamber is kept at a pressure of about 10^{-6} torr to minimize collisions between particles and interference from ionized air. Molecules that wander into the path of the *electron beam* are ionized to molecular ions, some of which undergo fragmentation to yield daughter ions. These ions are pushed toward a slit by a positively charged *repeller plate*, and then accelerated to a high velocity by electrically charged *accelerator plates* and directed into a *magnetic separator*. In the magnetic separator, a powerful magnetic field deflects each kind of ion into a curved path whose radius depends on the ion's mass-to-charge ratio. Lighter (lower m/e) ions are deflected more than heavier (higher m/e) ions. At a given magnetic-field strength, only ions of a given mass (or m/e value) can pass through the slit that leads to the *ion collector*. As the field strength is increased, ions of progressively higher mass reach the ion collector and are detected. When a beam of ions strikes the ion collector, an electrical signal is generated with an intensity proportional to the number of ions in the beam—that is, to their abundance. The signals are amplified and displayed by a monitor screen or recorder to produce a mass spectrum.

In a single-focusing mass spectrometer, variations in the kinetic energies of ions having the same mass cause their ion beam to broaden as it passes through the magnetic separator, reducing the instrument's resolving power. In a high-resolution *double-focusing* mass spectrometer, the ions are initially directed along a curved path by an electrostatic field, which allows only particles of the same kinetic energy to pass through a slit leading to the magnetic separator. Double-focusing instruments can often resolve such

In mass spectrometry, two adjacent peaks of equal amplitude are said to be resolved when the height of the valley between them is no more than 10% of their height.

ions as CH_2N^+ and N_2^+, whose mass numbers (28.0187 and 28.0061) differ by less than 0.05%.

Other kinds of mass spectrometers separate ions by very different methods. In a *quadrupole* mass spectrometer, the ions are introduced between four parallel metal rods that create a rapidly oscillating magnetic field between them. Ions whose mass-to-charge ratio is compatible with the frequency of the field will oscillate along a straight path toward the ion collector, while other ions will follow a different path and be removed. The frequency or intensity of the oscillating field is varied so that ions of all mass-to-charge ratios eventually reach the ion collector.

In a *time-of-flight* mass spectrometer, ion beams are produced by brief pulses of electrons and accelerated by an electrical field pulse that gives all ions, regardless of mass, the same kinetic energy. Ions of a given mass (or m/e value) then pass through a *drift tube* at a speed that is inversely proportional to their mass, so lighter ions reach the ion collector sooner than heavier ions.

In a *Fourier-transform* mass spectometer (sometimes called an *ion trap* mass spectrometer), ions are forced into a circular path by a strong magnetic field and subjected to a radio-frequency (RF) pulse. If the frequency of the radiation is equal to the frequency at which ions of a given m/e value move around their circular path (their *cyclotron frequency*), the ions will be accelerated and spiral outward. At the end of the RF pulse, they will be moving around a larger circle in a coherent packet. This packet of revolving ions generates an *image current* that decays with time after the pulse ends, producing a signal that is similar to the free induction decay signal generated by an FT-NMR spectrometer. The frequency of the RF pulse is varied to match the cyclotron frequencies of all ions in a sample, causing each kind of ion to produce its own image current. The resulting image currents are then detected and converted to a conventional mass spectrum by Fourier-transform analysis. Some Fourier-transform mass spectrometers are capable of resolving ions whose masses differ by only 0.001% or so.

Gas Chromatography/Mass Spectrometry

Rapid-scan mass spectrometers—including some quadrupole, Fourier-transform, and time-of-flight instruments—can be interfaced with gas chromatographs and used to analyze the components of a mixture as they emerge from the GC column. This combination of instruments is, not surprisingly, called a *gas chromatograph/mass spectrometer (GC/MS)*. The output of a GC capillary column [OP-37] can often be introduced directly into the ionization chamber of a mass spectrometer. With a packed column, however, most of the carrier gas must be removed to maintain a sufficiently low pressure in the evacuated ionizing chamber. The mass spectra of the components can be displayed by a screen or recorder in "real time," as each component exits the gas chromatogram, or stored to be printed out later.

Instruments in which mass spectrometers are interfaced with high-performance liquid chromatographs (LC/MS) and even with other mass spectrometers (MS/MS) are also available.

Gas chromatograph/mass spectrometers are particularly useful for identifying the components of natural products, biological systems, and ecosystems. For example, flavor and odor components of essential oils, physiologically active components of plants, and chemical pollutants in the environment can be characterized by GC/MS. Forensic chemists can use GC/MS to identify drug metabolites in the body fluids of a suspect, and physicians can diagnose certain illnesses on the basis of a GC/MS analysis of a patient's breath.

Experimental Considerations

Samples prepared for mass spectrometry should be very pure, because traces of impurities can make interpretation of a mass spectrum difficult. Sample sizes range from less than a microgram to several milligrams, and no special sample preparation is required. Liquids are inserted directly into the sample inlet with a syringe, micropipet, or break-off device, and solids can be introduced by means of a melting-point capillary. The samples vaporize in the sample inlet system, after which their molecules flow through a molecular leak into the evacuated ionization chamber.

A typical single-focus mass spectrometer is prepared for operation by turning on the magnet current and adjusting controls that set the potential of the repeller plates, the accelerating voltage, the ionizing current (the number of electrons in the electron beam), and the energy of the electron beam. Increasing the repeller potential reduces the time that the ions spend in the ionization chamber, while increasing the accelerating voltage increases the speed that the ions attain by the time they enter the mass separator. Increasing the ionizing current increases the number of ions that are formed, and increasing the energy of the electron beam increases the amount of fragmentation. Since the settings of these controls determine the appearance of a mass spectrum, it is important to set them within ranges that are appropriate for a given analysis. The user also selects the range of masses to be scanned and the scan time. The mass spectrum is then scanned and recorded as a chart or computer printout.

Other types of instruments may operate quite differently, so no operating procedures will be given here. These must be learned by special instruction and by studying the manufacturer's operating manual. Computer programs that simulate the operation of specific mass spectrometers may also be available for your use.

Interpretation of Mass Spectra

Each peak recorded on a mass spectrum is characterized by the mass-to-charge ratio of the ion that gave rise to it and the ion-beam intensity (also called relative abundance). This information can be displayed by a variety of output devices, including oscilloscopes, strip-chart recorders, and computer printers. The height of an ion's peak is proportional to its ion-beam intensity. The most intense peak in a mass spectrum, called the *base peak*, is assigned a relative intensity of 100, and the intensities of all other peaks are reported as percentages of the base-peak intensity. The molecular-ion peak may be the base peak, but often it is not.

Molecular Weight and Molecular Formula

For most organic compounds, the molecular weight of the compound is virtually equal to the mass of its strongest molecular-ion peak. This molecular-ion peak is usually the last strong peak on the spectrum, since no daughter ion should have a higher mass than the original molecular ion. However, the molecular-ion peak is usually followed by at least two low-intensity peaks corresponding to isotopic variations of the molecular ion, because most of the elements in organic compounds have at least one isotope of higher mass number than the common form.

Just over one carbon atom in a hundred (1.08%) is a carbon-13 atom; the rest are carbon-12 atoms. Since benzene, for example, contains six carbon atoms, the chance that any one of the six will be carbon-13 is $6 \times 1.08\%$, or 6.48%. Thus, the molecular-ion peak of benzene (C_6H_6, M.W. = 78) should be followed by a peak for a "heavy" form of benzene ($C_5{}^{13}CH_6$, M.W. = 79) with an intensity that is 6.48% of the parent peak intensity. Actually, the $m/e = 79$ peak, called the $M + 1$ *peak*, has a slightly higher intensity than this because benzene contains minute quantities of benzene-d_1 (C_6H_5D), which also has an approximate molecular weight of 79. Likewise, oxygen-18 occurs naturally to the extent of about 0.20 atoms for every 100 atoms of oxygen-16, so formaldehyde shows an $M + 2$ *peak* (corresponding to $CH_2{}^{18}O$) with an intensity that is 0.20% of the molecular-ion peak's intensity.

Intensities of the M + 1 and M + 2 peaks (relative to the molecular-ion peak's intensity) for a compound $C_wH_xN_yO_z$ can be calculated using Equations **1** and **2**:

$$\%(M + 1) = 1.08w + 0.015x + 0.37y + 0.037z \tag{1}$$

$$\%(M + 2) = 0.006w(w - 1) + 0.0002wx + 0.004wy + 0.20z \tag{2}$$

For example, quinine (as the hydrate) has the molecular formula $C_{20}H_{30}N_2O_3$. Using Equations **1** and **2** gives the intensity of its M + 1 and M + 2 peaks as 22.9% and 3.16%, respectively, of the M peak intensity.

$$\%(M + 1) = 1.08(20) + 0.015(30) + 0.37(2) + 0.037(3) = 22.9$$

$$\%(M + 2) = 0.006(20)(19) + 0.0002(20)(30)$$

$$+ 0.004(20)(2) + 0.20(3) = 3.16$$

No nonidentical sets of atoms are likely to yield M + 1 and M + 2 peaks of exactly the same relative intensity. Therefore, if the intensities of these peaks in the mass spectrum of an unknown compound can be measured to two or more decimal places, its molecular formula (or several possible formulas) can be determined using published formula mass tables, such as those in *Spectrometric Identification of Organic Compounds* [Bibliography, F20]. Here is a general procedure for determining molecular formulas.

1. Locate the molecular-ion peak and determine its mass.
2. Measure the intensities of the M, M + 1, and M + 2 peaks, and express the latter two as a percentage of the intensity of the M peak.
3. Find the formula (or formulas) listed under its M value that give M + 1 and M + 2 intensities close to the experimental values and that make sense from a chemical standpoint.

Some formulas can be eliminated immediately because they do not correspond to stable molecules or because compounds with those formulas would be impossible to obtain from a given reaction or source. Others can be eliminated because they do not have the expected *index of hydrogen deficiency* (*IHD*), where the IHD of a compound is equal to the number of rings plus the number of pi bonds (or their aromatic equivalent) in a molecule of the compound. For compounds with the general formula $C_wH_xN_yO_z$, the IHD can be calculated using Equation **3**.

$$IHD = \tfrac{1}{2}(2w - x + y + 2) \qquad \textbf{(3)}$$

For example, the calculated IHD of Compound **1** is 7, which is consistent with its molecular structure—one ring and six pi bonds.

$$IHD = \tfrac{1}{2}(18 - 7 + 1 + 2) = 7$$

A useful generalization that applies to most organic compounds is the *nitrogen rule*, which states that a stable compound whose molecular weight is an even number can have only zero or an even number of nitrogen atoms, whereas one whose molecular weight is an odd number can have only an odd number of nitrogen atoms.

Fragmentation Patterns

The use of fragmentation patterns to determine molecular structures is a broad subject covered in detail elsewhere (refer to Section F of the Bibliography), so only a few generalizations will be given here. A molecular ion or daughter ion often breaks down by eliminating small neutral molecules or free radicals such as CO, H_2O, HCN, C_2H_2, $H\cdot$, or $\cdot CH_3$, yielding ions with masses equal to $M - X$, where M represents the mass of the molecular ion or a daughter ion undergoing fragmentation, and X is the mass of the neutral species. Table G6 gives the masses and postulated structural formulas of some neutral species that are often lost by fragmentation.

Table G6 Formulas of some neutral species lost from molecular ions

Mass of neutral species	Mass of resulting ion	Possible formulas
1	$M - 1$	$H\cdot$
15	$M - 15$	$\cdot CH_3$
16	$M - 16$	$\cdot NH_2$
17	$M - 17$	$\cdot OH$, NH_3
18	$M - 18$	H_2O
26	$M - 26$	C_2H_2, $\cdot CN$
27	$M - 27$	$\cdot C_2H_3$, HCN
28	$M - 28$	CO, C_2H_4
29	$M - 29$	$\cdot CHO$, $\cdot C_2H_5$
30	$M - 30$	H_2CO, NO

continued

Table G6 *Continued*

Mass of neutral species	Mass of resulting ion	Possible formulas
31	$M - 31$	$CH_3O\cdot$, $\cdot CH_2OH$
35	$M - 35$	$Cl\cdot$ (also 37)
36	$M - 36$	HCl
42	$M - 42$	CH_2CO
43	$M - 43$	$CH_3CO\cdot$, $\cdot C_3H_7$
44	$M - 44$	CO_2, $\cdot CONH_2$
45	$M - 45$	$\cdot CO_2H$, $C_2H_5O\cdot$
46	$M - 46$	$\cdot NO_2$
49	$M - 49$	$\cdot CH_2Cl$
57	$M - 57$	$CH_3COCH_2\cdot$
59	$M - 59$	$\cdot CO_2CH_3$
77	$M - 77$	$C_6H_5\cdot$
79	$M - 79$	$Br\cdot$ (also 81)

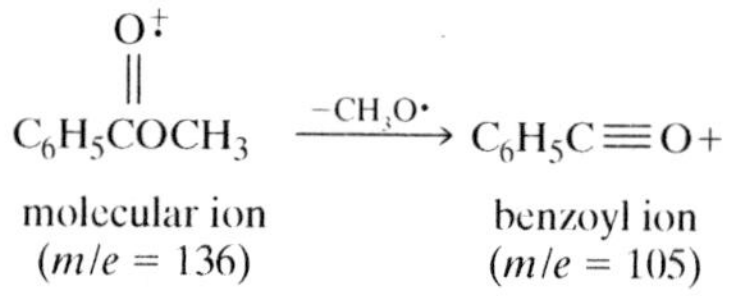

$$C_6H_5COCH_3 \xrightarrow{-CH_3O\cdot} C_6H_5C\equiv O+$$

molecular ion ($m/e = 136$)   benzoyl ion ($m/e = 105$)

$$C_6H_5C\equiv O+ \xrightarrow{-CO} C_6H_5^+$$

benzoyl ion ($m/e = 105$)   phenyl ion ($m/e = 77$)

$$C_6H_5^+ \xrightarrow{-C_2H_2} C_4H_3^+$$

phenyl ion ($m/e = 77$)   ($m/e = 51$)

To see how such data can be used to interpret mass spectra, consider the mass spectrum of methyl benzoate in Figure G47. The m/e value of the molecular-ion peak in this spectrum is 136, that of the base peak is 105, and there are other strong peaks at $m/e = 77$ and 51. The molecular ion is believed to be a radical cation with the structure shown in the margin. Since the mass numbers of the base peak and molecular-ion peak differ by 31 mass units, a neutral species with a mass number of 31 must have been lost from the molecular ion to produce the base-peak ion. Table G6 shows two neutral fragments, CH_3O and $\cdot CH_2OH$, with that mass number. Since the molecular ion has a methoxyl group, it must have lost CH_3O to form the base-peak ion, which must therefore be a benzoyl cation. The difference between the mass number of the benzoyl ion and that of the next major peak is $105 - 77 = 28$. The two neutral species with mass numbers of 28 are CO and C_2H_4; loss of CO from the benzoyl cation should yield a phenyl cation with the expected mass number of 77. Formation of the next major species ($m/e = 51$) requires a loss of 26 mass units from the phenyl ion. From Table G6, two species with that mass are $\cdot CN$ and C_2H_2 (acetylene), of which only the latter is a possibility. Loss of acetylene from the phenyl ion yields an ion with the formula $C_4H_3^+$; this species is often encountered in the mass spectra of aromatic compounds. There are a number of smaller peaks in the methyl benzoate spectrum that may provide additional structural information, but usually it is not possible (or necessary) to characterize all of the peaks in a mass spectrum.

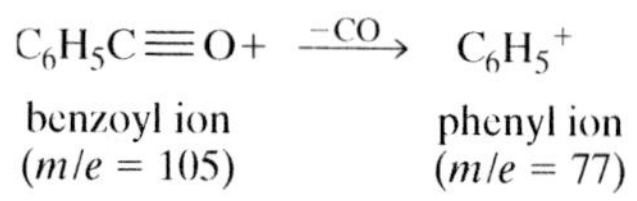

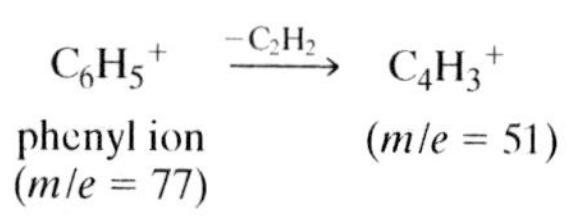

In the previous example, the interpretation was simplified because each major ion was produced from the preceding one along a single reaction path. Often there are fragmentation paths leading directly from the molecular ion

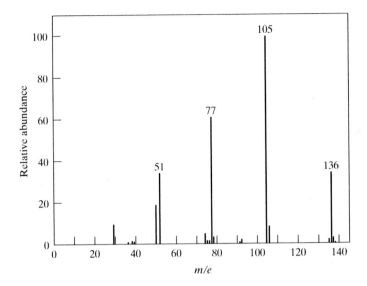

Figure G47 Mass spectrum of methyl benzoate

Table G7 Some common fragment ions

m/e	Possible formulas
15	CH_3^+
17	OH^+
18	H_2O^+, NH_4^+
26	$C_2H_2^+$
27	$C_2H_3^+$
28	$CO^+, C_2H_4^+$
29	$CHO^+, C_2H_5^+$
30	$CH_2NH_2^+, NO^+$
31	CH_2OH^+, CH_3O^+
35, 37	Cl^+
39	$C_3H_3^+$
41	$C_3H_5^+$
43	$CH_3CO^+, C_3H_7^+$
44	$CO_2^+, C_3H_8^+$
45	$CH_3OCH_2^+, CO_2H^+$
46	NO_2^+
49	CH_2Cl^+
51	$C_4H_3^+$
57	$C_4H_9^+, C_2H_5CO^+$
59	$COOCH_3^+$
65	$C_5H_5^+$
66	$C_5H_6^+$
71	$C_5H_{11}^+, C_3H_7CO^+$
76	$C_6H_4^+$
77	$C_6H_5^+$
78	$C_6H_6^+$
79, 81	Br^+
91	$C_7H_7^+, C_6H_5N^+$
93	$C_6H_5O^+$
94	$C_6H_6O^+$
105	$C_6H_5CO^+$

to a number of different daughter ions. It is therefore a common practice to compare the mass number of the molecular ion with those of the significant daughter ions before attempting to compare the mass numbers of individual daughter ions.

The fragment ions described previously and other common fragment ions are listed in Table G7. Structures have been determined for some (but not all) of these cations. For example, the $C_7H_7^+$ ion with mass number 91, which results from the cleavage of alkylbenzenes, has been formulated as either a benzyl cation or a tropylium ion; in most cases it appears to have the latter structure. Before you can interpret a mass spectrum proficiently using data like that in Tables G6 and G7, you need to learn about the characteristic fragmentation patterns and mechanisms for different classes of organic compounds. This kind of information and some general rules for interpretation of mass spectra are given in references on mass spectrometry listed in Category F of the Bibliography.

Appendixes and Bibliography

Laboratory Equipment

Chemical Glassware

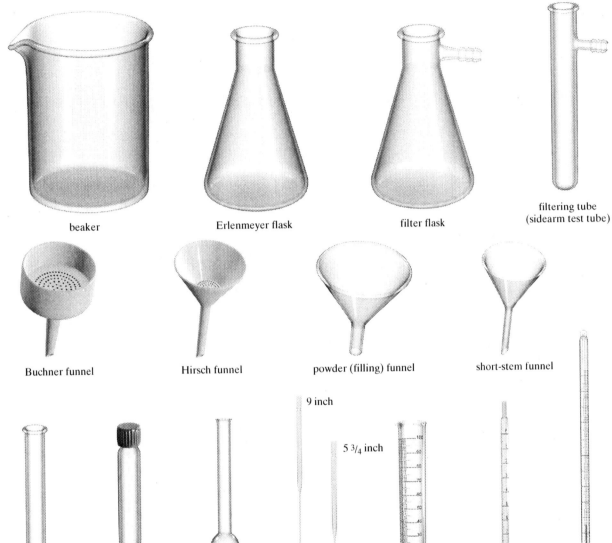

beaker

Erlenmeyer flask

filter flask

filtering tube
(sidearm test tube)

Buchner funnel

Hirsch funnel

powder (filling) funnel

short-stem funnel

9 inch

5 3/4 inch

test tube

centrifuge tube
with screw cap

drying tube

Pasteur pipets

graduated cylinder

measuring
pipet

thermometer

watch glass

evaporating dish

stirring rod

Lab Kit Components, Standard Scale

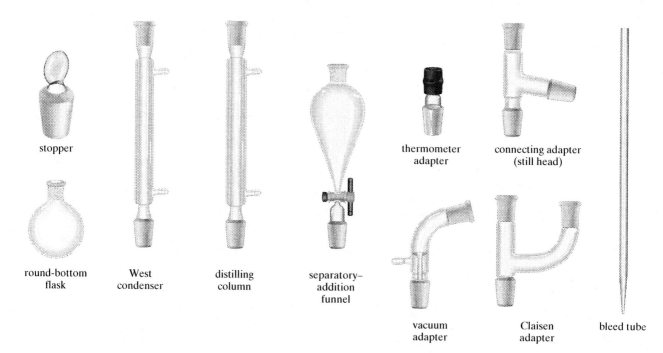

stopper

round-bottom
flask

West
condenser

distilling
column

separatory–
addition
funnel

thermometer
adapter

connecting adapter
(still head)

vacuum
adapter

Claisen
adapter

bleed tube

Lab Kit Components, Microscale

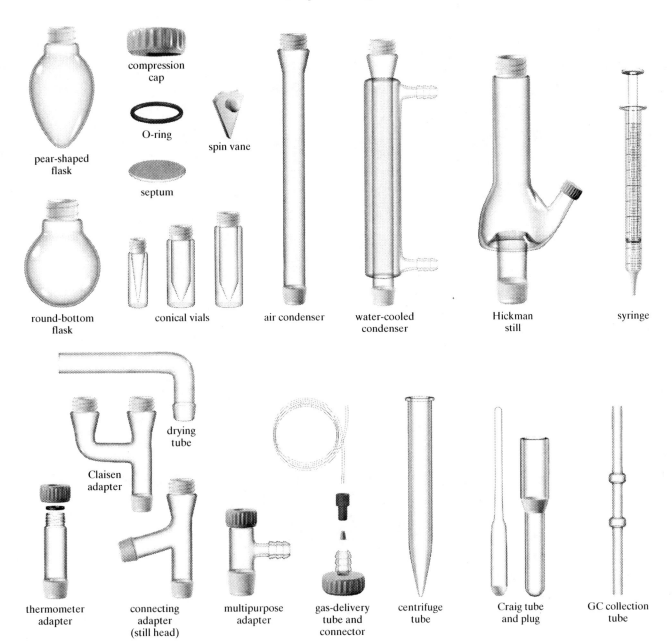

pear-shaped
flask

compression
cap

O-ring

septum

spin vane

round-bottom
flask

conical vials

air condenser

water-cooled
condenser

Hickman
still

syringe

drying
tube

Claisen
adapter

thermometer
adapter

connecting
adapter
(still head)

multipurpose
adapter

gas-delivery
tube and
connector

centrifuge
tube

Craig tube
and plug

GC collection
tube

Hardware

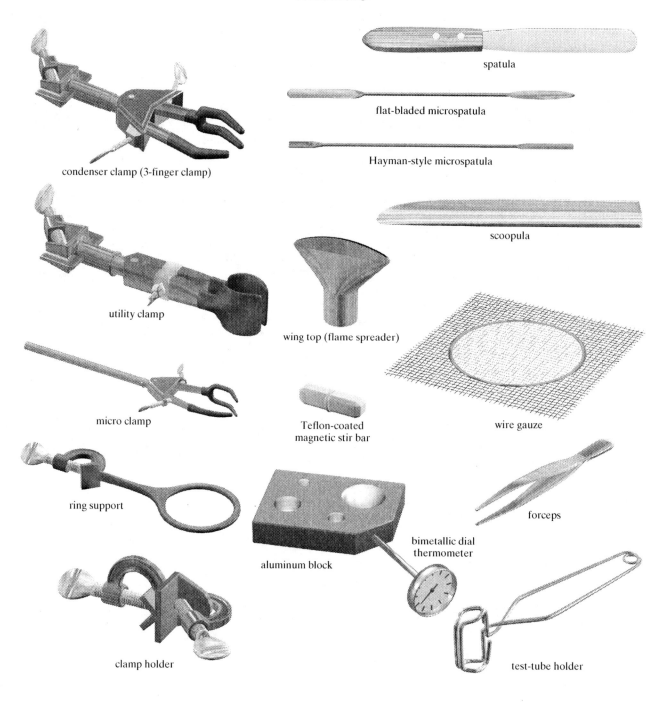

spatula

flat-bladed microspatula

Hayman-style microspatula

condenser clamp (3-finger clamp)

scoopula

utility clamp

wing top (flame spreader)

wire gauze

micro clamp

Teflon-coated
magnetic stir bar

ring support

aluminum block

bimetallic dial
thermometer

forceps

clamp holder

test-tube holder

Keeping a Laboratory Notebook

APPENDIX II

Your instructor may require that you maintain a laboratory notebook, write formal laboratory reports, or both. In either case, you should write an experimental plan before you begin most of the experiments in this book. This appendix tells you how to keep a laboratory notebook. Information about writing lab reports and experimental plans will be found in Appendix III and Appendix V, respectively.

A laboratory notebook is essentially a factual account of work performed in the laboratory. It may also include the writer's interpretation of the results. Your laboratory work may not require the documentation expected of a research chemist, but you should at least be aware of the characteristics of a good laboratory notebook. Although any kind of notebook can be used to record data and information, a notebook suitable for organic chemistry labs should be bound and have quadrilled (square-ruled) pages. If your instructor requires you to turn in a copy of your notes for each experiment, your notebook should also have duplicate pages. The first page or two should be reserved for a table of contents, and the pages should be numbered sequentially so that you can find information quickly. Whenever possible, each entry should be written immediately after the work is performed, and it should be dated and signed by the experimenter (notebooks of research chemists are usually signed by a witness as well). Each section of the notebook should have a clear, descriptive heading, and the writing should be grammatically correct and sufficiently legible to be read and understood by any knowledgeable individual.

Before each experiment, you should write an experimental plan (see Appendix V) in your lab notebook, summarizing what you expect to do in the laboratory and how you intend to go about it, and providing other relevant information. Your prelab writeup should include items from the following list:

- The experiment number and title.
- For a preparation, balanced equations for all significant reactions, including possible side reactions that might reduce the yield.
- A table listing relevant properties (mol wt, mp, solubility, hazardous characteristics, etc.) of the reactants, products, solvents and any other chemicals involved in the experiment.
- A calculation of the theoretical yield of product and any other necessary stoichiometric calculations, when applicable.
- A list of the materials (chemicals, supplies, equipment) needed for the experiment.
- A checklist, flow diagram (see Appendix V), or other kind of outline summarizing the experimental procedure.

Some of these items may not apply to certain kinds of experiments, such as kinetic studies or qualitative analysis experiments. At your instructor's request, you may also include the following:

- A clear, concise statement of the scientific problem the experiment is designed to solve
- A brief statement telling how you will attempt to solve the problem
- When appropriate, a working hypothesis predicting the outcome of the experiment.

During the experiment, you should keep a detailed account of your work, reporting everything of importance that you actually did and saw. Your notes should not simply restate the textbook procedure, but should describe in your own words how you carried out the experiment. You should include all relevant data, such as the quantities of materials that you actually used (not the quantities calculated or given in the procedure, unless they are exactly the same) and the results of any analyses you performed. Raw data should be recorded with particular care; if you forget to record data at the time you measure it, or if you record it incorrectly or illegibly, the results of an entire experiment may be invalidated. As you gather evidence pertaining to any scientific problem that the experiment is designed to solve, you can write down one or more tentative hypotheses and describe how you tested them. You should also report any observations that might have a bearing on the problem, such as those providing clues to the nature of the reaction or the identity of the product.

You may find it helpful to think of your lab notebook as telling a story about your accomplishments (or misadventures) in the laboratory. Although most professional journal articles are written in a dry, impersonal style, this need not be true of a lab notebook. The American Chemical Society's publication *Writing the Laboratory Notebook* [Bibliography, L46] suggests a more personal approach, stating that "the use of the active voice in the first person tells the story and clearly indicates who did the work." (For example, "I recorded the infrared spectrum of benzaldehyde ... "). Other scientists prefer to write in the passive voice, avoiding the use of personal pronouns. (For example, "The infrared spectrum of benzaldehyde was recorded. . . . ") If your instructor has a strong preference either way, you should follow his or her recommendation.

After you have finished an experiment, summarize your major results, write down any conclusion(s) you can draw from them, and explain how you arrived at your conclusion. If your notebook pages are to be turned in as your laboratory report, include answers to assigned exercises and any other information requested by your instructor. See Appendix III for additional suggestions about reporting and interpreting your observations and results.

APPENDIX III   Writing a Laboratory Report

You will be expected to submit a report for each experiment, either (1) from your laboratory notebook, (2) on a report form provided, or (3) on blank pages, bound or stapled together. Your instructor should tell you what kind of report he or she prefers. Handwritten reports are usually acceptable if they are written legibly in ink. Your report should include information under some or all of the following headings:

1. *Prelab Assignments:* Your experimental plan (see Appendix V), calculations, and any other material requested by your instructor. This may also include a prelab write-up of the kind described in Appendix II.
2. *Observations:* Any significant observations made during the course of the experiment. You should record all observations that might be of

help to you (or another experimenter) if you were to repeat the experiment at a later time. These include quantities of solvents or drying agents used, reaction times, distillation ranges, and a description of any experimental difficulties you encounter. You should also make note of phenomena that might provide clues about the nature of chemical or physical transformations taking place during the experiment, such as color changes, phase separations, tar formation, and gas evolution. If you keep a detailed record of your observations in a laboratory notebook, it may not be necessary to include them in a separate report.

3. *Raw Data:* All numerical data obtained directly from an experiment, before it is graphed, used in calculations, or otherwise processed. This can include quantities of reactants and products, titration volumes, kinetic data, gas chromatographic (GC) retention times, integrated GC peak areas, and spectrometric parameters.

4. *Calculations:* Yield and stoichiometry calculations and any other calculations based on the raw data. If a number of repetitive calculations are required, one or two sample calculations of each type may suffice.

5. *Results:* A list, graph, tabulation, or verbal description of the significant results of the experiment. For a preparation, this should include a physical description of the product (color, physical state, evidence of purity, etc.), the percent yield, and all significant physical constants, spectral data, and analytical data obtained for the product. For the qualitative analysis of an unknown compound, the results of all tests and derivative preparations should be included, along with spectral data and physical constants. The results of calculations based on the raw data should be reported in this section as well.

6. *Discussion:* In this section you should interpret your results and discuss possible sources of error. For example, if you obtain an impure or unexpected product, or a low yield of an expected product, you should propose reasons for these outcomes. At the instructor's request, your discussion can include a description of the scientific problem that the experiment was designed to solve, your interpretation of the results as they pertain to the problem, and a discussion of the significance of the results. You may also be asked to include statements of any working hypotheses you formulated and describe how they were tested, reporting and interpreting the relevant experimental evidence. When appropriate, this section may include your interpretation of spectra or chromatograms, which should be attached to the report.

7. *Conclusions:* A statement of your final conclusion (or conclusions) relating to the problem and an explanation of how you arrived at your conclusion, showing clearly how it is supported by the experimental evidence.

8. *Exercises:* Your answers to all exercises (questions, problems, etc.) assigned by your instructor, showing your calculations and describing your reasoning when applicable.

Each report should also include (on the first page or cover) the name and number of the experiment, your name, and the date when the report was submitted to the instructor. Your instructor should tell you what kind of report he or she prefers. The following example illustrates the kind of report that might be prepared by a conscientious student. Note that the experimental plan is not included here; see Appendix V for directions on writing experimental plans.

Experiment 60: Preparation of Tetrahedranol

Name: Cynthia Sizer
Date: Nov. 25, 2008

(*Attached*: Experimental plan, IR spectrum, gas chromatograph, and worked exercises)

Prelab Calculations

Tetrahedryl acetate required:

$$\text{mass} = 0.0150 \text{ mol} \times \frac{110 \text{ g}}{1 \text{ mol}} = 1.65 \text{ g}$$

$$\text{volume} = 1.65 \text{ g} \times \frac{0.951 \text{ g}}{1 \text{ mL}} = 1.57 \text{ mL}$$

6.0 *M* sodium hydroxide required:

$$\text{volume} = 0.060 \text{ mol} \times \frac{1 \text{ L}}{6.0 \text{ mol}} \times \frac{1000 \text{ mL}}{1 \text{ L}} = 10 \text{ mL}$$

Observations

The reaction of 1.644 g of the ester (tetrahedryl acetate) with 10 mL of 6.0 *M* NaOH was carried out under reflux for 60 minutes, during which time the organic (top) layer dissolved slowly to give a homogeneous solution and the "fruity" odor of the ester disappeared. The organic layer was no longer visible after 45 minutes of heating. The acidified reaction mixture was extracted with two 10-mL portions of diethyl ether, the ether extracts were washed with two 10-mL portions of saturated aqueous sodium chloride, and the ether solution was dried over 0.70 g of anhydrous magnesium sulfate. The residue was distilled over a 4°C boiling range and the distillate solidified on cooling. After the product dried for 1 week in a desiccator, its mass was 0.854 g. Approximately 0.10 g of the product was dissolved in 0.50 mL of dichloromethane, and gas chromatograms were recorded for (1) the original solution and (2) the solution spiked with an authentic sample of tetrahedryl acetate. The relative area of the second (72 s) product peak was considerably larger in the second chromatogram than in the first. The gas chromatograms were obtained using a 2-meter, $\frac{1}{8}$ inch i.d., OV-101/Chromosorb W packed column. The infrared spectrum of the product was recorded using a thin film between heated silver chloride plates.

Data

Mass of tetrahedryl acetate: 1.644 g
Mass of dry product: 0.854 g
Distillation boiling range of product: 78–82°C
Micro boiling point of product: 81°C
Gas chromatography data:

    Column temperature: 110°C

    Injector temperature: 150°C

Detector temperature: 150°C

Helium flow rate: 25 cm³/minute

Component	Retention time	Peak area
tetrahedranol	47 s	320 mm²
tetrahedryl acetate	72 s	12 mm²

Calculations

Percentage of tetrahedranol in product:

$$\frac{320 \text{ mm}^2}{332 \text{ mm}^2} \times 100\% = 96.4\%$$

Mass of tetrahedranol in product:

$$0.854 \text{ g} \times \frac{96.4\%}{100\%} = 0.823 \text{ g}$$

Theoretical yield of tetrahedranol (TetOAc = tetrahedryl acetate; TetOH = tetrahedranol):

$$1.644 \text{ g TetOAc} \times \frac{1 \text{ mol TetOAc}}{110.1 \text{ g TetOAc}} \times \frac{68.1 \text{ g TetOH}}{1 \text{ mol TetOH}} = 1.017 \text{ g TetOH}$$

Percent yield of tetrahedranol:

$$\frac{0.823 \text{ g}}{1.017 \text{ g}} \times 100\% = 80.9\%$$

Results

Tetrahedranol was obtained in 80.9% yield from the alkaline hydrolysis of tetrahedryl acetate. At room temperature, tetrahedranol was a colorless, almost transparent solid with a mild spirituous odor. It could easily be melted on a steam bath to form a clear, colorless liquid. It had a distillation boiling range of 78–82°C and a micro boiling point of 81°C. Its infrared spectrum contained significant absorption bands at 3370, 2975, 2885, and 1194 cm⁻¹.

Discussion

Statement of the problem: Can tetrahedranol with a purity of 95% or more be prepared by the alkaline hydrolysis of tetrahedryl acetate?

Working Hypothesis 1: The alkaline hydrolysis of tetrahedryl acetate *will* yield tetrahedranol as one of the products. I based this hypothesis on (1) our textbook's statement that the alkaline hydrolysis of an ester yields the corresponding hydroxy compound (alcohol or phenol) and the salt of the corresponding carboxylic acid, and (2) the results of a previous experiment in which I obtained salicylic acid (a phenol) from the alkaline hydrolysis of methyl salicylate. Observations in support of Hypothesis 1 include the following:

1. The slow disappearance of the organic layer during the reaction suggests that the water-insoluble ester was being converted to a water-soluble product. According to Table 60.1 in the lab textbook, tetrahedranol is soluble in water.
2. The disappearance of the "fruity" odor of tetrahedryl acetate also suggests that the ester was reacting.

3. The solidification of the distillate and easy melting of the solid between heated AgCl plates are consistent with the hypothesis, since Table 60.1 indicates that tetrahedranol has a melting point of 27°C.

Hypothesis 1 was tested by obtaining a micro boiling point and infrared spectrum of the product, with the following results:

1. The observed micro boiling point of 81°C is consistent with the hypothesis, since tetrahedranol has a reported boiling point of 82°C.
2. The infrared spectrum, suggesting a tertiary alcohol, is consistent with the hypothesis.

Wave number/cm^{-1}	Assignment	Interpretation
3370 (3368.4)	O—H stretch	alcohol or phenol
2975, 2885 (2975.1, 2884.2)	C—H stretch	absence of C—H absorption above 3000 cm^{-1} eliminates a phenol as a possibility
1194 (1193.6)	C—O stretch	consistent with a tertiary alcohol

The wave numbers of these bands are nearly identical to those on a published FTIR spectrum of tetrahedranol (shown in parentheses), and the band shapes and positions in the fingerprint region matched those of the published spectrum. My spectrum did have a weak band at 1739 cm^{-1} that was not present on the published spectrum, but this is probably the carbonyl (C=O) band of the tetrahedryl acetate impurity that was detected by gas chromatography.

Working Hypothesis 2: The purity of the product *will not* be 95% or better. I based this hypothesis on (1) the statement in the lecture textbook that some ester hydrolysis reactions, especially those involving esters of bulky alcohols, take more than an hour to reach completion; (2) the fact that the ester and alcohol boiling points are only 15°C apart, suggesting that simple distillation will not remove all of the impurity.

Hypothesis 2 was tested by obtaining a gas chromatogram of the product, identifying the peaks, and calculating the percentage of tetrahedranol in the product from the peak areas. Since the small 72 s peak became larger when the sample was spiked with tetrahedryl acetate, it must be the tetrahedryl acetate peak. Since the IR spectrum showed that the major product was tetrahedranol, the much larger peak at 47 s must be the tetrahedranol peak. The calculated mass percentage of tetrahedranol, 96.4%, is not consistent with my hypothesis. I am not disappointed with this result, since it was nice to know that the experiment came out better than expected! My assumptions that the reaction would not go to completion in an hour and that the impurity would not be completely removed from the product by simple distillation were both correct, but the reaction was more nearly complete than I had guessed.

The yield of tetrahedranol (0.823 g) was 0.194 g less than the theoretical value. Of this at least 0.019 g resulted from incomplete reaction of tetrahedryl acetate, based on the 0.031 g (0.854 g−0.823 g) of ester in the distillate. The remaining losses could have arisen from (1) losses during transfers, (2) losses during the extraction and washing operations, and (3) losses during the distillation. During each transfer I used additional solvent (ether or water) to rinse out the vessel from which the product was transferred, so losses during transfers should not be a major factor. About 0.1 mL of residue remained in the conical vial used for distillation; this accounts for less than 0.1 g of the loss, since some of the residue must have been tetrahedryl acetate. Most of the remaining losses must have occurred during the extraction and washing operations, due to the partial solubility of tetrahedranol in water. Such losses might have been reduced by saturating the aqueous layer with potassium carbonate to salt out the alcohol or by carrying out several more ether extractions.

Conclusion

I conclude that tetrahedranol with a purity of 95% or more can be prepared by the hydrolysis of tetrahedryl acetate. I base this conclusion primarily on the infrared spectrum of the product, whose resemblance to the published spectrum leaves little doubt that the product is tetrahedranol, and on the gas chromatographic analysis, which indicates a purity of 96.4 %. My conclusion that the product is tetrahedranol is supported by the other evidence cited in the discussion. My conclusion that the purity of the product is greater than 95% might possibly be in error, since the GC peak areas were not corrected by applying detector response factors.

Calculations for Organic Synthesis APPENDIX **IV**

Most organic chemistry lab procedures give the quantities of reactants and other chemicals in grams, milliliters, and similar units. But it is the relationship between *chemical* quantities—moles and millimoles—that is significant in a chemical reaction, not the relationship between such *physical* quantities as mass and volume. Because there are no "mole meters" that measure molar amounts directly, we are forced to use balances and volumetric glassware for that purpose. That should not obscure the fact that the chemical units are fundamental; only by knowing the chemical amounts of reactants involved in a preparation, for example, can you recognize the stoichiometric relationships between them or predict the yield of the expected product. Therefore you need to know how to convert physical quantities of reactants or products to chemical quantities, and vice versa.

It is helpful to regard a chemical calculation as a process by which a given quantity is "converted" to the required quantity. This can be accomplished by multiplying the given quantity by a series of ratios used as unit or dimensional *conversion factors*. Unit conversions are carried out by using conversion factors, such as 454 g/lb, that are written as ratios between two quantities whose quotient is unity (for example, 454 g = 1 lb, so 454 g/1 lb = 1). Dimensional conversions are carried out by using conversion factors that are ratios of quantities in different *dimensions*, such as mass, volume, and chemical amount (amount of substance). For example, the density of a substance can be regarded as a conversion factor linking the two dimensions of mass and volume, so it is used to convert the mass of a given quantity of the substance to units of volume, and vice versa. A conversion factor can be inverted when necessary. For example, the molar mass of butyl acetate, written as 116 g/1 mol, will convert moles of butyl acetate to grams; the inverse ratio, 1 mol/116 g, will convert grams of butyl acetate to moles. All calculations should be checked by making sure that the units involved cancel to yield the correct units in the answer. This does not ensure that your answer is correct, but if the units do *not* cancel, the answer is almost certainly wrong.

The following examples illustrate some fundamental types of calculations that you can expect to encounter in an organic chemistry lab course.

Chemical Amount and Mass. The chemical amount (in moles or millimoles) of a substance is converted to its mass by multiplying by its molar mass. Remember that the molar mass of a substance is obtained by simply appending the units g/mol to its molecular weight, which is a dimensionless quantity. For example, the mass of 15.0 mmol of butyl acetate (M.W. = 116) is 1.74 g.

$$15.0 \text{ mmol} \times \frac{1 \text{ mol}}{1000 \text{ mmol}} \times \frac{116 \text{ g}}{1 \text{ mol}} = 1.74 \text{ g}$$

Note that the chemical amount in millimoles must be converted to moles before the conversion factor is applied; otherwise, the units will not cancel. Mass can be converted to chemical amount by inverting the conversion factor before multiplying.

Chemical Amount and Volume. The chemical amount (in moles or millimoles) of a pure liquid is converted to volume by multiplying by the liquid substance's molar mass and by the inverse of its density. For example, the volume of 2.50

mmol of acetic acid (M.W. $= 60.1$; $d = 1.049$ g/mL) is 0.143 mL.

$$2.50 \text{ mmol} \times \frac{1 \text{ mol}}{1000 \text{ mmol}} \times \frac{60.1 \text{ g}}{1 \text{ mol}} \times \frac{1 \text{ mL}}{1.049 \text{ g}} = 0.143 \text{ mL}$$

The volume of a solution needed to provide a specified chemical amount of solute is calculated by multiplying the number of moles required by the inverse of the solution's molar concentration. For example, the volume of 6.0 M HCl (which contains 6.0 mol of HCl per liter of solution) needed to provide 18 mmol of HCl is 3.0 mL.

$$18 \text{ mmol} \times \frac{1 \text{ mol}}{1000 \text{ mmol}} \times \frac{1 \text{ L}}{6.0 \text{ mol}} \times \frac{1000 \text{ mL}}{1 \text{ L}} = 3.0 \text{ mL}$$

Note that concentrations expressed in mol/L and mmol/mL have the same numerical value. Thus a 6.0 M solution also has a concentration of 6.0 mmol/mL; using these units simplifies the previous calculation considerably:

$$18 \text{ mmol} \times \frac{1 \text{ mL}}{6.0 \text{ mmol}} = 3.0 \text{ mL}$$

Theoretical Yield. The maximum quantity of a product (usually expressed in mass units) that could be attained from a reaction is called the *theoretical yield* of the product. Theoretical yields can be calculated using *stoichiometric factors*— ratios derived from the coefficients (expressed in moles) of the products and reactants in a balanced equation for the reaction. For example, the stoichiometric factors relating the chemical amount of the organic product to the chemical amounts of the two reactants in the following reaction are (1 mol dibenzalacetone)/(1 mol acetone) and (1 mol dibenzalacetone)/(2 mol benzaldehyde).

$$\underset{\text{benzaldehyde (B)}}{2\text{PhCHO}} + \underset{\text{acetone (A)}}{\text{CH}_3\overset{\displaystyle\text{O}}{\overset{\|}{\text{C}}}\text{CH}_3} \xrightarrow{\text{NaOH}} \underset{\text{dibenzalacetone (DBA)}}{\text{PhCH}=\text{CHC}\overset{\displaystyle\text{O}}{\overset{\|}{\text{C}}}\text{CH}=\text{CHPh}} + 2\text{H}_2\text{O}$$

Suppose you were trying to prepare dibenzalacetone (M.W. $= 234.3$) starting with 0.500 g of benzaldehyde (M.W. $= 106.1$) and 0.150 g of acetone (M.W. $= 58.1$). (For convenience, we will abbreviate the names as dibenzalacetone $=$ DBA, benzaldehyde $=$ B, and acetone $=$ A.) You can calculate the maximum chemical amount of product that could be formed from each reactant by converting the given quantity to moles and then applying the appropriate stoichiometric factor:

$$0.500 \text{ g B} \times \frac{1 \text{ mol}}{106.1 \text{ g B}} \times \frac{1 \text{ mol DBA}}{2 \text{ mol B}} = 2.36 \times 10^{-3} \text{ mol DBA}$$

$$0.150 \text{ g A} \times \frac{1 \text{ mol A}}{58.1 \text{ g A}} \times \frac{1 \text{ mol DBA}}{1 \text{ mol A}} = 2.58 \times 10^{-3} \text{ mol DBA}$$

Since there is only enough benzaldehyde to produce 2.36×10^{-3} mol of dibenzalacetone, it is impossible to obtain more than that from the specified quantities of reactants. Once that much product has been formed, the reaction mixture will have run out of benzaldehyde, and the *excess* (leftover) acetone will have nothing to react with. Therefore, benzaldehyde is the *limiting reactant* upon which the yield calculations must be based. The theoretical yield of dibenzalacetone, in grams, is then

$$2.36 \times 10^{-3} \text{ mol DBA} \times \frac{234.3 \text{ g DBA}}{1 \text{ mol DBA}} = 0.553 \text{ g DBA}$$

Remember that the limiting reactant is always the one that would produce the least amount of product, which is not necessarily the one present in the lowest amount. In this example, benzaldehyde is the limiting reactant even though the mass and chemical amount of benzaldehyde are much greater than the mass and chemical amount of acetone.

Percent Yield. It is seldom, if ever, possible to attain the theoretical yield of product from an organic preparation. The reaction may not go to completion during the designated reaction period, leaving unreacted starting materials. There may be side reactions that reduce the yield of product, as in the reaction of benzaldehyde with acetone, where some benzalacetone ($PhCH{=}CHCOCH_3$) is formed as a by-product. And there are invariably material losses when the product is separated from the reaction mixture and purified. The *percent yield* of a preparation compares the actual yield to the theoretical yield as defined here:

$$\text{Percent yield} = \frac{\text{actual yield}}{\text{theoretical yield}} \times 100\%$$

For example, if you prepared 0.409 g of dibenzalacetone from 0.500 g of benzaldehyde and 0.150 g of acetone (theoretical yield = 0.553 g), the percent yield of your synthesis would be

$$\text{Percent yield} = \frac{0.409 \text{ g DBA}}{0.553 \text{ g DBA}} \times 100\% = 74.0\%$$

As a rule, you should estimate the theoretical yield of a preparation based on the amounts of reactants given in the experiment. This will help you assess your performance by comparing your actual yield with an estimate of the "ideal" yield. But the percent yield that you *report* should be based on the amounts of reactants that you actually used in the synthesis, not on the amounts given (unless they are exactly the same).

Mass Percentage and Percent Recovery. The mass percentage of a substance present in a particular mixture, such as the percentage of the compound eugenol in a sample of cloves, is calculated by dividing the mass of the component by the mass of the mixture and multiplying by 100%.

$$\text{Mass percentage} = \frac{\text{mass of component of a mixture}}{\text{mass of mixture}} \times 100\%$$

For example, if 5.00 g of cloves contains 0.850 g of eugenol the mass percentage of eugenol is 17.0%.

$$\text{Mass percentage} = \frac{0.850 \text{ g}}{5.00 \text{ g}} \times 100\% = 17.0\%$$

The percent recovery of a process that is used to isolate one or more component from a mixture is calculated similarly, but since the process is unlikely to isolate all of a component it is not the same as the component's actual mass percentage.

$$\text{Percent recovery} = \frac{\text{mass of component recovered}}{\text{mass of mixture}} \times 100\%$$

For example, if you isolate 0.625 g of eugenol by steam-distilling 5.00 g of cloves, the percent recovery of eugenol is 12.5%.

$$\text{Percent recovery} = \frac{0.625 \text{ g}}{5.00 \text{ g}} \times 100\% = 12.5\%$$

Preparation of Solutions. Suppose you need to prepare 75 mL of ~2.0 M sodium carbonate (Na_2CO_3) from solid sodium carbonate. First calculate the chemical amount of sodium carbonate (mol wt = 106) contained in 75 mL of 2.0 M sodium carbonate.

$$75 \text{ mL} \times \frac{1 \text{ L}}{1000 \text{ mL}} \times \frac{2.0 \text{ mol}}{1 \text{ L}} = 0.15 \text{ mol}$$

Now calculate the mass of that number of moles.

$$0.15 \text{ mol} \times \frac{106 \text{ g}}{1 \text{ mol}} = 16 \text{ g}$$

You can, of course, combine these calculations.

$$75 \text{ mL} \times \frac{1 \text{ L}}{1000 \text{ mL}} \times \frac{2.0 \text{ mol}}{1 \text{ L}} \times \frac{106 \text{ g}}{1 \text{ mol}} = 16 \text{ g}$$

For most purposes, you could prepare the solution by dissolving 16 g of sodium carbonate in distilled water, transferring it to a 100-mL graduated cylinder, and adding more distilled water to the 75 mL mark. If the concentration must be more accurate, you could use a 100-mL volumetric flask to prepare 100 mL of 2.00 M solution, which will require 21.2 g of sodium carbonate (do that calculation yourself).

Now suppose you want to prepare 9.0 mL of 3.0 M sulfuric acid by diluting concentrated sulfuric acid, which has a concentration of 18 mol/L. You need to know what volume of the concentrated acid must be diluted to yield the desired volume of 3.0 M sulfuric acid. The easiest way to do such a calculation is to use the dilution equation $M_1 \times V_1 = M_2 \times V_2$, where the terms on the left are the molar concentration and volume of the undiluted solution and the terms on the right are the molar concentration and volume of the diluted solution. Since V_1 is unknown,

$$18 \, M \times V_1 = 3.0 \, M \times 9.0 \text{ mL}$$

Solving for V_1 then yields

$$V_1 = \frac{3.0 \, M \times 9.0 \text{ mL}}{18 \, M} = 1.5 \text{ mL}$$

So to prepare the 3.0 M sulfuric acid you could *carefully* pour 1.5 mL of concentrated sulfuric acid into an amount of distilled water that is less than the final volume (5 mL, for example), wait for the solution to cool down (dilution of concentrated acids generates heat), and then transfer it to a 10-mL graduated cylinder and add more distilled water to the 9.0 mL mark. If it is important to control the concentration more accurately, you should calculate the volume of concentrated solution to at least three significant figures and use volumetric glassware to prepare the diluted solution. See the front endpaper for a table giving the concentrations of commercial acids and bases.

Planning an Experiment

Writing an Experimental Plan

Before starting any project, whether you are making a bookshelf, duck à l'orange, or isopentyl acetate, you must have a plan. An *experimental plan* should summarize what you expect to do in the laboratory during an experiment and how you intend to go about it. You should state how the work is to be done in short phrases, without excessive detail. When you carry out the experiment, you can refer to the procedure in your lab text for any necessary details, but as you become more proficient in the laboratory, you should find yourself relying less on the lab text and more on your experimental plan. A good plan should give you quick access to the essential information you will need while performing the experiment. Quantities of chemicals (including wash solvents, drying agents, etc.), reaction times, physical properties, hazard warnings, and other useful data should be included. You can list the supplies and equipment you will need for each operation (these are specified in most operation descriptions) so that you can have them cleaned and ready when you need them. You may also wish to sketch the apparatus you will be using so that you can assemble it quickly in the laboratory.

An experimental plan should help you organize your time efficiently by listing tasks in the approximate order in which you expect to accomplish them. For example, whenever a reflux period is specified in a procedure, you will have some free time to set up the apparatus for the next step, reorganize your work area, review an operation, take a melting point, record the spectrum of a previous product, or tie up other loose ends. Your plan should be flexible enough that you can alter it or deviate from it during the experiment, if there is good reason to do so.

One simple and effective way of organizing your time is to use a laboratory checklist, such as the one shown here for an experiment involving the synthesis of salicylic acid from methyl salicylate.

Sample Lab Checklist

- Collect and clean supplies for reaction
- Assemble reaction apparatus
- Measure methyl salicylate and 6 M NaOH solution
- Heat reactants under reflux for 30 min
- During reaction, collect supplies for H_2SO_4 addition, vacuum filtration, and recrystallization
- Measure 3 M H_2SO_4 solution
- Precipitate product with 3 M H_2SO_4
- Filter and wash product; air dry on filter
- Measure and boil water for recrystallization
- Recrystallize product from boiling water
- Filter and wash product
- Dry product
- Weigh and turn in product
- Measure melting point
- Clean up

Refer to the procedure and the descriptions of lab techniques to be used as you prepare your checklist. Leave enough space between the items on your checklist so that you can add new ones, as necessary, during the experiment. As you complete each task in the laboratory, check it off the list and go on to the next one.

Creating a Flow Diagram

A flow diagram can help you organize your time by giving you a quick overview of the procedure, showing the purpose of each step. To create such a flow diagram, first list all the substances that you know to be present in the reaction mixture before the reaction starts (reactants, solvents, catalysts), as shown in the following general flow diagram:

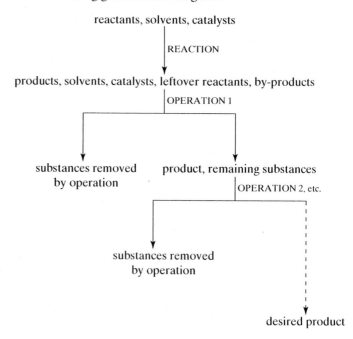

The reaction equation for the experiment tells you what substances will form as a result of the reaction. At the end of the reaction period, the reaction mixture will contain these products as well as the reaction solvent (if any), catalyst (if any), leftover reactants, and usually some by-products formed by various side reactions. The flow diagram should show how all of the unwanted substances in the reaction mixture are separated from the desired product. Each separation or purification operation is represented by a branch in the flow diagram, with the substance(s) being removed on one side and the desired product, along with any remaining substances, on the other side. After the last operation, the product stands alone, with all of the impurities eliminated—on paper, at least!

Planning an Organic Synthesis

(*Note:* In this section, numbers in brackets refer to literature sources from the Bibliography; thus B28 is the 28th source in category B.)

If you need to prepare an organic compound (which we will call your *target compound*) without reference to a standard lab textbook, you must plan your synthesis carefully and thoroughly. If the synthesis requires several steps, you will first have to outline a synthetic pathway for making it from some readily available starting material. To help you work out a synthetic pathway, you can consult an organic chemistry lecture textbook or such sources as *Organic Synthesis: The Disconnection Approach* [B30] and *Principles of Organic Synthesis* [B18]. After you develop one or more promising synthetic pathways, you should learn more about the reactions required for the synthesis. *Modern Synthetic Reactions* [B12], *Advanced Organic Chemistry* [H1], and *March's Advanced Organic Chemistry* [H7] describe a number of synthetic reactions and also give literature references to specific synthetic procedures. *Organic Reactions* [B19] is a comprehensive source of information about specific reactions and also includes some representative synthetic procedures, as well as numerous references to literature procedures. It describes the applications and limitations of many synthetic reactions and may help you tailor a synthetic procedure to fit the particular starting materials you will be working with. Many reactions and compound types are described in depth in review articles from such journals as *Chemical Reviews* and *Angewandte Chemie (International Edition in English)*.

There are many useful collections of detailed synthetic procedures. One of the best (and the first you may want to consult) is *Organic Syntheses* [B20], which provides a representative selection of carefully tested procedures. Other useful if less comprehensive sources include *Vogel's Textbook of Practical Organic Chemistry* [B31], *Organic Functional Group Preparations* [B27], and other works from Category B of the Bibliography, as well as a variety of basic and advanced organic chemistry lab textbooks. References to literature procedures for organic syntheses are given in many of the sources mentioned previously, such as *Organic Reactions* and the two advanced organic chemistry textbooks. *Theilheimer's Synthetic Methods of Organic Chemistry* [B32] is an extensive guide to synthetic procedures that is updated each year. Each entry gives a brief outline of the procedure and cites its source. Other useful if less complete sources include *Synthetic Organic Chemistry* [B34], which covers many of the older reactions, *Comprehensive Organic Transformations* [B13], and *Compendium of Organic Synthetic Methods* [B11].

The most comprehensive guides to the chemical literature of organic chemistry are *Chemical Abstracts (CA)* [J3] and *Beilstein* [A3]. With some practice, you can use them to find information about every aspect of your synthesis. For example, to find out more about your target compound, you should first find its *CA* index name and registry number in the *Index Guide*, which appears with the collective indexes that are published periodically by *Chemical Abstracts*. Then you can conduct a search of *Chemical Abstracts*—from an on-line database, if one is available—to locate articles that refer to your compound. See Appendix VII for more information about the use of *CA*, *Beilstein*, and on-line searches. For more information about these and other useful sources on organic reactions and syntheses, as well as suggested literature searching strategies, read the three-part series "Information Sources for Organic Chemistry" [K7].

If you locate a specific procedure for the synthesis of your target compound in the chemical literature, you should not assume that it will be suitable for your purposes as written. It may call for unavailable or excessively hazardous chemicals or suggest techniques and apparatus that are not appropriate to your laboratory situation. Primary literature procedures tend to be rather condensed, so you will probably have to "fill in the blanks" from your own experience or by consulting other sources. If you want to prepare more or less of a compound than the amount specified in a literature procedure, you will have to scale up or scale down the reaction as described in the next section.

If you can't locate a synthetic procedure for a specific synthetic conversion you plan to carry out, you may have to adapt a procedure for the same reaction of a closely related compound. Adapting a procedure is often just a matter of recalculating the masses of the reactants. For example, suppose you want to adapt a literature procedure for the $NaBH_4$ reduction of cyclohexanone to the same reaction of 4-methylcyclohexanone. If the original procedure used 3.00 g (30.5 mmol) of cyclohexanone (mol wt = 98.2), you will need to use about 3.43 g (30.5 mmol) of 4-methylcyclohexanone (mol wt = 112.2) while keeping the quantities of other reactants the same. Such an adaptation may also require changes in such things as the size of the glassware you use, the amount or kind of recrystallization solvent needed, and the boiling range for a distillation, so it is important to identify changes in the properties of the reactants and products that necessitate procedural changes.

An organic synthesis ordinarily requires the use of one more reagents, where a *reagent* is a chemical designed to bring about a specific kind of molecular transformation, such as the conversion of a carbonyl group to a hydroxyl group. Two excellent sources of detailed information about chemical reagents are *Fiesers' Reagents for Organic Synthesis* [B7] and *Encyclopedia of Reagents for Organic Synthesis* [B21]. General reference books, such as *The Merck Index* [A11], *Lange's Handbook of Chemistry* [A6], the *CRC Handbook of Chemistry and Physics* [A15], the *Dictionary of Organic Compounds* [A12], and the *Aldrich Catalog* [A1], can be consulted for information about the physical properties of the reagents, as well as other chemicals (including solvents) involved in your synthesis.

Before you carry out any chemical synthesis, you must learn how to handle the necessary chemicals safely. Detailed information about chemical hazards can be found in *Sax's Dangerous Properties of Industrial Materials* [C5], the *Sigma–Aldrich Library of Regulatory and Safety Data* [C4], and other sources from Category C of the Bibliography. You should also obtain and read the Material Safety Data Sheets (MSDS) for any hazardous chemicals you will be working with (see "Finding and Using Chemical Safety Information" in the Laboratory Safety section). Information about safe laboratory practices may be found in *Prudent Practices in the Laboratory* [C7], which also provides information about dealing with laboratory wastes.

If you are unfamiliar with some of the more advanced laboratory techniques required for a synthesis, consult *Guide for the Perplexed Organic Experimentalist* [D3], the appropriate volume in Weissberger's *Technique of Organic Chemistry* [D7], *Encyclopedia of Separation Science* [D10], or other references from Categories D and E of the Bibliography. If you need to purify a solvent or another common chemical, see *Purification of Laboratory Chemicals* [D1].

Once you have synthesized your target compound (or what you think is your target compound), you will have to confirm its identity by measuring one or more physical properties, such as its melting point or boiling point, and obtaining and interpreting at least one kind of spectrum. Refer to the appropriate sources in Category F if you need information about spectral analysis and the interpretation of spectra. *Spectrometric Identification of Organic Compounds* [F20] is a good place to start. There are many sources of published spectra, such as the *Aldrich Library of FT-IR Spectra* [F17] and the *Aldrich Library of ^{13}C and ^1H FT-NMR Spectra* [F15], and electronic libraries of spectra are available at most universities. Additional information about spectrometric and "wet" chemical methods for identifying organic compounds can be found in sources from Category G. You may also need

to assess the purity of your product by gas chromatography, TLC, HPLC, or another method. Refer to Category E of the Bibliography for appropriate sources. For additional help in carrying out a literature search, consult Appendix VII and appropriate sources listed in Category K of the Bibliography.

Scaling

If you want to prepare a target compound using a certain amount of starting material but can only find procedures that require more or less starting material than you expect to use, you will have to scale down or scale up the procedure. For example, if a standard-scale literature procedure calls for 12 g of acetophenone and you want to use only 1.8 grams, you should multiply the quantities of all reactants, reagents, catalysts, solvents, or other chemicals by the *scaling factor* 0.15, which is the ratio of 1.8 g to 12 g. You will also need to reduce the size of the apparatus accordingly, possibly switching to microscale glassware, and you may be able to reduce the reaction time as well. (Of course it can't hurt to use the reaction time specified.) When you are scaling down a very large-scale procedure, such as one published in *Organic Syntheses* [B20], you should use somewhat larger volumes of most organic solvents (such as extraction and reaction solvents) than the scaling factor would indicate, because evaporation of small amounts of a solvent reduces its volume by a greater percentage than evaporation of large amounts of the same solvent. You may then have to use larger glassware than the scaling factor would indicate to accommodate the larger solvent volumes. On the other hand, don't use glassware that is too large for the volume of the materials you are using, since larger glassware has more surface area, which leads to greater losses.

Most scaled-down syntheses do not give as high a percent yield as the original procedure. For example, losing 0.1 g of product during a transfer has a greater effect on the percent yield of a small-scale experiment than it does when you have more product to work with. Other factors, such as differences in surface-contact area or the rate of heat transfer, may also cause differences in the outcome of a scaled-down experiment, but most procedures can be scaled down to a moderate extent without difficulty. Before you attempt to convert a standard-scale procedure to a microscale procedure or vice versa, you should compare the standard scale and microscale versions of the operations you will be using.

Scaling up a procedure will ordinarily require larger glassware (you might have to switch from microscale to standard-scale glassware) and it may be advisable to increase the reaction time by 25–50%. If the scaling factor is quite large, you may want to increase the solvent volumes and container capacities somewhat less than the scaling factor would indicate.

Properties of Organic Compounds   APPENDIX **VI**

The tables in this appendix are intended for use during the identification of unknown organic compounds by qualitative organic analysis. Compounds that melt below ordinary ambient temperature (about 25°C) are listed in order of increasing boiling point (bp); those that (when pure) are generally solid at room temperature are listed in order of increasing melting point (mp). Both melting and boiling points are specified for some borderline cases; if either value is out of sequence, it is shown in italics.

Procedures for derivative preparations may be provided by your instructor or obtained from a lab textbook; otherwise refer to one or more appropriate books from Section G of the Bibliography. Melting points in parentheses are for derivatives that exist in more than one crystalline form or for which significantly different melting points have been reported in the literature. Sometimes the recrystallization solvent will determine the form in which a derivative crystallizes, so a significant deviation from a listed melting point should not be considered conclusive proof that a product is not the expected derivative. If a given compound can form more than one product (for example, mononitro and dinitro derivatives of an aromatic hydrocarbon), the reaction conditions for the preparation may determine the derivative isolated (a mixture of derivatives may also result). A dash ($-$) in a derivative column indicates either that the derivative has not been reported in the literature or that it is not suitable for identification (it may be a liquid, for example). See references from Category G in the Bibliography for physical constants and derivative melting points not listed in these tables.

Most compounds are listed by their systematic (IUPAC) names, except when those names would be too lengthy.

Abbreviations Used in Tables

d = decomposes on melting
s = sublimes at or below melting point
m = monosubstituted derivative (such as a mononitrated aromatic hydrocarbon)
di = disubstituted derivative
t = trisubstituted derivative
tet = tetrasubstituted derivative

List of Tables

Table 1 Alcohols

Compound	bp	mp	3,5-Dinitro-benzoate	4-Nitro-benzoate	1-Naphthyl-urethane	Phenyl-urethane
methanol	65		108	96	124	47
ethanol	78		93	57	79	52
2-propanol	82		123	110	106	88
2-methyl-2-propanol*	83	*26*	142	—	—	136

continued

Table 1 *Continued*

Compound	bp	mp	3,5-Dinitro-benzoate	4-Nitro-benzoate	1-Naphthyl-urethane	Phenyl-urethane
2-propen-1-ol	97		49	28	108	70
1-propanol	97		74	35	105	57
2-butanol	99		76	26	97	65
2-methyl-2-butanol	102		116	85	72	42
2-methyl-1-propanol	108		87	69	104	86
3-pentanol	116		101	17	95	48
1-butanol	118		64	36	71	61
2-pentanol	120		62	24	74	—
3-methyl-3-pentanol	123		94 (62)	69	104	43
3-methyl-1-butanol	132		61	21	68	57
4-methyl-2-pentanol	132		65	26	88	143
1-pentanol	138		46	11	68	46
cyclopentanol	141		115	62	118	132
2-ethyl-l-butanol	148		51	—	60	—
1-hexanol	157		58	5	59	42
cyclohexanol*	161	*25*	113	50	129	82
furfuryl alcohol	172		80	76	130	45
1-heptanol	177		47	10	62	60
2-octanol	179		32	28	63	114
1-octanol	195		61	12	67	74
1-phenylethanol	202		95	43	106	92
benzyl alcohol	205		113	85	134	77
2-phenylethanol	219		108	62	119	78
1-decanol	231		57	30	73	60
3-phenyl-propanol	236		45	47	—	92
1-dodecanol*	259	*24*	60	45 (42)	80	74
1-tetradecanol		39	67	51	82	74
(−)-menthol		44	153	62	119	111
1-hexadecanol		49	66	58	82	73
1-octadecanol		59	77	64	89	79
diphenyl-methanol		68	141	132	136	139
cholesterol		148	—	185	176	168
(+)-borneol		208	154	137 (153)	132 (127)	138

*May be solid at or just below room temperature.

Note: All temperatures are in °C. Italicized values refer to melting or boiling points that are out of sequence.

Table 2 Aldehydes

Compound	bp	mp	2,4-Dinitrophenyl-hydrazone	Semicarbazone	Oxime
ethanal	21		168 (157)	162	47
propanal	48		148 (155)	154	40
propenal	52		165	171	—
2-methylpropanal	64		187 (183)	125 (119)	—
butanal	75		123	106	—
3-methylbutanal	92		123	107	48
pentanal	103		106 (98)	—	52
2-butenal	104		190	199	119
2-ethylbutanal	117		95 (30)	99	—
hexanal	130		104	106	51
heptanal	153		106	109	57
2-furaldehyde	162		212 (230)	202	91
2-ethylhexanal	163		114 (120)	254d	—
octanal	171		106	101	60
benzaldehyde	179		239	222	35
4-methylbenzaldehyde	204		234	234 (215)	80
3,7-dimethyl-6-octenal	207		77	84 (91)	—
2-chlorobenzaldehyde	213		213 (209)	229 (146)	76 (101)
4-methoxy-benzaldehyde	248		253d	210	133
phenylethanal	*195*	33	121 (110)	153 (156)	99
2-methoxybenzaldehyde		38	254	215	92
4-chlorobenzaldehyde		48	265	230	110 (146)
3-nitrobenzaldehyde		58	290	246	120
4-nitrobenzaldehyde		106	320	221 (211)	133 (182)

Note: All temperatures are in °C. Italics designate melting or boiling points that are out of sequence.

Table 3 Ketones

Compound	bp	mp	2,4-Dinitrophenyl-hydrazone	Semicarbazone	Oxime
acetone	56		126	187	59
2-butanone	80		118	146	—
3-methyl-2-butanone	94		124	113	—
2-pentanone	102		143	112 (106)	58

continued

Table 3 *Continued*

Compound	bp	mp	2,4-Dinitrophenyl-hydrazone	Semicarbazone	Oxime
3-pentanone	102		156	138	69
3,3-dimethyl-2-butanone	106		125	157	75 (79)
4-methyl-2-pentanone	117		95 (81)	132	58
2,4-dimethyl-3-pentanone	124		95 (88)	160	34
2-hexanone	128		110	125	49
4-methyl-3-penten-2-one	130		205	164 (133)	48
cyclopentanone	131		146	210 (203)	56
4-heptanone	144		75	132	—
2-heptanone	151		89	123	—
cyclohexanone	156		162	166	91
2,6-dimethyl-4-heptanone	168		92	122	210
2-octanone	173		58	124	—
cycloheptanone	181		148	163	23
2,5-hexanedione	194		257 (di)	185 (m) 224 (di)	137 (di)
acetophenone	202	*20*	238	198 (203)	60
2-methylacetophenone	214		159	205	61
propiophenone	218	*21*	191	182 (174)	54
3-methylacetophenone	220		207	203	57
2-undecanone	228		63	122	44
4-phenyl-2-butanone	235		127	142	87
3-methoxyacetophenone	240		—	196	—
2-methoxyacetophenone	245		—	183	83 (96)
4-methylacetophenone	*226*	28	258	205	88
4-methoxyacetophenone		38	228	198	87
4-phenyl-3-buten-2-one		42	227 (223)	187	117
benzophenone		48	238	167	144
2-acetonaphthone		54	262d	235	145
3-nitroacetophenone		80	228	257	132
9-fluorenone		83	283	234	195
camphor		179	177	237	118

Note: All temperatures are in °C. Italics designate melting or boiling points that are out of sequence.

Table 4 Amides

Compound	bp	mp	Carboxylic acid	N-Xanthylamide
formamide	195d		—	184
propanamide		81	—	211
ethanamide		82	17	240
heptanamide		96	—	155
nonanamide		99	12	148
hexanamide		100	—	160
hexadecanamide		106	63	142
pentanamide		106	—	167
octadecanamide		109	69	141
butanamide		115	—	187
chloroacetamide		120	61 (53)	209
4-methylpentanamide		121	—	160
succinimide		126	185	247
2-methylpropanamide		129	—	211
benzamide		130	122	223
3-methylbutanamide		136	—	183
o-toluamide		143	104 (108)	200
furamide		143	133	210
phenylacetamide		156	77	195
p-toluamide		159	180	225
4-nitrobenzamide		201	240	233
phthalimide		238	210d	177

Note: All temperatures are in °C.

Table 5 Primary and secondary amines

Compound	bp	mp	Benzamide	p-Toluene-sulfonamide	Phenylthio-urea	Picrate
t-butylamine	44		134	—	120	198
propylamine	48		84	52	63	135
diethylamine	56		42	60	34	155
sec-butylamine	63		76	55	101	140
2-methylpropylamine	69		57	78	82	150
butylamine	77		42	—	65	151
diisopropylamine	84		—	—	—	140
pyrrolidine	89		—	123	—	112 (164)

continued

Table 5 *Continued*

Compound	bp	mp	Benzamide	p-Toluene-sulfonamide	Phenylthio-urea	Picrate
3-methylbutylamine	95		—	65	102	138
pentylamine	104		—	—	69	139
piperidine	106		48	96	101	152
dipropylamine	109		—	—	69	75
morpholine	128		75	147	136	146
pyrrole	131		—	—	143	69d
hexylamine	132		40	—	77	126
cyclohexylamine	134		149	—	148	—
diisobutylamine	139		—	—	113	121
N-methylcyclohexylamine	147		86	—	—	170
dibutylamine	159		—	—	86	59
N-ethylbenzylamine	181		—	95	—	118
aniline	184		163	103	154	198 (180)
benzylamine	185		105	116 (185)	156	199 (194)
N-methylaniline	196		63	94	87	145
o-toluidine	200		144	108	136	213
m-toluidine	203		125	114	104 (92)	200
N-ethylaniline	205		60	87	89	138
2-chloroaniline	209		99	105 (193)	156	134
2,6-dimethylaniline	215		168	212	204	180
2-methoxyaniline	225		60	127	136	200
2-ethoxyaniline	229		104	164	137	—
3-chloroaniline	230		120	138 (210)	124 (116)	177
4-ethoxyaniline	250		173	106	136	69
dicyclohexylamine	255d		153 (57)	—	—	173
N-benzylaniline		37	107	149	103	48
p-toluidine		44	158	118	141	182d
diphenylamine		54	180 (109)	141	152	182
4-methoxyaniline		58	154	114	157 (171)	170
4-bromoaniline		66	204	101	148	180
2-nitroaniline		71	110 (98)	142	—	73
4-chloroaniline		72	192	95 (119)	152	178
1,2-diaminobenzene		102	301 (di)	260 (di)	—	208

continued

Table 5 *Continued*

Compound	bp	mp	Benzamide	p-Toluene-sulfonamide	Phenylthio-urea	Picrate
3-nitroaniline		114	155 (di)	138	160	143
1,4-diaminobenzene		142	300 (di)	266 (di)	—	—
4-nitroaniline		147	199 (di)	191	—	100

Note: All temperatures are in °C. See other references in the Bibliography (Category G) for melting points of benzenesulfonamides and α-naphthylthioureas.

Table 6 Tertiary amines

Compound	bp	mp	Methiodide	Picrate
triethylamine	89		280	173
pyridine	115		117	167
2-methylpyridine	129		230	169
3-methylpyridine	143		92	150
4-methylpyridine	143		152	167
tripropylamine	157		207	116
2,4-dimethylpyridine	159		113	183 (169)
N,N-dimethylbenzylamine	183		179	93
N,N-dimethylaniline	193		228d	163
tributylamine	216 (211)		186	105
N,N-diethylaniline	217		102	142
quinoline	237		133 (72)	203
isoquinoline	243	*26*	159	222
tribenzylamine	—	91	184	190
acridine	—	111	224	208

Note: All temperatures are in °C. Italics designate melting or boiling points that are out of sequence.

Table 7 Carboxylic acids

Compound	bp	mp	Amide	p-Toluidide	Anilide	p-Nitrobenzyl Ester
formic acid	101		43	53	50	31
acetic acid	118		82	148	114	78
propenoic acid	139		85	141	104	—
propanoic acid	141		81	124	103	31
2-methylpropanoic acid	154		128	107	105	—
Butanoic acid	164		115	75	95	35

continued

Table 7 *Continued*

Compound	bp	mp	Amide	p-Toluidide	Anilide	p-Nitrobenzyl Ester
3-methylbutanoic acid	176		135	107	109	—
pentanoic acid	186		106	74	63	—
2-chloropropanoic acid	186		80	124	92	—
dichloroacetic acid	194		98	153	118	—
2-methylpentanoic acid	196		79	80	95	—
hexanoic acid	205		101	75	95	—
2-bromopropanoic acid	205d	*24*	123	125	99	—
octanoic acid	239		107	70	57	—
nonanoic acid	254		99	84	57	—
decanoic acid		32	108	78	70	—
2,2-dimethylpropanoic acid	*164*	35	178 (154)	—	129 (133)	—
dodecanoic acid		44	110 (99)	87	78	—
3-phenylpropanoic acid		48	105	135	98	36
tetradecanoic acid		54	103	93	84	—
hexadecanoic acid		62	106	98	90	42
chloroacetic acid		63	120	162	137	—
octadecanoic acid		70	109	102	95	—
trans-2-butenoic acid		72	160	132	118	67
phenylacetic acid		77	156	136	118	65
2-methoxybenzoic acid		101	129	—	131	113
oxalic acid (dihydrate)		101	419d (di)	268 (di)	254 (di)	204(di)
2-methylbenzoic acid		104	142	144	125	91
nonanedioic acid		106	175 (di)	201 (di)	186 (di)	44
3-methylbenzoic acid		112	94	118	126	87
benzoic acid		122	130	158	163	89
maleic acid		130	181 (m) 266 (di)	142 (di)	187 (di)	91
decanedioic acid		133	170 (m) 210 (di)	201 (di)	122 (m) 200 (di)	73 (di)
cinnamic acid		133	147	168	153	117
propanedioic acid		135	50 (m) 170 (di)	86 (m) 253 (di)	132 (m) 230 (di)	86
2-chlorobenzoic acid		140	140	131	118	106
3-nitrobenzoic acid		140	143	162	155	141
diphenylacetic acid		148	167	172	180	—
2-bromobenzoic acid		150	155	—	141	110

continued

Table 7 *Continued*

Compound	bp	mp	Amide	*p*-Toluidide	Anilide	*p*-Nitrobenzyl Ester
hexanedioic acid		152	125 (m) 224 (di)	238	151 (m) 241 (di)	106
4-methylbenzoic acid		180s	160	160 (165)	145	104
4-methoxybenzoic acid		184	167 (163)	186	169	132
butanedioic acid		188	157 (m) 260 (di)	180 (m) 255 (di)	143 (m) 230 (di)	—
3,5-dinitrobenzoic acid		205	183	—	234	157
phthalic acid		210d	220 (di)	201 (di)	253 (di)	155
4-nitrobenzoic acid		240	201	204	211	168
4-chlorobenzoic acid		242	179	—	194	129
terephthalic acid		>300s	—	—	337	263 (di)

Note: All temperatures are in °C. Italics designate melting or boiling points that are out of sequence.

Table 8 Esters

Compound	bp	mp	Carboxylic acid	Alcohol or phenol	*N*-Benzylamide	3,5-Dinitro-benzoate
ethyl formate	54		8	—	60	93
methyl acetate	57		17	—	61	108
ethyl acetate	77		17	—	61	93
methyl propanoate	80		—	—	43	108
methyl acrylate	80		13	—	237	108
isopropyl acetate	91		17	—	61	123
tert-butyl acetate	98		17	26	61	142
ethyl propanoate	99		—	—	43	93
methyl 2,2-dimethylpropanoate	101		35	—	—	108
propyl acetate	102		17	—	61	74
methyl butanoate	102		—	—	38	108
ethyl 2-methylpropanoate	111		—	—	87	93
sec-butyl acetate	112		17	—	61	76
methyl 3-methylbutanoate	117		—	—	54	108
isobutyl acetate	117		17	—	61	87
ethyl butanoate	122		—	—	38	93
butyl acetate	126		17	—	61	64
methyl pentanoate	128		—	—	43	108
ethyl 3-methylbutanoate	135		—	—	54	93
3-methylbutyl acetate	142		17	—	61	61

continued

Table 8 *Continued*

Compound	bp	mp	Carboxylic acid	Alcohol or phenol	N-Benzylamide	3,5-Dinitro-benzoate
ethyl chloroacetate	145		63	—	—	93
pentyl acetate	149		17	—	61	46
ethyl hexanoate	168		—	—	53	93
hexyl acetate	172		17	—	61	58
cyclohexyl acetate	175		17	25	61	113
dimethyl malonate	182		135	—	142	108
diethyl oxalate	185		101*	—	223	93
heptyl acetate	192		17	—	61	47
phenyl acetate	197		17	42	61	146
methyl benzoate	199		122	—	105	108
diethyl malonate	199		135	—	142	93
o-tolyl acetate	208		17	31	61	135
m-tolyl acetate	212		17	12	61	165
ethyl benzoate	213		122	—	105	93
p-tolyl acetate	213		17	36	61	189
methyl o-toluate	215		104	—	—	108
benzyl acetate	217		17	—	61	113
diethyl succinate	218		188	—	206	93
isopropyl benzoate	218		122	—	105	123
methyl phenylacetate	220		77s	—	122	108
diethyl maleate	223		137	—	150	93
ethyl phenylacetate	228		77s	—	122	93
propyl benzoate	230		122	—	105	74
diethyl adipate	245		152	—	189	93
butyl benzoate	250		122	—	105	64
ethyl cinnamate	271		133	—	226	93
dimethyl phthalate	284		210d	—	179	108
(+)-bornyl acetate	226	27	17	208	61	154
methyl p-toluate		33	180s	—	133	108
methyl cinnamate		36	133	—	226	108
benzyl cinnamate		39	133	—	226	113
1-naphthyl acetate		49	17	94	61	217
ethyl p-nitrobenzoate		56	240	—	—	93
phenyl benzoate		69	122	42	105	146

continued

Table 8 *Continued*

Compound	bp	mp	Carboxylic acid	Alcohol or phenol	N-Benzylamide	3,5-Dinitro-benzoate
2-naphthyl acetate		71	17	123	61	210
p-tolyl benzoate		71	122	36	105	189
methyl m-nitrobenzoate		78	140	—	101	108
methyl p-nitrobenzoate		96	240	—	142	108

Note: All temperatures are in °C. Italics designate melting or boiling points that are out of sequence. Additional derivatives of the acid and alcohol portions of most esters can be found in Tables 1 and 7.

* Dihydrate; the anhydrous acid melts at 190°C.

Table 9 Alkyl halides

Compound	bp	Density, d_4^{20}	S-Alkylthiuronium picrate
bromoethane	38	1.461	188
2-bromopropane	60	1.314	196
1-chloro-2-methylpropane	69	0.879	167 (174)
3-bromopropene	71	1.398	155
1-bromopropane	71	1.354	177
iodoethane	72	1.936	188
1-chlorobutane	78	0.884	177
2-iodopropane	89	1.703	196
1-bromo-2-methylpropane	93	1.264	167 (174)
1-chloro-3-methylbutane	100	0.875	173
1-bromobutane	101	1.274	177
I-iodopropane	102	1.749	177
3-iodopropene	102	1.848	155
1-chloropentane	108	0.882	154
I-bromo-3-methylbutane	119	1.207	173 (179)
2-iodobutane	119	1.595	166
1-iodo-2-methylpropane	120	1.606	167 (174)
1-bromopentane	129	1.218	154
1-iodobutane	131	1.617	177
1-chlorohexane	134	0.876	157
1-iodo-3-methylbutane	148	1.503	173
1-iodopentane	155	1.516	154
1-bromohexane	155	1.173	157
1-iodohexane	181	1.439	157
1-bromooctane	201	1.112	134
1-iodooctane	225	1.330	134

Note: All temperatures are in °C; density is in g/mL.

Table 10 Aryl halides

Compound	bp	mp	Density, d_4^{20}	Nitro derivative	Carboxylic acid
chlorobenzene	132		1.106	52	—
bromobenzene	156		1.495	75 (70)	—
2-chlorotoluene	159		1.083	63	140
3-chlorotoluene	162		1.072	91	158
4-chlorotoluene	162		1.071	38 (m)	242
1,3-dichlorobenzene	173		1.288	103	—
1,2-dichlorobenzene	181		1.306	110	—
2-bromotoluene	182		1.423	82	150
3-bromotoluene	184		1.410	103	155
iodobenzene	188		1.831	171 (m)	—
2,6-dichlorotoluene	199		1.269	50 (m)	139
2,4-dichlorotoluene	200		1.249	104	164
3-iodotoluene	204		1.698	108	187
2-iodotoluene	211		1.698	103 (m)	162
1-chloronaphthalene	259		1.191	180	—
4-bromotoluene	*184*	28	—	—	251
4-iodotoluene		35	—	—	270
1,4-dichlorobenzene		53	—	106, 54 (m)	—
2-chloronaphthalene		56	—	175	—
1,4-dibromobenzene		89	—	84	—

Note: All temperatures are in °C; density is in g/mL. Italics designate melting or boiling points that are out of sequence. Nitro derivatives signified by (m) are mononitro compounds; all others are dinitro derivatives.

Table 11 Aromatic hydrocarbons

Compound	bp	mp	Nitro derivative	Carboxylic acid	Picrate
benzene	80		89 (di)	—	84u
toluene	111		70 (di)	122	88u
ethylbenzene	136		37 (t)	122	96u
1,4-xylene	138		139 (t)	300s	90u
1,3-xylene	139		183 (t)	330s	9lu
1,2-xylene	144		118 (di)	210d	88u
isopropylbenzene	152		109 (t)	122	—
propylbenzene	159		—	122	103u

continued

Table 11 *Continued*

Compound	bp	mp	Nitro derivative	Carboxylic acid	Picrate
1,3,5-trimethylbenzene	165		86 (di) 235 (t)	350 (t)	97u
t-butylbenzene	169		62 (di)	122	—
4-isopropyltoluene	177		54 (di)	300s	—
1,3-diethylbenzene	181		62 (t)	330s	—
1,2,3,4-tetrahydronaphthalene	206		96 (di)	210d	—
diphenylmethane	262	26	172 (tet)	—	—
1,2-diphenylethane		53	180 (di) 169 (tet)	—	—
naphthalene		80	61 (m)	—	149
triphenylmethane		92	206 (t)	—	—
acenaphthene		96	101 (m)	—	161
fluorene		114	199 (di) 156 (m)	—	87 (77)
anthracene		216	—	—	138u

Note: All temperatures are in °C. Picrates designated u are unstable and cannot easily be purified by recrystallization.

Table 12 Phenols

Compound	bp	mp	Aryloxyacetic acid	Bromo Derivative	1-Naphthyl-urethane
2-chlorophenol	176		145	49 (m)	120
				76 (di)	
3-methylphenol	202		—	84 (t)	128
2-methylphenol	*192*	31	152	56 (di)	142
4-methylphenol	*232*	36	135	49 (di)	146
				108 (tet)	
phenol	*182*	42	99	95 (t)	133
4-chlorophenol		43	156	33 (m)	166
				90 (di)	
2-nitrophenol		45	158	117 (di)	113
4-ethylphenol		47	97	—	128
5-methyl-2-isopropylphenol		50	149	55 (m)	160
3,4-dimethylphenol		63	163	171 (t)	142
4-bromophenol		64	157	95 (t)	169
2,5-dimethylphenol		75	118	178 (t)	173

continued

Table 12 *Continued*

Compound	bp	mp	Aryloxyacetic acid	Bromo Derivative	1-Naphthyl-urethane
1-naphthol		94	194	105 (di)	152
3-nitrophenol		97	156	91 (di)	167
4-*t*-butylphenol		100	86	50 (m)	110
1,2-dihydroxybenzene		105	—	193 (tet)	175
1,3-dihydroxybenzene		110	195	112 (di)	206
4-nitrophenol		114	187	142 (di)	150
2-naphthol		123	154	84 (m)	157
1,2,3-trihydroxybenzene		133	198	158 (di)	—
1,4-dihydroxybenzene		172	250	186 (di)	—

Note: All temperatures are in °C. Italics designate melting or boiling points that are out of sequence.

The Chemical Literature

The literature of chemistry consists of *primary*, *secondary*, and *tertiary* sources. Most primary sources in chemistry contain descriptions of original research carried out by professional chemists. They include scientific periodicals (such as the *Journal of Organic Chemistry*), patents, dissertations, technical reports, and government bulletins. Secondary sources contain material from the primary literature that has been systematically organized, condensed, or restated to make it more accessible and understandable to users. Secondary sources include most monographs, textbooks, dictionaries, encyclopedias, reference works, review publications, and abstracting journals dealing with chemistry. Tertiary sources are intended to aid users of the primary and secondary sources or to provide facts about chemists and their work. Tertiary sources include guides to the chemical literature, directories of scientists and scientific organizations, bibliographies, trade catalogs, and publications devoted to the financial and professional aspects of chemistry. Some sources may combine several different functions; for example, *Chemical & Engineering News* prints articles about chemical research as well as financial and professional information, and thus serves as both a secondary and a tertiary source.

Although some secondary sources critically evaluate the primary material they review, primary sources are generally used when it is important to obtain the most accurate and detailed information available on a topic. Errors can always occur when primary material reappears in a secondary source, and important information may be left out. Because primary sources are not organized in any systematic way, it is generally necessary to refer to other sources to determine where the desired information can be found.

The Bibliography following this Appendix lists a number of secondary and tertiary sources that can be used to obtain information directly, to gain access to the primary literature, or both. For example, *Beilstein's Handbook of Organic Chemistry* (hereafter referred to as *Beilstein*) gives detailed, reliable information about organic compounds and also provides citations to the literature in which the information was first reported. References to the

Bibliography will be given in the form (C7), where the letter indicates a category (such as "Laboratory Safety") and the number indicates a specific work in that category.

Using *Chemical Abstracts* and *Beilstein*

Most of the reference works cited in the Bibliography are limited in scope; they make no attempt to cover the entire field of chemistry or to list all of the known organic compounds. The two major works that do attempt that kind of coverage are *Beilstein* (A3) and *Chemical Abstracts* (J3). *Beilstein* summarizes all of the important information published about specific compounds, but it is many years behind the current literature in most areas. The earlier volumes of *Beilstein* are available only in German, so at least a rudimentary knowledge of that language is necessary to make good use of this resource. *Chemical Abstracts (CA)* prints *abstracts* (brief summaries) of scientific papers, patents, and other printed material related to chemistry shortly after publication.

Before using *Chemical Abstracts* for the first time, read the introduction that appears in Issue 1 of each volume (two volumes are published each year), which describes the layout of the abstracts. Each abstract of a scientific paper (or other article) contains an abstract number, the title and author of the paper, a citation that tells where the original paper can be located, and a concise summary of the important information in the paper. The contents of *CA* can be searched by computer (as described in "On-Line Searches") or by consulting the indexes. Although each weekly issue of *CA* has its own indexes, the semiannual and collective indexes are far more useful for literature searches. A Collective Index is published every 5 years; prior to 1957, these indexes came out every 10 years. In 1972, the Subject Index was divided into two parts: the Chemical Substances Index and the General Subject Index. Author, formula, and patent indexes are also published, along with ancillary materials such as the Ring Systems Handbook, Registry Handbook, Index Guide, and Service Source Index, which are updated periodically.

The first thing you must do before searching *Chemical Abstracts* for information about a particular compound or subject is to find the *CA index name* of the compound or the *CA index heading* for the subject. In some cases, it may be possible to derive (or guess) the index name, but that is not always easy because *CA* follows its own rules of nomenclature, which often differ from IUPAC rules. The Index Guide, which now appears with each Collective Index and at intervals in between, gives cross-references from alternative names of substances to the *CA* index name. Thus the entry under "aniline" lists the index name benzeneamine, followed by the *CA registry number* [62-53-3]. The Index Guide does not list every compound indexed or give every synonym for the compounds it does list, so you may have to try different approaches to find what you are looking for. If you can't locate a specific compound, try looking up a possible parent (unsubstituted) compound under its trivial name; the index name of this parent compound should begin with a root name under which you will find your compound listed in the Chemical Substances Index (or the Subject Index, prior to 1972). For example, suppose you are searching for the following compound:

The unsubstituted compound (PhCH=CHCOPh) is known by such names as chalcone and benzalacetophenone. Looking up "chalcone" in a recent Index Guide provides the *CA* index name "2-propen-1-one, 1,3-diphenyl," so you will find the substituted compound listed—under the same root name—as "2-propen-1-one, 3-(4-methoxy)phenyl-1-phenyl." Keep in mind that the index name of a compound may change from time to time. This compound was listed under "chalcone, 4-methoxy" during the 8th Collective Index period (1967–1971) and before. Between the 8th and 9th Collective Index periods some major changes were made in the *CA* nomenclature rules, which now require rigorously systematic names for most chemical substances.

If you have trouble finding the *CA* index name for a compound using the Index Guide, you might look for it in another source such as *The Merck Index* (A11) or the *Dictionary of Organic Compounds* (A12). If you can locate the registry number for a compound, you can easily find its index name in the Registry Handbook. If you have a fairly good idea what the index name for a compound might be, you may be able to locate it in the formula indexes.

Once you locate the index name in use during a particular index period, you can locate abstracts listed under that name in the Chemical Substances Index or Subject Index for that period. Abstract citations in indexes from 1967 on are given in the form **80**:12175e, where the first number is the *CA* volume number and the second is the abstract number. In an index prior to 1967, an abstract citation such as **51**:4321^b refers to the volume and column number (there are two columns on each page) in which the abstract appears; the superscript (either a letter from "a" through "i" or a number from 1 through 9) indicates the location of the abstract in that column. The information you are looking for may appear in the abstract itself, or you may have to read the original article cited in the abstract. Citations for such articles now appear in the form *Tetrahedron Lett.* **1996**, *37*(37), 6767–6770 (Eng.), where the abbreviated name of the publication appears first, followed by the date, volume and issue number, page numbers, and language in which the paper is written. (Earlier citations were given with the volume number first, followed by the pages and year.) The full name of the publication will be found in the *Chemical Abstracts Service Source Index* (*CASSI*, A5), which also provides a brief publication history of each source and a list of the libraries that carry it.

The letter "e" in this citation is a check letter. If you looked up abstract number 12715 by mistake, you would find that its check letter is "b."

To carry out a thorough index search of the print version of *Chemical Abstracts*, it is best to start with the most recent Collective Index and all semiannual indexes published since then and work your way back through the previous Collective Indexes, using the index guides or other sources to locate the appropriate index names. If you are looking for information about a specific compound, you may find it easier to search *Beilstein* through the most recent supplemental series that lists the compound, and *Chemical Abstracts* from that time to the present. For further information about the use of *Chemical Abstracts*, refer to one or more of the literature guides in Category K of the Bibliography.

Beilstein's Handbook of Organic Chemistry (*Handbuch der organischen Chemie*) is by far the most comprehensive source of organized information about organic compounds. *Beilstein* provides information on the structure, characterization, natural occurrence, preparation, purification, energy parameters, physical properties, and chemical properties of organic compounds. It also cites the primary sources from which the information was obtained. *Beilstein*, unlike *Chemical Abstracts*, evaluates its sources critically and corrects errors that appeared in previous series. Beginning with the fifth supplemental series, covering the period from 1960 to 1979, Beilstein is available in an English-language edition. The basic series (*Hauptwerke*) and four previous supplemental series (*Erganzungswerke* I–IV, abbreviated E I–IV) are available only in German and cover the literature through 1959.

To obtain all of the information about a particular compound in *Beilstein*, you must search the basic series and all of the available supplemental series. The enormous size of this so-called "handbook"—along with the language barrier—could make that seem a monumental task, but there is really no reason to be intimidated by *Beilstein*. The amount of German you need to know is quite limited and can be learned with the help of the *Beilstein Dictionary* (A2), a slim dictionary written specifically for *Beilstein* users. And *Beilstein* is so well organized that it is not difficult to find the information you seek. If a compound has been around for some time, you can locate its *Beilstein* entries by the following procedure:

1. Write the molecular formula of the compound with C and H first, followed by other elements in alphabetical order.
2. Locate the formula in volume (*Band*) 29 of the second supplemental series (*General–Formelregister, Zweites Erganzungswerke*) and look for the name of the compound among those listed under that formula. Although the names are in German, many are similar or identical to the English names. (If you need help, use a German–English dictionary.)
3. Write down the volume number and pages on which information about the compound appears in the basic series (H) and the first and second supplemental series (E I and E II), and look up the appropriate entries in those volumes. (Each volume may include several individually bound subvolumes.)
4. Once you know the index name of the compound and its page number in the basic series, you can locate its entry in the corresponding volume of any later series. Alternatively, you can look it up in the cumulative subject index for that volume. Each compound is also assigned a system number, which can be used to locate its entries in the same way.

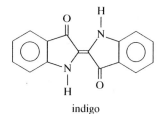

indigo

For example, the notation for indigo ($C_{16}H_{10}N_2O_2$) in volume 29 of E II reads "Indigo **24**, 417, I 370, II 233." So you will find entries for indigo on page 417 of volume 24 in the basic series and on pages 370 and 233 of volume 24 in the first and second supplemental series, respectively. The page number in the basic series, written as "**H**, 417," is called its *coordinating reference*; to find indigo's entry in volume 24 of a later series, you can locate the pages with **H**, 417 printed at the top and leaf through them until you find the entry for indigo. Knowing the E II index name of a compound may also help you locate it in the current cumulative subject index (*Sachsregister*) for the appropriate volume. The cumulative indexes for some volumes are combined; thus the listing for indigo is found in the volume 23–25 subject index, and it reads "**24** 417 d, I 370 d, II 233 e, IV 469." The letters refer to the location of an entry on the page; for example, "II 233 e" means that information about indigo will be found under the fifth entry on page 233 of E II. There is no listing for E III because supplementary series III and IV were issued jointly for volumes 17–27. Entries in the joint series are designated by E IV rather than by E III/IV.

Locating entries for a compound such as adamantane, which does not appear in the E II formula indexes, may take a little more time. All *Beilstein* entries are organized according to a detailed system, and learning that system is the best way to get complete access to the information contained in this work. However, you can usually locate such entries by either (1) finding the volume number in which a structurally similar compound appears and searching the cumulative indexes of that volume or (2) locating a *Beilstein* reference from another source. The E II indexes indicate that cyclohexane appears in volume 5, which contains all cyclic compounds lacking functional groups, so you will find adamantane listed in the cumulative indexes for that volume. You can also find *Beilstein* references for many compounds in certain reference books (A1, A6, A11, and A15, for example). The entry for adamantane in the *CRC Handbook* gives the notation "B5^4, 469," referring to a *Beilstein* entry on page 469 in volume 5 of E IV. If you look up that entry, you will find a back reference (E III 393) and a coordinating reference (**H**, 165) that will help you find information about adamantane in the other series. For more detailed information about searching *Beilstein*, see reference K2 or other sources in Category K of the Bibliography.

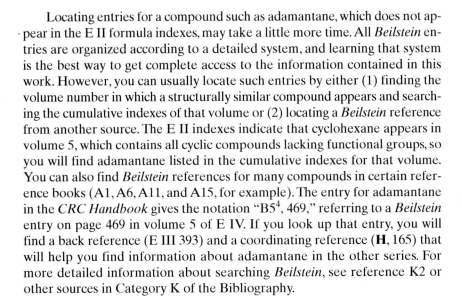

adamantane

The Beilstein *system is described in the "Notes for Users" (in English) at the beginning of each volume in the recent supplemental series.*

On-Line Searches

Both *Chemical Abstracts* and *Beilstein* can be searched electronically using a variety of on-line search services. Limited *Chemical Abstracts* searching capabilities are available through FirstSearch CA Student Edition, with coverage of the most commonly held journals at academic libraries. More sophisticated search options are available using the on-line service STN (Scientific & Technical Information Network) International, which is operated by the Chemical Abstracts Service (CAS) and provides a variety of scientific and technical databases. The fundamental CAS database, called *CAplus*, includes all entries from the printed *Chemical Abstracts* since 1970 (and some back to 1967), plus additional bibliographic information. Other CAS databases available through STN International include *CAOLD,* which includes abstracts from *CA* prior to 1967; the *REGISTRY* File, a list of virtually all currently known chemicals with their *CA* registry numbers; *CJACS,* which gives the complete texts of articles published in selected American Chemical Society journals since 1982; *CASREACT,* a database of recent organic reactions; *CHEMSOURCES,* a database of information on chemical products and their suppliers; and *CHEMLIST,* a listing of regulated and other hazardous substances. STN databases can be accessed with any computer that is connected to a telecommunications network, and are often accessible through university libraries. A simplified search option called STN*Easy* provides access to a variety of STN databases through the World Wide Web and features a graphical interface that doesn't require special training to use. Basic and advanced searches are available on STN*Easy*, and on-line help is provided when needed. For a basic search, the user simply enters a category that determines the databases to be searched, types in the words to be searched, and selects a search strategy. Search strategies include "any of these terms," which retrieves references that contain any or all of the words listed, and "all of these words," which only retrieves references that contain all of them.

Access to *Beilstein* is available through the BEILSTEIN on-line database (provided by STN International), DIALOG, and other vendors. The

BEILSTEIN database is intended to cover not only the contents of the printed work, but also information from Beilstein file cards and primary literature up to the current date. For detailed information about on-line searching of *Beilstein* or *CA*, see reference K3 or K6 in the Bibliography.

Using the Bibliography

A number of books, articles, and other literature sources in organic chemistry are listed in the following Bibliography under 12 general categories:

A. Reference Works
B. Organic Reactions and Syntheses
C. Laboratory Safety
D. General Laboratory Techniques
E. Chromatography
F. Spectrometry and Structure Analysis
G. Qualitative Organic Analysis
H. Reaction Mechanisms and Advanced Topics
J. Reports of Chemical Research
K. Guides to the Chemical Literature
L. Sources on Selected Topics
M. Software for Organic Chemistry

Each source is referred to here by the category letter and its number within the category.

Category A: Reference Works

While *Beilstein* attempts to provide all of the important information about the millions of organic compounds mentioned in the chemical literature, the reference books that follow provide selected information about a much smaller number of compounds, usually numbering in the tens of thousands. The *CRC Handbook of Chemistry and Physics* (A15), *Lange's Handbook of Chemistry* (A6), and *Dean's Handbook of Organic Chemistry* (A7) tabulate physical properties and other data for many common organic compounds and contain a large amount of useful information about chemistry. *The Merck Index* (A11) is an excellent source of information on approximately 10,000 organic and inorganic compounds. It describes their uses and hazardous properties, provides detailed physical and structural data, and gives literature references for the isolation and synthesis of many compounds. The *CRC Handbook of Data on Organic Compounds (HODOC)* (A9) contains data and references to published spectra for more than 27,000 organic compounds. The *Aldrich Catalog* (A1) lists the many chemicals manufactured by the Aldrich Chemical Company and gives their physical properties, hazard warnings, procedures for safe disposal, references to published Aldrich spectra, and references to listings in *The Merck Index*, *Beilstein*, and *Fieser* (B7). The *Dictionary of Organic Compounds (DOC)* (A12) is an important multivolume set, updated by annual supplements, that gives structures, physical constants, hazard descriptions, sources, uses, derivatives, and bibliographic references for more than 145,000 organic compounds. It is also available on CD-ROM. Figure 1 shows the level of information provided by three of these reference works.

Before you use such a reference work to find information about organic compounds, always read the introduction or explanatory material at

No.	Name	Formula	Formula weight	Beilstein reference	Density, g/mL	Refractive index	Melting point, °C	Boiling point, °C	Flash point, °C	Solubility in 100 parts solvent
b44	Benzoic acid	C_6H_5COOH	122.12	9, 92	1.321		122.4	249	121 (CC)	0.29 aq^{25}; 43 alc; 10 bz; 22 chl; 33 eth; 33 acet; 30 CS_2

A. *Lange's Handbook of Chemistry.* (Reprinted with permission from *Lange's Handbook of Chemistry*, 15th ed., by N. A. Lange, edited by J. A. Dean. Copyright McGraw-Hill, Inc., New York, 1999.)

1092. Benzoic Acid. [65-85-0] Benzenecarboxylic acid; phenylformic acid; dracylic acid. $C_7H_6O_2$; mol wt 122.12. C 68.85%, H 4.95%, O 26.20%. Occurs in nature in free and combined forms. Gum benzoin may contain as much as 20%. Most berries contain appreciable amounts (around 0.05%). Excreted mainly as hippuric acid by almost all vertebrates, except fowl. Mfg processes include the air oxidation of toluene, the hydrolysis of benzotrichloride, and the decarboxylation of phthalic anhydride: *Faith, Keyes & Clark's Industrial Chemicals,* F. A. Lowenheim, M. K. Moran, Eds. (Wiley-Interscience, New York, 4th ed., 1975) pp 138-144. Lab prepn from benzyl chloride: A. I. Vogel, *Practical Organic Chemistry* (Longmans, London, 3rd ed, 1959) p 755; from benzaldehyde: Gattermann-Wieland, *Praxis des organischen Chemikers* (de Gruyter, Berlin, 40th ed, 1961) p 193. Prepn of ultra-pure benzoic acid for use as titrimetric and calorimetric standard: Schwab, Wicher, *J. Res. Nat. Bur. Standards* 25, 747 (1940). Review: A. E. Williams in *Kirk-Othmer Encyclopedia of Chemical Technology* vol. 3 (Wiley-Interscience, New York, 3rd ed., 1978) pp 778-792.

Monoclinic tablets, plates, leaflets. d 1.321 (also reported as 1.266). mp 122.4°. Begins to sublime at ~100°. bp$_{760}$ 249.2°; bp$_{400}$ 227°; bp$_{200}$ 205.8°; bp$_{100}$ 186.2°; bp$_{60}$ 172.8°; bp$_{40}$ 162.6°; bp$_{20}$ 146.7°; bp$_{10}$ 132.1°. Volatile with steam. Flash pt 121°C. pK (25°) 4.19, pH of satd soln at 25°: 2.8. Soly in water (g/l) at 0° = 1.7; at 10° = 2.1; at 20° = 2.9; at 25° = 3.4; at 30° = 4.2; at 40° = 6.0; at 50° = 9.5; at 60° = 12.0; at 70° = 17.7; at 80° = 27.5; at 90° = 45.5; at 95° = 68.0. Mixtures of excess benzoic acid and water form two liquid phases beginning at 89.7°. The two liquid phases unite at the critical soln temp of 117.2°. Composition of critical mixture: 32.34% benzoic acid, 67.66% water: *see* Ward, Cooper, *J. Phys. Chem.* 34, 1484 (1930). One gram dissolves in 2.3 ml cold alc, 1.5 ml boiling alc, 4.5 ml chloroform, 3 ml ether, 3 ml acetone, 30 ml carbon tetrachloride, 10 ml benzene, 30 ml carbon disulfide, 23 ml oil of turpentine; also sol in volatile and fixed oils, slightly in petr ether. The soly in water is increased by alkaline substances, such as borax or trisodium phosphate, *see also* Sodium Benzoate.

Barium salt dihydrate. Barium benzoate. $C_{14}H_{10}$-$BaO_4.2H_2O$. Nacreous leaflets. *Poisonous!* Soluble in about 20 parts water; slightly sol in alc.

Calcium salt trihydrate. Calcium benzoate. $C_{14}H_{10}$-$CaO_4.3H_2O$. Orthorhombic crystals or powder. d 1.44. Soluble in 25 parts water; very sol in boiling water.

Cerium salt trihydrate. Cerous benzoate. $C_{21}H_{15}$-$CeO_6.3H_2O$. White to reddish-white powder. Sol in hot water or hot alc.

Copper salt dihydrate. Cupric benzoate. $C_{14}H_{10}$-$CuO_4.2H_2O$. Light blue, cryst powder. Slightly soluble in cold water, more in hot water; sol in alc or in dil acids with separation of benzoic acid.

Lead salt dihydrate. Lead benzoate. $C_{14}H_{10}O_4Pb.2H_2O$. Cryst powder. *Poisonous!* Slightly sol in water.

Manganese salt tetrahydrate. Manganese benzoate. C_{14}-$H_{10}MnO_4.4H_2O$. Pale-red powder. Sol in water, alc. Also occurs with $3H_2O$.

Nickel salt trihydrate. Nickel benzoate. $C_{14}H_{10}$-$NiO_4.3H_2O$. Light-green odorless powder. Slightly sol in water; sol in ammonia; dec by acids.

Potassium salt trihydrate. Potassium benzoate. C_7H_5-$KO_2.3H_2O$. Crystalline powder. Sol in water, alc.

Silver salt. Silver benzoate. $C_7H_5AgO_2$. Light-sensitive powder. Sol in 385 parts cold water, more sol in hot water; very slightly sol in alc.

Uranium salt. Uranium benzoate; uranyl benzoate. $C_{14}H_{10}$-O_6U. Yellow powder. Slightly sol in water, alc.

Caution: Mild irritant to skin, eyes, mucous membranes.

USE: Preserving foods, fats, fruit juices, alkaloidal solns, etc; manuf benzoates and benzoyl compds, dyes; as a mordant in calico printing; for curing tobacco. As standard in volumetric and calorimetric analysis. Pharmaceutic aid (antifungal).

THERAP CAT (VET): Has been used with salicylic acid as a topical antifungal.

B. *The Merck Index: An Encyclopedia of Chemicals, Drugs, and Biologicals,* 13th Edition, Maryadele J. O'Neil, Ann Smith, Patricia E. Heckelman, John R. Obenchain Jr., Eds. (Reproduced with permission from *The Merck Index*, 13th Edition. Copyright © 2001 by Merck & Co., Inc., Whitehouse Station, NJ, USA. All rights reserved.)

Figure 1 Entries for benzoic acid from some reference works

Benzoic acid, 9CI   **B-0-00650**

Benzenecarboxylic acid

[65-85-0]

PhCOOH

$C_7H_6O_2$ M 122.1

Widespread in plants esp. in essential oils, mostly in esterified form. Obt. in 17th Century by sublimation of *Styrax* spp. resin. Produced industrially mainly by oxidation of toluene. Preservative in the food industry. Used in manuf. of preservatives, plasticisers, alkyd resin coatings and caprolactam. Antiseptic and expectorant. Used as alkalimetric standard; in photometric detn. of U and Zr (anionic complexes associated with basic dyes). Reference material used in elemental microanalysis. Leaflets or needles (H_2O). V. spar. sol. H_2O. Mp 122°. Bp 249°, Bp_{10} 133°. Subl. *ca.* 100°. Steam-volatile.

▶ Fl. p. 121°, autoignition temp. 570°. Eye, skin and mucous membrane irritant. Hypersensitivity reactions reported. Low systemic toxicity. DG0875000.

Na salt: [532-32-1].
Used as food preservative, anticorrosion agent. Cryst.
▶ DH6650000.

K salt: [582-25-2].
Cryst.

Me ester: [93-58-3]. *Methyl benzoate*
$C_8H_8O_2$ M 136.1   Used in perfumery and flavourings. Liq. d_{23}^{23} 1.09. Fp −12.3°. Bp 199.6°, Bp_{24} 96-98°.
▶ Fl. p. 83°. Skin and eye irritant. LD_{50} (rat, orl) 1350 mg/kg. DH3850000.

Et ester: [93-89-0]. *Ethyl benzoate*
$C_9H_{10}O_2$ M 150.1   Polymerisation catalyst. Used in perfumery and flavourings. Liq. d^{25} 1.04. Fp −34°. Bp 212.9°, Bp_{10} 87.2°.
▶ Fl. p. 88°, autoignition temp. 490°. Skin and eye irritant. LD_{50} (rat, orl) 2100 mg/kg. DH0200000.

Vinyl ester: see Ethenol, E-0-00320
Propyl ester: [2315-68-6]. *Propyl benzoate*
$C_{10}H_{12}O_2$ M 164.2   Flavour ingredient. d^{15}_{15} 1.03. Bp 230°.

Isopropyl ester: [939-48-0]. *Isopropyl benzoate*
Polymerisation catalyst, flavour ingredient. d^{15}_{15} 1.02. Bp 218-219°.
▶ Fl. p. 89/99°. Skin and eye irritant. LD_{50} (rat, orl) 3730 mg/kg. DH3150000.

Butyl ester: [136-60-7]. *Butyl benzoate*
$C_{11}H_{14}O_2$ M 178.2   Dye carrier; used in perfumery. d^{15}_{15} 1.01. Bp 248-249°.
▶ Fl. p. 107° (oc). Eye and skin irritant. DG4925000.

tert-Butyl ester: [774-65-2]. tert-*Butyl benzoate*
$C_{11}H_{14}O_2$ M 178.2 Bp_2 96°.

Benzyl ester: [120-51-4]. *Benzyl benzoate*, USAN. Ascabin. Benylate. Vanzoate. Many other names

$C_{14}H_{12}O_2$ M 212.2   Contained in Peru balsam. Isol. from other plants e.g. *Jasminum* spp., ylang-ylang oil. Insect repellant component. Acaricide and pediculicide. Used in perfumery as fixative and in food flavouring. Leaflets. d^{18} 1.11. Mp 21° (19.5°). Bp 323-324° (316-317°), $Bp_{0.1}$ 80-82°. Spar. steam-volatile.
▶ Fl. p. 148°, autoignition temp. 480°. Eye, mucous membrane, and possible skin irritant. Hypersensitivity reactions reported. LD_{50} (rat, orl) 500 mg/kg. DG4200000.

Ph ester: see Phenyl benzoate, P-0-01360
Fluoride: [455-32-3]. *Benzoyl fluoride*
C_7H_5FO M 124.1   Fuming liq. Bp 159-161°. Hydrolysed by hot H_2O.
▶ Highly irritant, causes burns, violent reaction with DMSO. Fl. p. 72/102°.

Chloride: [98-88-4]. *Benzoyl chloride*
C_7H_5ClO M 140.5   Polymerisation catalyst, benzoylating agent. Can be used for synth. of aliphatic acid chlorides. Used to derivatise steroids and carbohydrates for chromatog. Fuming liq. d^{15}_{15} 1.22. Fp −1°. Bp 197°.
▶ Fl. p. 72/102°. Violent reaction with DMSO. Corrosive and irritating to all tissues. Potent lachrymator. DM6600000.

Bromide: [618-32-6]. *Benzoyl bromide*
C_7H_5BrO M 185.0   Fuming liq. d^{15} 1.57. Fp −24°. Bp 218-219°, $Bp_{0.05}$ 48-50°.

Iodide: [618-38-2]. *Benzoyl iodide*
C_7H_5IO M 232.0   Needles. Mp 3°. Bp_{20} 128°.

Amide: see Benzamide, B-0-00069
Anilide: see Benzanilide, B-0-00074
Azide: Benzazide. Benzoylazimide
$C_7H_5N_3O$ M 147.1   Plates. Mp 32°.
▶ Explodes on heating.

Hydrazide: [613-94-5]. *Benzoylhydrazine*
$C_7H_8N_2O$ M 136.1   Used as 0.2M aq. soln. for photometric detn. of V (λ_{max} 400 nm, ε 9000); as 0.1M aq. soln. for photometric detn. of $IO_4^\ominus$. Cryst. (H_2O). Sol. H_2O, acids, EtOH, C_6H_6, Me_2CO. Mp 112.5°.
▶ DH1575000.

Hydroxamate: see N-Hydroxybenzamide, H-0-01671
Nitrile: see Benzonitrile, B-0-00747
Anhydride: [93-97-0]. *Benzoic anhydride*
$C_{14}H_{10}O_3$ M 226.2   Cross-linking agent for polymers. Acylation and decarboxylating agent, can be used in polymer-linked form. Can be used to prep. derivs. of e.g. glycosphingolipids for hplc. Rhombic prisms. d^{15} 1.99. Mp 42°. Bp 360°.
▶ Mild irritant and allergen.

Aldrich Library of ^{13}C and ^1H FT NMR Spectra, 2, 1063B, 1199A, 1240A, 1240B, 1241A, 1241B, 1244A, 1337C, 1411A (*nmr*)
Aldrich Library of FT-IR Spectra, 1st edn., 2, 186A, 271B, 291B, 291C, 292D, 340A, 340B, 340C, 380D (*ir*)

Aldrich Library of FT-IR Spectra: Vapor Phase, 3, 181D, 1322C, 1357D, 1358A, 1358B, 1358C, 1360A, 1389D, 1390A, 1390B (*ir*)
Org. Synth., Coll. Vol., 1, 1932, 75, 361 (*synth, deriv*)
Jesson, J.P. *et al*, *Proc. R. Soc. London, A*, 1962, **268**, 68 (*Raman*)
Beynon, J.H. *et al*, *Z. Naturforsch., A*, 1965, **20**, 883 (*ms*)
Moeken, H.H. *et al*, *Anal. Chim. Acta*, 1967, **37**, 480 (*detn, U*)
Fieser and Fieser's Reagents for Organic Synthesis, Wiley, 1967, 1, 49, 1004; 1975, 5, 23, 24, 249; 1979, 7, 405 (*use*)
Evans, H.B. *et al*, *J. Phys. Chem.*, 1968, **72**, 2552 (*pmr*)
Escarrilla, A.M. *et al*, *Anal. Chim. Acta*, 1969, **45**, 199 (*use*)
Analyst (London), 1972, **97**, 740 (*microanal*)
Bel'tyukova, S.V. *et al*, *Zh. Anal. Khim.*, 1972, **27**, 191 (*detn, Zr*)
Fitzpatrick, F.A. *et al*, *Anal. Chem.*, 1973, **45**, 2310 (*chloride, use*)
Morris, W.W., *J. Assoc. Off. Anal. Chem.*, 1973, **56**, 1037 (*ir*)
Dubey, S.C. *et al*, *Talanta*, 1977, **24**, 266 (*detn, U*)
Pilipenko, A.T. *et al*, *Zh. Anal. Khim.*, 1977, **32**, 1369 (*hydrazide, detn, V*)
Fauvet, G. *et al*, *Acta Cryst. B*, 1978, **34**, 1376 (*cryst struct, nitrile*)
White, C.A. *et al*, *Carbohydr. Res.*, 1979, 76, 1 (*chloride, use*)
Opdyke, D.L.J., *Food Cosmet. Toxicol.*, 1979, 17, 715 (*rev, tox*)
Hassan, M.M.A. *et al*, *Anal. Profiles Drug Subst.*, 1981, 10, 55 (*rev, benzyl ester*)
Rama Rao, A.V. *et al*, *Chem. Ind. (London)*, 1984, 270 (*synth*)
Ullmann's Encycl. Ind. Chem., 5th Ed, VCH, Weinheim, 1985, A3, 555 (*rev*)
Lewandowski, W., *Can. J. Spectrosc.*, 1987, 32, 41 (*salts, ir*)
Ullman, M.D. *et al*, *Methods Enzymol.*, 1987, **138**, 117 (*use, anhydride*)
Negwer, M., *Organic-Chemical Drugs and their Synonyms*, 6th edn., Akademie-Verlag, Berlin, 1987, 3041.
Cook, I.B., *Aust. J. Chem.*, 1989, 42, 1493 (*cmr*)
Lewis, R.J., *Food Additives Handbook*, Van Nostrand Reinhold International, New York, 1989, BCL750, BCM000, EGR000, MHA500.
Merck Index, 11th edn., 1989, No. 1107, No. 1141 (*nitrile, benzyl ester*)
Kirk-Othmer Encycl. Chem. Technol., 4th edn., Wiley, New York, 1991, 4, 103 (*rev*)
Martindale, The Extra Pharmacopoeia, 30th edn., Pharmaceutical Press, London, 1993, 1124, 1132.
Lewis, R.J., *Sax's Dangerous Properties of Industrial Materials*, 8th edn., Van Nostrand Reinhold, 1992, BBV250, BCL750, BCM000, BCQ250, BDM500, BQK250, EGR000, IOD000, MHA750, PKW760, SFB000.
Bretherick, L., *Handbook of Reactive Chemical Hazards*, 4th edn., Butterworth, London and Boston, 1990, 2511.
Luxon, S.G., *Hazards in the Chemical Laboratory*, 5th edn., Royal Society of Chemistry, Cambridge, 1992, 117.
Chemical Hazards of the Workplace, (eds. Proctor, N.H. *et al*), 3rd edn., VNR, 1991, 107.

C. Dictionary of Organic Compounds. (Reprinted with permission from *Dictionary of Organic Compounds*, 6th ed., edited by P. H. Rhodes. Copyright Chapman & Hall, London, 1995.)

Figure 1 (*continued*)

the beginning of the work or preceding the table you intend to use. The introductory section of a reference work will usually (1) describe the content and organization of the material, (2) list symbols and abbreviations, and (3) describe the system of nomenclature used. Different sources may use very different naming systems. For example, *The Merck Index* emphasizes therapeutic uses of compounds, so it lists aspirin under that name; but in *Lange's Handbook* you will find aspirin listed as acetylsalicylic acid, and in some editions of the *CRC Handbook of Chemistry and Physics* it appears as salicylic acid acetate. Often, the index of a reference work provides the quickest and most reliable access to a given entry. When using *The Merck Index* or the *Dictionary of Organic Compounds*, you should first consult the name index to locate the entry for a given compound. If you can't find the compound in the name index, you may be able to locate it in a formula index. In most formula indexes, carbon and hydrogen are listed first, followed by the other elements in alphabetical order.

Other useful reference works in organic chemistry include the *Ring Systems Handbook* (A4), which (with its supplements) records all known organic ring systems and provides information allowing users to locate compounds having a particular ring system; and *CASSI* (A5), which provides bibliographic information for the journals and other sources indexed by *Chemical Abstracts* and lists the libraries holding each source. The *Beilstein Dictionary* (A2) is an invaluable aid to understanding the parts of *Beilstein* (A3) that are available only in German. *The Organic Chemist's Desk Reference* (A13) includes a user's guide to the *Dictionary of Organic Chemistry* as well as a discussion of nomenclature in *Chemical Abstracts*, a list of reference works in organic chemistry, and other useful information. *The Chemist's Ready Reference Handbook* (A14) provides practical information about a variety of theoretical and experimental topics. *Organic Chemistry: An Alphabetical Guide* (A10) discusses and defines many terms used in organic chemistry. *Kirk–Othmer* (A8) is an excellent source of comprehensive, up-to-date articles on a variety of chemical topics. It offers particularly good coverage of industrial chemistry and commercial products, but also includes entries on natural products, such as coffee, terpenoids, and vitamins.

Category B: Organic Reactions and Syntheses

The works in this category are intended primarily to help chemists and chemistry students design and carry out organic syntheses. Although you may not be required to work out synthetic procedures on your own, an understanding of the strategy and techniques of organic synthesis will help you perform better in your lab and lecture courses and get more out of them. The article by Nicolaou (B17) gives an interesting account of the evolution of organic synthesis, especially as it applies to the synthesis of natural products. *Modern Organic Synthesis* (B35) and *Organic Synthesis: Concepts and Methods* (B9) each provide a general overview of organic synthesis. *Organic Synthesis: The Disconnection Approach* (B30) describes strategies for planning an organic synthesis. At a more advanced level, *The Logic of Chemical Synthesis* (B6) deals with the analysis of complex synthetic problems, while *Principles of Organic Synthesis* (B18) discusses such topics as thermodynamics, kinetics, and stereochemistry as they apply to organic synthesis.

A number of works can help the experimenter select the type of reaction that will best accomplish a given synthetic transformation. *Synthetic Organic Chemistry* (B34) and *Modern Synthetic Reactions* (B12) survey many important synthetic reactions, with references to the earlier literature. More comprehensive coverage of organic transformations, including the more recent synthetic reactions, can be found in *Modern Methods of Organic Synthesis* (B4), *Compendium of Organic Synthetic Methods* (B11), *Comprehensive Organic Transformations* (B13), *Comprehensive Organic Synthesis* (B33), and *Theilheimer's Synthetic Methods of Organic Chemistry* (B32). Reaction files from Theilheimer and other works can be searched on-line using the *REACCS* database from Molecular Design, Ltd.

General works on organic synthetic reactions range from the *Reaction Guide for Organic Chemistry* (B16), a basic compilation of most reactions covered in the sophomore-level organic chemistry course, to *Organic Reactions* (B19), a multivolume set containing very comprehensive monographs on a large variety of reactions. *Organic Reactions* describes each reaction's mechanism, scope, limitations, and experimental conditions. It also provides detailed experimental procedures and a table that lists examples of each reaction with references to the original sources. *Named Organic Reactions* (B14) describes those synthetic reactions that—like the Hell–Volhard–Zelinskii reaction—are identified by the names of one or more discoverers. *Rodd's Chemistry of Carbon Compounds* (B5) describes the reactions of different classes of organic compounds in considerable depth. Each volume of *Chemistry of Functional Groups* (B22) covers the chemical reactions of a different functional group. *Asymmetric Synthesis* (B25) deals with the synthesis of chiral compounds and *Stereoselective Synthesis* (B1) with the use of stereoselective reactions in organic synthesis. Many books cover only one type of synthetic reaction, such as the Diels–Alder reaction and related cycloaddition reactions (B3, B8); a search of your library's catalog should reveal similar works about other types of reactions. *Protective Groups in Organic Synthesis* (B10) describes the use of protective groups in syntheses involving multifunctional reactants. *Microwaves in Organic Synthesis* (B15) covers the theory and applications of microwave-assisted synthesis. In addition to the works dedicated to organic reactions, several advanced textbooks, such as March (H7) and Carey-Sundberg (H1), describe the most important synthetic reactions and give literature references to specific synthetic procedures.

Three multivolume sources of information about chemical reagents are *Fiesers' Reagents for Organic Syntheses* (B7), *Encyclopedia of Reagents for Organic Synthesis* (B21), and *Handbook of Reagents for Organic Synthesis* (B23). *Fiesers'* provides information about the preparation, purification, handling, and hazards of many chemical reagents, as well as examples of their use, with literature citations. Only the individual volumes are indexed, but if you locate the entry for a reagent in a recent volume, it will provide back references to the previous volumes. The *Encyclopedia of Reagents* reviews nearly 3500 reagents, listed alphabetically, and gives a critical assessment of each reagent. The four-volume *Handbook of Reagents* (B23) covers some 500 reagents selected from this encyclopedia. *Borane Reagents* (B24) deals with the applications of boranes in organic synthesis. *Organic Solvents* (B26) gives physical properties and purification methods for many solvents used in organic synthesis.

Although the primary chemical literature is the most important source of experimental procedures, a number of secondary sources (in addition to

Organic Reactions) provide relatively detailed, reliable procedures. *Organic Syntheses* (B20) is a continuing series that contains an excellent selection of carefully tested synthetic procedures. Other works that provide experimental procedures include *Organicum* (B2), several works by Sandler and Karo (B27, B28, B29), and *Vogel's Textbook of Practical Organic Chemistry* (B31). Vogel is also a good source of information about laboratory techniques. *Houben-Weyl* (D5), which is listed under "General Laboratory Techniques," includes many synthetic procedures (in German) as well.

Category C: Laboratory Safety

Accidents can happen in the organic chemistry lab, so it is important to know how to prevent accidents and what to do in case of an accident. *Working Safely with Chemicals in the Laboratory* (C2) is a short booklet about laboratory safety written for students. *Prudent Practices in the Laboratory* (C7) is an authoritative guide to safe laboratory practices; it also includes procedures for the safe handling and disposal of chemicals. *Hazards in the Chemical Laboratory* (C6) describes the toxic effects of hazardous substances and reviews recent developments in the safe design and operation of chemical laboratories. The *CRC Handbook of Laboratory Safety* (C1) deals with the recognition and control of hazards and compliance with safety regulations, and includes a chapter on responding to laboratory emergencies. The *First Aid Manual for Chemical Accidents* (C3) gives first aid procedures for accidents caused by specific chemicals and classes of chemicals. The *Sigma–Aldrich Library of Regulatory and Safety Data* (C4) and *Sax's Dangerous Properties of Industrial Materials* (C5) provide detailed health and safety data for many common chemicals. *Bretherick's Handbook of Reactive Chemical Hazards* (C8) describes the properties of chemicals that are hazardous by virtue of their instability or their tendency to react with other chemicals. The ninth edition of *The Merck Index* (see A11) contains a section on first aid for poisoning and chemical burns; this section is not included in the more recent editions.

Category D: General Laboratory Techniques

Although *The Student's Laboratory Companion* covers the techniques you are most likely to use in your organic chemistry lab course, sources from this category and the following two categories may provide more detailed practical and theoretical information about specific lab techniques, information about more advanced techniques, or a different approach to the methods described here. *The Organic Chem Lab Survival Manual* (D11) describes many of the lab techniques used by organic chemistry students and tells you what things *not* to do, such as plugging a heating mantle directly into a wall socket. *Guide for the Perplexed Organic Experimentalist* (D3) deals with the practical aspects of laboratory work for anyone intending to do research in organic chemistry. Weissberger's *Technique of Organic Chemistry* (D7) and *Techniques of Chemistry* (D8) are multivolume sets that cover a wide variety of experimental methods. Information on classical laboratory techniques, such as distillation and recrystallization, can be found in Volume I of reference D7, which is subtitled *Physical Methods of Organic Chemistry*. The first four volumes of *Houben–Weyl* (D5) describe many laboratory methods for organic chemistry, in German. Microscale techniques based on Mayo–Pike and Williamson type glassware, respectively, are described in *Microscale Techniques for the Organic Laboratory* (D4) and *Macroscale and Microscale Organic Experiments*

(D9). The *Encyclopedia of Separation Technology* (D6) and *Encyclopedia of Separation Science* (D10) provide comprehensive up-to-date descriptions of separation techniques, including modern microscale techniques. *Purification of Laboratory Chemicals* (D1) provides methods for the purification of more than 4000 common chemicals. *Natural Products* (D2) describes laboratory techniques and gives specific procedures for the isolation and structure determination of natural products.

Category E: Chromatography

During your organic chemistry lab course you may use a variety of chromatographic methods for the separation and analysis of organic compounds. These methods include thin-layer chromatography (TLC), paper chromatography (PC), gas chromatography (GC) and high-performance liquid chromatography (HPLC). *Chromatography Today* (E6) and *Principles and Practice of Chromatography* (E7) are good general sources of information on the theory and practice of all types of chromatography. *Gas Chromatography* (E1) and *High Performance Liquid Chromatography* (E3) are "open learning" texts designed for self-study. The remaining works provide up-to-date coverage of GC, HPLC, and TLC techniques and applications. *Principles of Instrumental Analysis* (F21) in the next section has chapters on instrumental chromatographic methods.

Category F: Spectrometry and Structure Analysis

During your organic chemistry lab course you will probably need to record and interpret various kinds of spectra of organic compounds, such as infrared (IR) spectra, nuclear magnetic resonance (NMR) spectra, ultraviolet–visible (UV–VIS) spectra, and mass spectra (MS). *Principles of Instrumental Analysis* (F21) is a good source of information about the principles and applications of all important kinds of spectrometric methods, as well as other instrumental methods of analysis. An article in the *Journal of Chemical Education* (F8) covers the basics of IR and NMR spectral interpretation. The works by Silverstein (F20), Feinstein (F5), Kemp (F9), Pavia (F13), Whittaker (F24), and Yadav (F25) are good one-volume introductions to the interpretation of spectra of organic compounds. More comprehensive coverage of specific spectrometric methods is provided for infrared spectrometry by references F4 and F22; for nuclear magnetic resonance spectrometry by F1, F2, F6, and F12; for mass spectrometry by F3, F7, F10, F11, F19, and F23; and for ultraviolet–visible spectrometry by F14. The *Sadtler Standard Spectra* (F18) series consists of a large number of IR, NMR, and UV–VIS spectra in ring binders; although they are not arranged systematically, individual spectra can be located by using the index volumes. Spectra in the Aldrich collections (F15–F17) are arranged by functional class and in order of increasing molecular complexity within a functional class, making it possible to observe the effect of various structural features on the spectra.

Category G: Qualitative Organic Analysis

During your organic chemistry lab course you may be required to identify one or more unknown organic compounds using either "wet-chemistry" methods (involving chemical tests and derivative preparations) or spectrometric methods, or both. *Organic Structure Determination* (G4), *The Systematic Identification of Organic Compounds* (G6), and *Spectral and Chemical Characterization of Organic Compounds* (G2) cover the traditional wet-chemistry

methods but include chapters on spectrometric methods as well. *Qualitative Organic Analysis* (G3) emphasizes spectral methods of identification, and *Organic Structure Analysis* (G1) focuses on the use of multiple spectrometric methods to identify a molecule's major structural elements. The *CRC Handbook of Tables for Organic Compound Identification* (G5) lists the properties and derivative melting points for many organic compounds in the most important functional classes.

Category H: Reaction Mechanisms and Advanced Topics

Although reaction mechanisms are more often explored in an organic chemistry lecture course than in the laboratory course, you may be expected to understand and write mechanisms for some of the reactions you perform in the lab. *Electron Flow in Organic Chemistry* (H11) teaches an intuitive approach to organic chemistry by breaking down reaction mechanisms into elementary electron-flow pathways. *A Guidebook to Mechanism in Organic Chemistry* (H12) by Sykes is an excellent survey of reaction mechanisms suitable for advanced students; his *Primer* (H13) is a more basic introduction to mechanisms based on a simplified classification scheme. Other how-to books include *The Art of Writing Reasonable Reaction Mechanisms* (H6), *Writing Reaction Mechanisms in Organic Chemistry* (H8) and *Reaction Mechanisms at a Glance* (H9). *Name Reactions* (H4) provides detailed mechanisms for reactions known familiarly by the names of their discoverers. *Mechanism and Theory in Organic Chemistry* (H5) and *Perspectives on Structure and Mechanism in Organic Chemistry* (H3) are advanced textbooks that present the theoretical aspects of organic chemistry and provide up-to-date information about important reaction mechanisms. *Advanced Organic Chemistry* (H1) and *March's Advanced Organic Chemistry* (H7) provide good coverage of the mechanisms and synthetic applications of a large number of organic reactions, giving numerous references to the primary literature. *Determination of Organic Reaction Mechanisms* (H2) describes experimental techniques for studying reaction mechanisms, and *Organic Reaction Mechanisms* (H10) is an annual survey of recent developments in the field.

Category J: Reports of Chemical Research

Virtually all professional chemists do *chemical research*—experimental or theoretical work designed to discover new facts about the various forms of matter, develop new techniques that can be used to study matter, or provide new insights about the fundamental nature of matter. Papers describing the results of their research are reported in such a large number of professional journals and other publications that it impossible for anyone to investigate them all. For that reason, scientists consult various reports of chemical research to locate the papers that deal with their own research interests. *Chemical Abstracts* (J3), previously described in detail, is the most comprehensive single source of information about research in chemistry. The *Science Citation Index* (J7) is an index of literature citations to papers, patents, and books published in the past. For example, if you find an interesting paper by Linus Pauling in a chemistry journal, you can look up the paper in the *Science Citation Index* to find later articles that were based, in part, on Pauling's original paper. In this way, you can sometimes trace the development of an idea or a method from its origin to the present day. A similar index for chemistry, the *Chemistry Citation Index*, is available on CD-ROM. *Chemical Titles* (J4) and *Current Contents* (J5) reproduce the current tables of contents of the most important chemistry journals to inform chemists quickly of recent research in their fields. *Index Chemicus* (J6), a weekly guide to new organic compounds and their chemistry, is available in print and on a searchable database. The other works in this category (J1, J2) provide annual summaries and reviews of research in organic chemistry.

Category K: Guides to the Chemical Literature

If you need specific information to complete a lab report or write a research paper, you have to know where to look for it. *Information Sources in Chemistry* (K1) and *How to Find Chemical Information* (K4) are general guides to the chemical literature that list and describe a large number of information sources. *Library Handbook for Organic Chemists* (K5) is an up-to-date guide to the use of library information resources. *The Beilstein System* (K2) and *The Beilstein Online Database* (K3) tell how to search and use *Beilstein's* print and on-line versions, respectively. *From CA to CAS Online* (K6) serves the same function for *Chemical Abstracts*. The three articles by Somerville (K7) describe the contents and uses of some major works on organic reactions and syntheses. A brief but useful guide to information sources for organic chemistry can be found in Appendix A of *March's Advanced Organic Chemistry* (H7).

Category L: Sources on Selected Topics

This category includes a number of books and articles that can be used as resources for library research papers. Others can simply be read for enjoyment and enlightenment, on everything from coffee and perfumes to the O. J. Simpson trial. Some works are listed here because they don't fit into any of the other categories. For example, L16 and L46 are guides for writing scientific papers and laboratory notebooks, L41 gives suggestions about presenting papers and posters, and L22 tells you how to name organic compounds.

Category M: Software for Organic Chemistry

A number of software titles are designed to be used in preparation for or during an organic chemistry laboratory. *Identification of Organic Compounds* (M8) and *MacSQUALOR* (M9) allow the user to identify simulated unknowns for qualitative organic analysis. *Introduction to Spectroscopy* (M4) helps the user analyze and interpret spectral data. *IR Simulator* (M10) and *NMR Simulator* (M11) simulate the operation of actual instruments to generate IR and NMR spectra. *MassSpec* (M5) helps the user identify the structural fragments that correspond to peaks on a mass spectrum. *SynTree* (M7) helps the user work out retrosynthetic pathways leading from a selected target compound back to a readily available starting material. *ChemDraw* (M2) allows the user to generate a variety of chemical structures. *Name It* (M6) provides practice in naming organic compounds and drawing structures from their names. *Beaker* (M1) predicts properties and generates spectra of organic molecules, as well as providing some structure-drawing tools. *Chem3D* (M3) and *Spartan Student Edition* (M12) are used for molecular modeling and computational chemistry applications.

Bibliography

A. Reference Works

1. *Aldrich Catalog Handbook of Fine Chemicals.* Milwaukee, WI: Aldrich Chemical Co., 2005–06 (and other years).
2. *Beilstein Dictionary: German–English: For the Users of the Beilstein Handbook of Organic Chemistry.* Ft. Worth, TX: W.B. Saunders, 1992.
3. *Beilstein's Handbook of Organic Chemistry.* New York: Springer-Verlag, 1918 to date.
4. *Chemical Abstracts Ring Systems Handbook.* Washington, D.C.: American Chemical Society, 1993 with cumulative supplements.
5. *Chemical Abstracts Service Source Index (CASSI).* Washington, D.C.: American Chemical Society, 1907–2004 with quarterly supplements.
6. Dean, J. A., ed., *Lange's Handbook of Chemistry*, 16th ed. New York: McGraw-Hill, 2004.
7. Gokel, G. W., *Dean's Handbook of Organic Chemistry.* New York: McGraw-Hill, 2004.
8. *Kirk–Othmer Encyclopedia of Chemical Technology*, 5th ed. New York: Wiley, 2004–.
9. Lide, D. R., and Milne, G. W. A., eds., *CRC Handbook of Data on Organic Compounds*, 3rd ed. Boca Raton, FL: CRC Press, 1994.
10. Mundy, B. P., and Ellerd, M. G., *Organic Chemistry: An Alphabetical Guide.* New York: Wiley, 1996.
11. O'Neil, M. J., et al., eds., *The Merck Index: An Encyclopedia of Chemicals, Drugs, and Biologicals*, 14th ed. Whitehouse Station, NJ: Merck & Co., 2006.
12. Rhodes, P. H., ed., *Dictionary of Organic Compounds*, 6th ed. London: Chapman & Hall, 1995.
13. Rhodes, P. H., *The Organic Chemist's Desk Reference: A Companion Volume to the Dictionary of Organic Compounds*, 6th ed. London: Chapman & Hall, 1995.
14. Shugar, G. J., and Dean, J. A., *The Chemist's Ready Reference Handbook.* New York: McGraw-Hill, 1990.
15. Weast, R. C., ed., *CRC Handbook of Chemistry and Physics*, new editions annually. Boca Raton, FL: CRC Press, 2007 (and other years).

B. Organic Reactions and Syntheses

1. Atkinson, R. S., *Stereoselective Synthesis.* New York: Wiley, 1995.
2. Becker, H., et al., *Organicum: Practical Handbook of Organic Chemistry*, trans. by B. J. Hazzard. Reading, MA: Addison-Wesley, 1973.
3. Carruthers, W., *Cycloaddition Reactions in Organic Synthesis.* New York: Pergamon Press, 1990.
4. Carruthers, W., *Modern Methods of Organic Synthesis*, 4th ed. Cambridge: University Press, 2004.
5. Coffey, S. (1964–1989), Ansell, M. F. (1973–), Sainsbury (1991–), eds., *Rodd's Chemistry of Carbon Compounds*, 2nd ed. and supplements. New York: Elsevier, 1964–.
6. Corey, E. J., and Cheng, X.-M., *The Logic of Chemical Synthesis.* New York, Wiley, 1995.
7. Fieser, L. F., et al., *Fiesers' Reagents for Organic Synthesis.* New York: Wiley, 1967–.
8. Fringuelli, F. and Taticchi, A. *The Diels-Alder Reaction: Selected Practical Methods.* New York: Wiley, 2002.
9. Fuhrhop, J. H., and Li, G., *Organic Synthesis: Concepts and Methods*, 3rd ed. New York: Wiley, 2003.
10. Greene, T. W., and Wuts, P. G. M., *Protective Groups in Organic Synthesis*, 3rd ed. New York: Wiley, 1999.
11. Harrison, I. T., and Harrison, S., *Compendium of Organic Synthetic Methods.* New York: Wiley, 1971–.
12. House, H. O., *Modern Synthetic Reactions*, 2nd ed. Menlo Park, CA: Benjamin, 1972.
13. Larock, R. C., *Comprehensive Organic Transformations: A Guide to Functional Group Preparations*, 2nd ed. New York: Wiley, 1999.
14. Laue, T., and Plagens, A., *Named Organic Reactions.* New York: Wiley, 2000.
15. Loupy, A., *Microwaves in Organic Synthesis.* New York: Wiley, 2003
16. Millam, M. J., *Reaction Guide for Organic Chemistry.* Lexington, MA: D.C. Heath, 1989.
17. Nicolaou, K. C., et al., "The Art and Science of Organic and Natural Product Synthesis." *J. Chem. Educ.* **1998**, 75, 1225.
18. Norman, R. O. C., and Coxon, J. M., *Principles of Organic Synthesis*, 3rd ed. Cheltenham, UK: Stanley Thornes, 1993.
19. *Organic Reactions.* New York: Wiley, 1942–.
20. *Organic Syntheses, Collective Volumes*, 2nd ed. New York: Wiley, 2004 (and previous years).
21. Paquette, L. A., ed., *Encyclopedia of Reagents for Organic Synthesis.* New York: Wiley, 1995.
22. Patai, S., ed., *Chemistry of Functional Groups.* New York: Wiley, 1964–.
23. Pearson, A. J., *Handbook of Reagents for Organic Synthesis.* New York: Wiley, 1999.
24. Pelter, A., Smith, K., and Brown, H. C., *Borane Reagents.* London: Academic Press, 1988.
25. Proctor, R. G., *Asymmetric Synthesis.* New York: Oxford University Press, 1996.

26. Riddick, J. A., and Bunger, W. M., *Organic Solvents: Physical Properties and Methods of Purification*, 4th ed. New York: Wiley, 1986.

27. Sandler, S. R., and Karo, W., *Organic Functional Group Preparations*, 2nd ed. Orlando, FL: Academic Press, 1983, 1986, 1989.

28. Sandler, S. R., and Karo, W., *Polymer Syntheses*, 2nd ed. Orlando, FL: Academic Press, 1997.

29. Sandler, S. R., and Karo, W., *Sourcebook of Advanced Organic Laboratory Preparations*. San Diego, CA: Academic Press, 1992.

30. Stuart, W., *Organic Synthesis: The Disconnection Approach*. New York: Wiley, 1982.

31. Tatchell, A. R., et al., *Vogel's Textbook of Practical Organic Chemistry*, 5th ed. New York: Wiley, 1989.

32. Theilheimer, W. (1948–81), Finch, A. F. (1982–), eds., *Theilheimer's Synthetic Methods of Organic Chemistry*. Basel: Karger, 1946–.

33. Trost, B. M., and Fleming, I. eds., *Comprehensive Organic Synthesis: Selectivity, Strategy & Efficiency in Modern Organic Chemistry*. Elmsford, NY: Pergamon Press, 1991.

34. Wagner, R. B., and Zook, H. D., *Synthetic Organic Chemistry*. New York: Wiley, 1953.

35. Zweifel, G. and Nantz, M., *Modern Organic Synthesis: An Introduction*. New York: Freeman, 2006.

C. Laboratory Safety

1. Furr, A. K., ed., *CRC Handbook of Laboratory Safety*, 5th ed. Boca Raton, FL: CRC Press, 2000.

2. Gorman, C. E., ed., *Working Safely with Chemicals in the Laboratory*, 2nd ed. Schnectady, NY: Genium, 1995.

3. Lefèvre, M. J., *First Aid Manual for Chemical Accidents*, 2nd ed. New York: Van Nostrand Reinhold, 1989.

4. Lenga, R. E., ed., *The Sigma–Aldrich Library of Regulatory and Safety Data*. Milwaukee, WI: Sigma–Aldrich, 1993.

5. Lewis Sr., R. J., *Sax's Dangerous Properties of Industrial Materials*, 10th ed. New York: Wiley, 2000.

6. Luxon, S. G., ed., *Hazards in the Chemical Laboratory*, 5th ed. Cambridge University Press: Royal Society of Chemistry, 1992.

7. National Research Council, *Prudent Practices in the Laboratory: Handling and Disposal of Chemicals*. Washington, D.C.: National Academies Press, 1995.

8. Urben, P. G., ed., *Bretherick's Handbook of Reactive Chemical Hazards*, 7th ed. Burlington, MA: Academic Press, 2006.

D. General Laboratory Techniques

1. Armarego, W. L. F., and Chai, C., *Purification of Laboratory Chemicals*, 5th ed. New York: Elsevier, 2003.

2. Ikan, R., *Natural Products: A Laboratory Guide*, 2nd ed. San Diego, CA: Academic Press, 1991.

3. Loewenthal, H. J. E., *Guide for the Perplexed Organic Experimentalist*, 2nd ed. New York: Wiley, 1992.

4. Mayo, D. W., et al., *Microscale Techniques for the Organic Laboratory*, 2nd ed. New York: Wiley, 2000.

5. *Methoden der Organischen Chemie, Houben-Weyl*, 4th ed. Stuttgart: Georg Thieme, 1952–.

6. Ruthven, D., *Encyclopedia of Separation Technology*. New York: Wiley, 1997.

7. Weissberger, A., ed., *Technique of Organic Chemistry*, 3rd ed. New York: Wiley, 1959–.

8. Weissberger, A., ed., *Techniques of Chemistry*. New York: Wiley, 1971–.

9. Williamson, K. L., *Macroscale and Microscale Organic Experiments*, 5th ed. Boston, MA: Houghton–Mifflin, 2007.

10. Wilson, I. D., et al., eds., *Encyclopedia of Separation Science*. Orlando, FL: Academic Press, 2000.

11. Zubrick, J. W., *The Organic Chem Lab Survival Manual: A Student's Guide to Techniques*, 6th ed. New York: Wiley, 2004.

E. Chromatography

1. Fowlis, I. A., *Gas Chromatography*, 2nd ed. New York: Wiley, 1995.

2. Grob, R. L. and Barry, E. F., eds., *Modern Practice of Gas Chromatography*, 4th ed. New York: Wiley, 2004.

3. Lindsay, S., *High Performance Liquid Chromatography*, 2nd ed. New York: Wiley, 1992.

4. McNair, H. M., and Miller, J. M., *Basic Gas Chromatography*. New York: Wiley, 1998.

5. Meyer, V. R., *Practical High-Performance Liquid Chromatography*, 3rd ed. Chichester, UK: Wiley, 1999.

6. Poole, C. F., and Poole, S. K., *Chromatography Today*. New York: Elsevier, 1991.

7. Ravindranath, B., *Principles and Practice of Chromatography*. New York: Halsted, 1989.

8. Schomburg, G., *Gas Chromatography: A Practical Course*. New York: VCH, 1990.

9. Sherma, J., and Fried, B., *Thin-Layer Chromatography: Techniques and Applications*, 3rd ed. New York: Marcel Dekker, 1996.

10. Touchstone, J. C., *Practice of Thin Layer Chromatography*, 3rd ed. New York: Wiley, 1992.

F. Spectrometry and Structure Analysis

1. Akitt, J. W., and Mann, B. E., *NMR and Chemistry: An Introduction to Modern NMR Spectroscopy*, 4th ed. Cheltenham, UK: Stanley Thornes, 2000.
2. Bovey, F. A., *Nuclear Magnetic Resonance Spectroscopy*, 2nd ed. San Diego, CA: Academic Press, 1988.
3. Chapman, J. R., *Practical Organic Mass Spectrometry: A Guide for Chemical and Biochemical Analysis*, 2nd ed. New York: Wiley, 1995.
4. Colthup, N. B., Daly, L. H., and Wiberley, S. E., *Introduction to Infrared and Raman Spectroscopy*, 3rd ed. Orlando, FL: Academic Press, 1990.
5. Feinstein, K., *Guide to Spectroscopic Identification of Organic Compounds*. Boca Raton, FL: CRC Press, 1995.
6. Günther, H., *NMR Spectroscopy: Basic Principles, Concepts, and Applications in Chemistry*, 2nd ed. New York: Wiley, 1995.
7. Hoffmann, Edmond de, *Mass Spectrometry: Principles and Applications*. New York: Wiley, 1996.
8. Ingham, A. M., and Henson, R. C., "Interpreting Infrared and Nuclear Magnetic Resonance Spectra of Simple Organic Compounds for the Beginner." *J. Chem. Educ.* **1984**, *61*, 704.
9. Kemp, W., *Organic Spectroscopy*, 3rd ed. New York: W. H. Freeman, 1991.
10. Lee, T. A., *A Beginners Guide to Mass Spectral Interpretation*. New York: Wiley, 1998.
11. McLafferty, F. W., and Turecek, F., *Interpretation of Mass Spectra*, 4th ed. Mill Valley, CA: University Science Books, 1993.
12. Nelson, J. H., *Nuclear Magnetic Resonance Spectroscopy*. Upper Saddle River, NJ: Prentice Hall, 2003.
13. Pavia, D. L., et al., *Introduction to Spectroscopy: A Guide for Students of Organic Chemistry*, 3rd ed. Ft. Worth, TX: W.B. Saunders, 2001.
14. Perkampus, H.-H., *UV–VIS Spectroscopy and its Applications*. New York: Springer-Verlag, 1992.
15. Pouchert, C. J., and Behnke, J., *The Aldrich Library of ^{13}C and ^1H FT-NMR Spectra*. Milwaukee, WI: Aldrich Chemical Co., 1992.
16. Pouchert, C. J., and Campbell, J. R., *The Aldrich Library of NMR Spectra*, 2nd ed. Milwaukee, WI: Aldrich Chemical Co., 1983.
17. Pouchert, C. J., *The Aldrich Library of FT-IR Spectra*, 2nd ed. Milwaukee, WI: Aldrich Chemical Co., 1997.
18. *Sadtler Standard Spectra* (Collections of infrared, ultraviolet, and NMR spectra). Philadelphia, PA: Sadtler Research Laboratories.
19. Herbert, C. G., and Johnstone, R. A. W., *Mass Spectrometry Basics*. Boca Raton, FL: CRC Press, 2002.
20. Silverstein, R. M., et al., *Spectrometric Identification of Organic Compounds*, 7th ed. New York: Wiley, 2003.
21. Skoog, D. A., et al., *Principles of Instrumental Analysis*, 5th ed. Belmont, CA: Brooks/Cole, 1997.
22. Smith, B. C., *Fundamentals of Fourier Transform Infrared Spectroscopy*. Boca Raton, FL: CRC Press, 1996.
23. Smith, R. M., *Understanding Mass Spectra, A Basic Approach*, 2nd ed. Hoboken, NJ: Wiley-Interscience, 2004.
24. Whittaker, D., *Interpreting Organic Spectra*. New York: Springer-Verlag, 2000.
25. Yadav L. D. S., *Organic Spectroscopy*. Norwell, MA: Kluwer Academic, 2003

G. Qualitative Organic Analysis

1. Crews, P., et al., *Organic Structure Analysis*. New York: Oxford University Press, 1998.
2. Criddle, W. J., *Spectral and Chemical Characterization of Organic Compounds: A Laboratory Handbook*, 3rd ed. New York: Wiley, 1990.
3. Kemp, W., *Qualitative Organic Analysis: Spectrochemical Techniques*, 2nd ed. New York: McGraw-Hill, 1986.
4. Pasto, D. J., and Johnson, C. R., *Organic Structure Determination*. Englewood Cliffs, NJ: Prentice Hall, 1969.
5. Rappoport, Z., ed., *CRC Handbook of Tables for Organic Compound Identification*, 3rd ed. Cleveland, OH: Chemical Rubber Co., 1967.
6. Shriner, R. L., et al., *The Systematic Identification of Organic Compounds*, 8th ed. New York: Wiley, 2003.

H. Reaction Mechanisms and Advanced Topics

1. Carey, F. A., and Sundberg, R. J., *Advanced Organic Chemistry*, 4th ed. New York: Springer, 2000.
2. Carpenter, B. K., *Determination of Organic Reaction Mechanisms*. New York: Wiley, 1984.
3. Carroll, F. A., *Perspectives on Structure and Mechanism in Organic Chemistry*. Belmont, CA: Brooks/Cole, 1998.
4. Li, J. J., *Name Reactions: A Collection of Detailed Reaction Mechanisms*. New York: Springer, 2002.
5. Lowry, T. H., and Richardson, K. S., *Mechanism and Theory in Organic Chemistry*, 3rd ed. New York: Harper & Row, 1987.
6. Grossman, R. B., *The Art of Writing Reasonable Organic Reaction Mechanisms*, 2nd ed. New York: Springer, 2003.
7. Smith, M., *March's Advanced Organic Chemistry: Reactions, Mechanisms and Structure*, 6th ed. New York: Wiley, 2006.

8. Miller, A., and Solomon, P. H., *Writing Reaction Mechanisms in Organic Chemistry*, 2nd ed. San Diego, CA: Harcourt/Academic Press, 2000.

9. Moloney, M. G., *Reaction Mechanisms at a Glance: A Stepwise Approach to Problem-Solving in Organic Chemistry*. Malden, MA: Blackwell Science, 2000.

10. *Organic Reaction Mechanisms*. New York: Wiley, 1965–.

11. Scudder, P. H., *Electron Flow in Organic Chemistry*. New York: Wiley, 1992.

12. Sykes, P., *A Guidebook to Mechanism in Organic Chemistry*, 6th ed. Upper Saddle River, NJ: Prentice Hall, 1996.

13. Sykes, P., *A Primer to Mechanism in Organic Chemistry*. Upper Saddle River, NJ: Prentice Hall, 1996

J. Reports of Chemical Research

1. *Annual Reports in Organic Synthesis*. New York: Academic Press, 1970–.

2. *Annual Reports on the Progress of Chemistry, Section B: Organic Chemistry*. London: Royal Society of Chemistry, 1904–.

3. *Chemical Abstracts*. Columbus, OH: CA Service, American Chemical Society, 1907–.

4. *Chemical Titles*. Columbus, OH: CA Service, American Chemical Society, 1961–.

5. *Current Contents: Physical, Chemical & Earth Sciences*. Philadelphia, PA: ISI Press, 1967–.

6. *Index Chemicus*. Philadelphia, PA: ISI Press, 1960–.

7. *Science Citation Index*. Philadelphia, PA: ISI Press, 1961–.

K. Guides to the Chemical Literature

1. Bottle, R. T., and Rowland, J. F. B., eds., *Information Sources in Chemistry*, 4th ed. New Providence, NJ: Bowker-Saur, 1993.

2. Heller, S. R., ed., *The Beilstein System: Strategies for Effective Searching*. New York: Oxford University Press, 1997.

3. Heller, S. R., ed., *The Beilstein Online Database: Implementation, Content, and Retrieval*. New York: Oxford University Press, 1990.

4. Maizell, R. E., *How to Find Chemical Information: A Guide for Practicing Chemists, Educators, and Students*, 3rd ed. New York: Wiley, 1998.

5. Poss, A. J., *Library Handbook for Organic Chemists*. New York: Chemical Publishing Co., 2000.

6. Schulz, H., and Georgy, U., *From CA to CAS Online: Databases in Chemistry*, 2nd ed. New York: Springer-Verlag, 1994.

7. Somerville, A. N., "Information Sources for Organic Chemistry." *J. Chem. Educ.* **1991**, *68*, 553, 843; **1992**, *69*, 379.

L. Sources on Selected Topics

1. Agosta, W. C., *Bombardier Beetles and Fever Trees: A Close-up Look at Chemical Warfare and Signals in Animals and Plants*. Reading, MA: Addison-Wesley, 1996.

2. Agosta, W. C., "Medicines and Drugs from Plants." *J. Chem. Educ.* **1997**, *74*, 857.

3. Agosta, W. C., *Chemical Communication: The Language of Pheromones*. New York: Scientific American Library, 1992.

4. Anastas, P. T., and Warner, J. C., *Green Chemistry: Theory and Practice*. New York: Oxford University Press, 2000.

5. Atkins, P. W., and Atkins, P., *Atkins' Molecules*. Cambridge: University Press, 2003.

6. Bauer, K., et al., *Common Fragrance and Flavor Materials: Preparation, Properties, and Uses*, 2nd ed. Deerfield Beach, FL: VCH, 1990.

7. Belitz, H. D., and Grosch, W., *Food Chemistry*, 2nd ed. New York: Springer-Verlag, 1999.

8. Benfey, O. T., *From Vital Force to Structural Formulas*. Philadelphia, PA: Beckman Center for the History of Chemistry, 1992.

9. Buxton, S. R., and Roberts, S. M., *Guide to Organic Stereochemistry: From Methane to Macromolecules*. Menlo Park, CA: Benjamin/Cummings, 1996.

10. Carraher Jr., C. E., *Polymer Chemistry*, 5th ed. New York: Marcel Dekker, 2000.

11. Cole, L. A., *The Eleventh Plague: The Politics of Biological and Chemical Warfare*. Darby, PA: Diane Publishing, 2001.

12. Coultate, T. P., *Food: The Chemistry of Its Components*, 4th ed. Cambridge: Royal Society of Chemistry, 2002.

13. Cresswell, S. L. and Haswell, S. J., "Microwave Ovens— Out of the Kitchen." *J. Chem. Educ.* **2001**, *78*, 900

14. Dehmlow, E. V., and Dehmlow, S. S., *Phase Transfer Catalysis*, 3rd ed. New York: VCH, 1993.

15. Djerassi, C., *From the Lab into the World: A Pill for People, Pets, and Bugs*. Washington, D.C.: American Chemical Society, 1994

16. Dodd, J. S., ed., *The ACS Style Guide: A Manual for Authors and Editors*, 3rd ed. Washington, D.C.: American Chemical Society, 2006.

17. Donnelly, T. H., "The Origins of the Use of Antioxidants in Foods." *J. Chem. Educ.* **1996**, *73*, 159.

18. DuPré, D. B., "Blood or Taco Sauce? The Chemistry behind Criminalists' Testimony in the O. J. Simson Murder Case." *J. Chem. Educ.* **1996**, *73*, 60.

19. Eliel, E. L, et al., *Basic Organic Stereochemistry*. New York: Wiley, 2001.

20. Ellis, J. W., et al., "Symposium: Sweeteners and Sweetness Theory." *J. Chem. Educ.* **1995**, *72*, 671, 676, 680.

21. Fossey, J., et al., *Free Radicals in Organic Chemistry*. New York: Wiley, 1995.

22. Fox, R. B., and Powell, W. H., *Nomenclature of Organic Compounds: Principles and Practice*, 2nd ed. Washington, D.C.: American Chemical Society, 2000.

23. French, L. G., "The Sassafrass Tree and Designer Drugs: from Herbal Tea to Ecstasy." *J. Chem. Educ.* **1995**, *72*, 479.

24. Foye, W. O., et al., *Principles of Medicinal Chemistry*, 4th ed. Baltimore, MD: Lippincott Williams and Wilkins, 1995.

25. Gerber, S. M., ed., *Chemistry and Crime: from Sherlock Holmes to Today's Courtroom*. New York: Oxford University Press, 1983.

26. Gerber, S. M., and Saferstein, R., eds., *More Chemistry and Crime: from Marsh Arsenic Test to DNA Profile*. New York: Oxford University Press, 1997.

27. G. A. Giffin, et al., "Modern Sport and Chemistry: What a Chemically Aware Sports Fanatic Should Know." *J. Chem. Educ.* **2002**, *79*, 813

28. Gilchrist, T. L., *Heterocyclic Chemistry*, 3rd ed. New York: Wiley, 1997.

29. Goldsmith, R. H., "A Tale of Two Sweeteners." *J. Chem. Educ.* **1987**, *64*, 954.

30. Goodman, J. M., *Chemical Applications of Molecular Modeling*, 2nd ed. Cambridge, UK: Royal Society of Chemistry, 2007.

31. Gribble, G. W., "Natural Organohalogens: A New Frontier for Medicinal Agents?" *J. Chem. Educ.* **2004**, *81*, 1441.

32. Gribble, G. W., "Natural Organohalogens: Many More than You Think!" *J. Chem. Educ.* **1994**, *71*, 907.

33. Gunstone, F. D., *Fatty Acid and Lipid Chemistry*. New York: Blackie, 1996.

34. Hammond, G. S., and Kuck, V. J., *Fullerenes: Synthesis, Properties, and Chemistry of Large Carbon Clusters*. Washington, D.C.: American Chemical Society, 1992.

35. Hirsch, A., et al., *Fullerenes: Chemistry and Reactions*. New York: Wiley, 2005.

36. Hocking, M. B., "Vanillin: Synthetic Flavoring from Spent Sulfite Liquor." *J. Chem. Educ.* **1997**, *74*, 1055.

37. Höltje, H.-D. et. al., *Molecular Modeling: Basic Principles and Applications*, 2nd ed. Weinheim, Germany: Wiley–VCH, 2003.

38. Honeybourne, C. L., "Organic Vapor Sensors for Food Quality Assessment," *J. Chem. Educ.* **2000**, *77*, 338.

39. Hosler, D. M., and Mikita, M. A., "Ethnobotany: The Chemist's Source for the Identification of Useful Natural Products." *J. Chem. Educ.* **1987**, *64*, 328.

40. Houghton, P. J., "Old Yet New—Pharmaceuticals from Plants." *J. Chem. Educ.* **2001**, *78*, 175

41. Huddle, P. A., "How to Present a Paper or Poster." *J. Chem. Educ.* **2000**, *77*, 1091

42. Hughes, P. "Was Markovnikov's Rule an Inspired Guess?" *J. Chem. Educ.* **2006**, *83*, 1152

43. Jacques, J., *The Molecule and Its Double*. New York: McGraw-Hill, 1993.

44. James, L. K., ed., *Nobel Laureates in Chemistry, 1901–1992*. Washington, D.C.: American Chemical Society, 1993.

45. Jandacek, R. J., "The Development of Olestra, a Noncaloric Substitute for Dietary Fat." *J. Chem. Educ.* **1991**, *68*, 476.

46. Kanare, H. M., *Writing the Laboratory Notebook*. Washington, D.C.: American Chemical Society, 1985.

47. Kauffman, G. B., "Wallace Hume Carothers and Nylon, the First Completely Synthetic Fiber." *J. Chem. Educ.* **1988**, *65*, 803.

48. Kauffman, G. B., and Seymour, R. B., "Elastomers: I. Natural Rubber." *J. Chem. Educ.* **1990**, *67*, 422.

49. Kent, J. A., ed., *Kent and Riegel's Handbook of Industrial Chemistry and Biotechnology*, 11th ed. New York: Springer, 2006

50. Kikuchi, S., "A History of the Structural Theory of Benzene—the Aromatic Sextet Rule." *J. Chem. Educ.* **1997**, *74*, 194.

51. Kimbrough, D. R., "Hot and Spicy vs. Cool and Minty as an Example of Organic Structure–Activity Relationships." *J. Chem. Educ.* **1997**, *74*, 861.

52. Kimbrough, D. R., "The Photochemistry of Sunscreens." *J. Chem. Educ.* **1997**, *74*, 51.

53. King, F. D., ed., *Medicinal Chemistry: Principles and Practice*, 2nd ed. Cambridge, UK: Royal Society of Chemistry, 2003.

54. Kopecky, J., *Organic Photochemistry: A Visual Approach*. New York: Wiley, 1992.

55. Laing, M., "Beware—Fertilizer can EXPLODE!" *J. Chem. Educ.* **1993**, *70*, 393.

56. Lancaster, M. *Green Chemistry: An Introductory Text*. Cambridge, UK: Royal Society of Chemistry, 2002.

57. Leung, A.Y., and Foster, S., *Encyclopedia of Common Natural Ingredients Used in Foods, Drugs, and Cosmetics*, 2nd ed. New York: Wiley, 2003.

58. Manahan, S. E., *Green Chemistry and the Ten Commandments of Sustainability*, 2nd ed. Columbia, MO: ChemChar Research, 2005

59. Mann, J., et al., *Natural Products: Their Chemistry and Biological Significance*. New York: Wiley, 1994.

60. Milgrom, L. R., *The Colours of Life: An Introduction to the Chemistry of Porphyrins and Related Compounds*. New York: Oxford University Press, 1997.

61. Morris, E. T., *Fragrance: The Story of Perfume from Cleopatra to Chanel*. Mineola, NY: Dover Publications, 2002.

62. Nicholson, J. W., and Anstice, H. M., "The Chemistry of Modern Dental Filling Materials." *J. Chem. Educ.* **1999**, *76*, 1497.

63. Ohloff, G., *Scent and Fragrances: The Fascination of Odors and Their Chemical Perspectives*. New York: Springer-Verlag, 1994.

64. Parsons, A. *An Introduction to Free-Radical Chemistry*. Malden, MA: Blackwell, 2000.

65. Pellenbarg, R. E., and Max, M. D., "Gas Hydrates: From Laboratory Curiosity to Potential Global Powerhouse." *J. Chem. Educ.* **2001**, *78*, 896.

66. Perrine, D. M., *The Chemistry of Mind-Altering Drugs: History, Pharmacology, and Cultural Context*. New York: Oxford University Press, 1996.

67. Petracco, M., "Our Everyday Cup of Coffee: The Chemistry behind Its Magic." *J. Chem. Educ.* **2005**, *82*, 1161.

68. Pinto, G., "Chemistry of Moth Repellents." *J. Chem. Educ.* **2005**, *82*, 1267

69. Robinson, M. J. T., *Organic Stereochemistry*. New York: Oxford University Press, 2000.

70. Schmid, G., *Nanoparticles: From Theory to Application*. New York: Wiley, 2004.

71. Seymour, R. B., and Kauffman, G. B., "The Ubiquity and Longevity of Fibers." *J. Chem. Educ.* **1993**, *70*, 449.

72. Selinger, B., *Chemistry in the Marketplace*, 5th ed. Orlando, FL: Academic Press, 1998.

73. Silverman, R. B., *The Organic Chemistry of Drug Design and Drug Action*, 2nd ed. San Diego, CA: Academic Press, 2004.

74. Sotheeswaran, S., "Herbal Medicine: The Scientific Evidence." *J. Chem. Educ.* **1992**, *69*, 444.

75. Spessard, G. O., and Miessler, G. L., *Organometallic Chemistry*. Upper Saddle River, NJ: Prentice Hall, 1997.

76. Starks, C. M., et al., *Phase Transfer Catalysis: Fundamentals, Applications, and Industrial Perspectives*. New York: Chapman & Hall, 1994.

77. Stick, R. V., *Carbohydrates: The Sweet Molecules of Life*. San Diego, CA: Academic Press, 2001.

78. Stocker, J. H., *Chemistry and Science Fiction*. New York: Oxford University Press, 1998.

79. Sundberg, R. J., *Indoles*. San Diego, CA: Academic Press, 1996.

80. Tannenbaum, G., "Chocolate: A Marvelous Natural Product of Chemistry." *J. Chem. Educ.* **2004**, *81*, 1131

81. Turro, N., *Modern Molecular Photochemistry of Organic Molecules*. Sausalito, CA: University Science Books, 2006.

82. Vartanian, P. F., "The Chemistry of Modern Petroleum Product Additives." *J. Chem. Educ.* **1991**, *68*, 1015.

83. Waddell, T. G., and Rybolt, T. R., "The Chemical Adventures of Sherlock Holmes: The Ghost of Gordon Square." *J. Chem. Educ.* **2000**, *77*, 471.

84. Waddell, T. G., et al., "Legendary Chemical Aphrodisiacs." *J. Chem. Educ.* **1980**, *57*, 341.

85. Walters, E. E., et al., eds., *Sweeteners: Discovery, Molecular Design, and Chemoreception*. Washington, D.C.: American Chemical Society, 1991.

86. Waring, D. R., and Hallas, G., eds., *The Chemistry and Application of Dyes*. New York: Plenum Press, 1990.

87. Weissermel, K., and Arpe, H.-J., *Industrial Organic Chemistry*, 4th ed. *New York*: Wiley-VCH, 2003.

88. White, M. A., "The Chemistry behind Carbonless Copy Paper." *J. Chem. Educ.* **1999**, *75*, 1061.

89. Wigfield, D. C., *Environmental Aspects of Organic Chemistry*. Winnipeg, Canada: Wuerz, 1997.

90. Yee, G. T., "Through the Looking Glass and What Alice Ate There." *J. Chem. Educ.* **2002**, *79*, 569

91. Zanger M., et al., "The Aromatic Substitution Game." *J. Chem. Educ.* **1993**, *70*, 985.

92. Zielinski, T. J., and Swift, M. L., eds., *Using Computers in Chemistry and Chemical Education*. New York: Oxford University Press, 1997.

93. Zimpleman, J. M., "Dioxin, Not Doomsday." *J. Chem. Educ.* **1999**, *76*, 1662.

M. Software for Organic Chemistry

1. Brockwell, J. C., et al., *Beaker: An Expert System for the Organic Chemistry Student. (Windows or Macintosh)*. Belmont, CA: Thompson Wadsworth, 1995.

2. *ChemDraw (Windows or Macintosh)*. CambridgeSoft.com

3. *Chem3D (Windows or Macintosh)*. CambridgeSoft.com

4. Clough, F. W., *Introduction to Spectroscopy: IR, NMR, CMR, and Mass Spec (Windows)*. Campton, NH: Trinity Software.

5. Figueras, J., *MassSpec: A Graphics-Based Mass Spectrum Analyzer (Windows or Macintosh)*. Campton, NH: Trinity Software.

6. Figueras, J., *Name It (Windows or Macintosh)*. Campton, NH: Trinity Software.

7. Figueras, J., *SynTree: A Program for Exploring Organic Synthesis (Windows or Macintosh)*. Campton, NH: Trinity Software.

8. Pavia, D. L., and Clough, F. W., *Identification of Organic Compounds (Windows)*. Campton, NH: Trinity Software.

9. Pavia, D. L., and Clough, F. W., *MacSQUALOR (Macintosh)*. Campton, NH: Trinity Software.

10. Schatz, P. F., *IR Simulator (Windows or Macintosh)*. Wentworth, NH: Falcon Software.

11. Schatz, P. F., *NMR Simulator (Windows or Macintosh)*. Wentworth, NH: Falcon Software.

12. *Spartan Student Edition for Macintosh* and *Spartan Student Edition for Windows*. Irvine, CA: Wavefunctions, Inc

Index